China in Family Photographs

A PEOPLE'S HISTORY OF REVOLUTION AND EVERYDAY LIFE

Edited and Translated
by Ed Krebs and Hanchao Lu

Bridge21 Publications | Encino

Published 2018 by Bridge21 Publications

Distributed exclusively by Routledge
2 Park Square, Milton Park, Abingdon, Oxon OX14 4RN
711 Third Avenue, New York, NY 10017, USA

Routledge is an imprint of the Taylor & Francis Group, an informa business

Cover Design by Linda Ronan
Layout Design by Linda Ronan

ISBN: 9781626430549

For information contact Bridge21 Publications, LLC
16705 Huerta Road, Encino, California 91436 or info@bridge21.us

Published in the United States

Contents

Old Photos: A Living Archive

by Ed Krebs

Translations in this book are taken from the Chinese periodical *Old Photos* (*Lao Zhaopian*), the best-known publication of its kind in China. Launched in 1996 by Shandong Pictorial Press (*Shandong huabao chubanshe*), *Old Photos* published issue Number 100 in April 2015. In 2009, this *Reader's Digest*-sized magazine was acclaimed "One of the 60 Most Influential Publications in the History of the [People's] Republic."[1]

Old Photos was the successor to Shandong Pictorial Press's pioneering three-volume photo history of modern China.[2] Feng Keli, editor of *Old Photos*, served as an editor of that ambitious work. Upon publication, *Old Photos* became a first of its kind in post-Mao China. Over the next several years, many other publications appeared that featured historic photos and memorabilia—and "old" in their titles.[3] A powerful nostalgia was in the air.

This broad communal recall did much to shape ongoing waves of nostalgia in Chinese society of the late 1990s. Shanghai played

1. This honor was presented by *China Readers' News* (*Zhonghua dushu bao*) in 2009, on the sixtieth anniversary of the People's Republic. *Old Photos* was also named "one of the most unforgettable publications of the period of 'Reform and Opening'" (*gaige kaifang*).
2. This was *A Century of Chinese History in Photographs* (*Tupian Zhongguo bainian shi*), published in 1995.
3. In 1998, a series on "Old Cities," quality photo-histories of a dozen Chinese cities, was produced by Jiangsu Fine Arts Publishers (*Jiangsu meishu chubanshe*). Beijing, Tianjin, Shanghai, Xi'an, and other major cities were included. In 1999 the Hundred Flowers Cultural Arts Press (*Baihua wenyi chubanshe*) of Tianjin brought out its richly illustrated *Old Stories of Ration Coupons* (*Piaozheng jiushi*), edited by Xue Yanwen and Wang Tongli.

a leading role in this "nostalgia fever." The city had been a beacon of modernity that produced countless photos of people, places, and events in a receding past. It was both a hub of avant-garde activities such as movie-making and a favorite backdrop for posters purveying cigarettes. Such ads featured stylish but "naughty" city women whose images also proved useful for selling ordinary things such as boxed matches and flashlight batteries. During the 1990s, reprints of these ads took people back to the days when all those things were new in Chinese life. Photos of historic Shanghai, retro calendars, and reprinted handbills helped to launch this rage for nostalgia. *Old Photos*, a new-style magazine, mirrored this nostalgia but also took it to a deeper level by stimulating thoughtful reflection on the past.

Old Photos published its first issue in December 1996. Reprinted twice, this inaugural issue was followed by a second in March 1997. It was first published as a series, not yet a periodical produced on a regular schedule. The editors created a format with subtle yet powerful appeal. The cover photo almost always featured a single individual, suggesting a respect for people as individuals that had broader implications. The cover design also featured a pair of autumn leaves, projecting a seasoned maturity. Readers could interpret this image as they saw fit. It could certainly suggest the memories of China's long struggles toward success as a nation. Both the cover and the pages of the magazine were printed on an off-white, slightly yellow-toned paper. This color gave the photos (virtually all of them in black-and-white) a sepia tone that accented their "oldness."

Essential to the appeal of *Old Photos* was the editors' invitation for readers to contribute their own old photos with accompanying texts. The response was enthusiastic from the beginning. The editors responded promptly to these contributions: If they wanted your piece, you would soon see it in print. These individual contributions were often written by people who had never had anything published before. Meanwhile, articles by more experienced and professional writers supplemented the amateur stories. Letters to the magazine from readers and contributors created a forum that further built *Old Photos*' popularity. As the magazine flourished, a community of enthusiastic readers quickly grew.

The magazine has made changes from time to time over the

years, especially as it expanded. The iconic pair of leaves on the cover was dropped after Number 10. Another change at that point was the addition of a pair of foldout pages in each issue, allowing for the presentation of wider-scale and panoramic views. This has been a part of every subsequent issue. The pair of leaves later reappeared in smaller form at the upper left corner of the cover but was subsequently dropped again. These periodic changes of dress have helped to keep *Old Photos* fresh. The most basic changes have been additions to the number of pages and modest price increases. One further change seems especially important: as the qualitative change in paper and photo sharpness was introduced with Number 70 in 2010, two phrases of four characters each were added alongside the banner: "Determining Principles of History, Gathering a Storehouse of Memory."

As *Old Photos* approached its 100th issue, the editors invited readers to send comments or essays on the theme of "*Old Photos* and Me." One reader, Ding Li, noted the changes outlined above. The first issue included 83 photos and about 70,000 words, she said, and was a bargain at 6.50 *yuan*. By February 2015 the magazine had reached 188 pages and included 132 photos and 120,000 words. While the price had risen to 20 *yuan*, Ding clearly thought this was still a bargain. Her comment on the relationship of *Old Photos* to her personal life reflects the experience of many readers. Newly married when the magazine began publication, she then became a mother; she and her daughter grew up together as avid readers of the magazine.

Ding Li offers this evaluation: "History itself is multi-faceted, and a democratic society should begin with memories that are multifaceted. With *Old Photos*, many years are like a day, and they have quietly preserved complex memories for later generations. At the start they broke new ground, and now the trees they planted have become a forest."[4]

Another contributor, Zhu Xindi, also responded to the invitation for comment: "In this era of such rapid change, our heartfelt thanks to *Old Photos* for providing a platform that gives countless ordinary people an opportunity to look back at the past and to be involved in

4. See "Greetings, Old Photos" (*Ni hao, Lao Zhaopian*), no. 102, pp. 181–86.

'written history.' . . . History always continues: ordinary people are part of it, and their stories of old photos will be told endlessly."[5]

• • •

This book will be of special interest to students in courses on China's history of the past century. Our selections include material on events that brought great change and upheaval to China. This book will also appeal to older readers as an introduction to China's recent history presented in human terms. The advantage to readers of this English version is the same as for Chinese readers. These individual experiences fit into the vast mosaic of a great country's history and offer insights into that history.

Some basic questions remain. When did photography begin to influence China's modern culture? The earliest photographs were made in the middle nineteenth century, almost all of them by foreigners. By the turn of the twentieth century, this new form of information and art was expanding into many aspects of the culture. Who had cameras? How could a person get a photograph without owning a camera? As one would suppose, camera ownership was mostly an upper-class phenomenon, but some less prosperous people also owned cameras. Local institutions, businesses, and military units commonly had cameras that were used to record and publicize their activities. Photo studios began to operate in large and medium-sized cities during the 1890s, and on-site photographers stood by to record people's visits to China's greater and lesser tourist sites.

A word about the choice of articles for translation. This process was actually rather subjective. The translators, both professional historians of modern China with a strong interest in everyday life, selected articles that looked interesting and would illustrate the experience of Chinese people over the past century. It is not comprehensive: one topic not included, for example, is military life and action. That was passed over in favor of stories about more everyday kinds of activities. We like our choices and hope our readers will like

---

5. "'Old Photos' That Speak Truth," (*Shuo zhenhua de 'Lao Zhaopian'*), no. 100, pp. 178–82. Zhu is the author of "Hard Times but Good Times"—the opening selection in this book.

them. The variety and volume of material in *Old Photos* is suggested above. What appears here is no more than one percent of this vast archive of China's recent history.

In preparing these translations for publication, we have written introductions to suggest connections to larger social and historical contexts. The notes offer information beyond the published text. The translations are organized into sections according to broad themes, but individual articles often have relevance beyond these themes. We have done some cross-referencing to provide connections and comparisons, especially in the notes.

Finally, for anyone beginning to study Chinese or for those doing graduate work and looking for research topics, *Old Photos* offers a rich lode of research possibilities. The *Lao Zhaopian* website is a great resource (www.lzp1996.com).

The editors, contributors, and readers of *Old Photos* welcome you into their community!

ACKNOWLEDGMENTS First I thank the circumstances that led me to buy the very first issue of *Old Photos* at a subway station bookstand in Beijing in the summer of 1997. I probably would have discovered this magazine in due time, but having found the first issue and enjoying it thoroughly as did countless Chinese readers, I sought issue after issue and have them all after subscribing to *Old Photos* through China International Bookstore. In Beijing, again in the summer of 2000, I contacted editor Feng Keli, and we met when he came to Beijing for a meeting.

I have visited Feng Keli in Jinan and met Zhang Jie and other assistants on the capable staff of *Old Photos*. Mr. Feng is an intensely hard worker who has built both a magazine and a community. He published a translation of my study first done for a conference on "Modern Chinese Historiography and Historical Thought" at Heidelberg University in 2001. This paper was eventually published as "Old in the Newest New China: Photographic History, Private Memories, and Individual Views of History" in *The Chinese Historical Review* (Spring 2004).

Feng Keli has said that the ideas offered in my paper arrived in a timely way for him and his staff and helped them to focus on their work from a new perspective. He has reprinted that translation in two collections of materials and comments on *Old Photos*. It is most pleasing to think my work might have had such an influence!

We are also indebted to Feng Keli for permission to use the photographs that appear in this book. He has supported us with ongoing cooperation.

Hanchao Lu invited me to submit that revised paper to the *Chinese Historical Review* and later joined me in doing the translations in this book. I am grateful to Hanchao for years of friendship and many enjoyable sessions comparing notes on our work. Thanks to Hanchao for his thorough review of the pieces I translated and for greatly improving the accuracy and quality of the completed book.

More recently I want to thank two friends of many years for looking over parts of the book: Douglas Reynolds, the East Asia specialist at Georgia State University in Atlanta and author of two outstanding studies of late Qing China. And Bill Rope, whose career in the Foreign Service included tours in China and Hong Kong. Thanks also to Vera Schwarcz, emerita professor at Wesleyan University, for thoughts and sharing of her work on Zhang Shenfu.

I met Gregory Kaplan, president of Bridge21 Publications, who joined members of the US-China Friendship Association at the national convention that we of the Atlanta chapter hosted in September 2015. We appreciate Gregory's enthusiasm for this collection of translations and hope his interest will be rewarded. Many thanks also to Mark Gonnerman for his editorial work done with great care, to Amanda Brown for her review of the manuscript, and to Linda Ronan for her great work in creating the final form of this book.

Finally, I can never express sufficient thanks to the editor who is always nearby, my wife Sylvia. She has read most of the translations I did for this book and her suggestions always help. She has forborne as I put off everyday matters to complete this project, especially. Sylvia has joined me in my interests in China and in our many travels and years of living there.

Shortcomings remain my responsibility as editor. Corrections and comments are welcome via e-mail at edkrebs@hotmail.com.

Ed Krebs (also published as Edward S. Krebs)

Inside Chinese Family Albums: Memories and Reflections of Everyday People

by Hanchao Lu

Photographs are a valuable resource for studying history and human society, and China is no exception. Indeed, to a great extent this resource is more precious in China than in Western nations. Although the technology of photography arrived in China no later than the 1850s, cameras remained a sheer luxury in Chinese lives until very recently. As late as 2003, only 20 percent of Chinese families owned a camera, and in early 2000 only 3 percent of Chinese families owned a digital camera.[6] Historical photographs of modern China available for researchers as well as casual viewers were provided mostly by Westerners who visited or sojourned in the country on various occasions. For instance, photographs taken by the Scottish traveler John Thomson (1837–1919), the American traveler James Ricalton (1844–1929), the American social activist Sidney D. Gamble (1890–1968), and American photographer Francis Stafford (1884–1938) are among some of the best known and most widely used photographic collections in Chinese studies.[7] Few foreigners

6. *China Economic Times* (*Zhongguo jingji shibao*), August 23, 2008. Statistics are based on a report by the State Council of the PRC on the consumption of digital products among Chinese urban families in 2002.

7. Duke University library has Sidney D. Gamble's collection of over 5,000 photographs, primarily of China in the period from 1908 to 1932. Francis Eugene Stafford's collection of 1,055 photographs taken in 1909–1915 and 1932–1934 is housed at Stanford University's Hoover Institution. For a glimpse of Stafford's photographs related to the era when China's imperial monarchy ended, see Hanchao Lu, *The Birth of a Republic: Francis Stafford's Photographs of China's 1911 Revolution and Beyond* (Seattle: University of Washington Press, 2010).

had the opportunity to visit Mao's China. Still, glimpses of the country at the time were caught in photographs taken by Westerners who had the chance to visit, such as John Brian Brake (1927–1988) of New Zealand and Marc Riboud (1923–2016) and Bruno Barbey (1941–) of France.[8]

Generally missing in the field, then, are photos taken by the Chinese themselves, especially by ordinary people. Photos taken by the Chinese before the establishment of the People's Republic are mostly in private family albums and rarely available in the public domain. Subsequently, most photos of Chinese lives were taken by state-controlled media and published in periodicals such as the *People's Pictorial* (*Renmin Huabao*, published since 1950), *China Today* (published since 1952, first as *China Reconstructs* and renamed in 1990), and various newspapers.[9] Autobiographies published in English since the late 1970s by Chinese who left their country typically contain a few family photos, but there are too few such illustrations to be of significant use for historians.[10]

Since the 1990s, along with the rapid changes brought about by reforms, a social phenomenon known as the "wind of nostalgia" (*huaijiu feng*) has popularized reminiscences about life in the past century.[11] As a result, there has been a torrent of publications about all things "old" in China: old trades, old customs, old news, old newspapers, old cartoons, old "ration coupons," old advertisements, and most of all, old photos.[12] The book series *Old Photos (Lao Zhaopian)*,

8. See Brian Brake, *China: The 1950s* (Te Aro: Te Papa Tongarewa Museum of New Zealand, 1995); Marc Riboud, *Marc Riboud in China: 40 Years of Photography* (New York: Harry N. Abrams, 1997); and Bruno Barbey, *China: From Mao to Modernity, 1973–2013* (Singapore: Didier Millet, 2015).

9. For a glimpse of Chinese photobooks published in the twentieth century, particularly photo brochures published after the founding of the People's Republic, see Martin Parr and Wassink Lundgren, eds., *The Chinese Photobook: From the 1900s to the Present* (New York: Aperture, 2015).

10. See, for example, Liang Heng and Judith Shapiro, *Son of the Revolution* (first published in 1983) and Jung Chang's *Wild Swans: Three Daughters of China* (first published in 1991) for this type of family photos.

11. Hanchao Lu, "Nostalgia for the Future: The Resurgence of an Alienated Culture in China." *Pacific Affairs*, vol. 75, no. 2 (Summer 2002): 169–186.

12. Edward S. Krebs, "Old in the Newest New China: Photographic History, Private Memories and Individual Views of History." *Chinese Historical Review*, vol. 11, no. 1 (Spring 2004): 87–116.

launched in 1996 by Shandong Pictorial Press, offered the Chinese public a medium for recognizing the experiences of ordinary people amidst momentous historical events. The series quickly became a publishing phenomenon. As of April 2015, one hundred issues had been published; some have sold more than 300,000 copies. The series was recognized as one of the top sixty publishing events in the People's Republic since 1949. It is also credited with reigniting China's "cultural fever for old photographs" at the turn of the twenty-first century.

A distinguishing feature of *Old Photos* is that the editors invite readers to send in their own photographs with accompanying narratives. The response has been strong. The "Call for Contributions" specifies that the photos must be at least twenty years old. Most of the photos published in the series, however, were taken at least half a century ago. Readers' contributions, including both the photos and accompanying essays, are the main source for the book series. This approach has offered alternatives to authorized presentations of China's modern historical experience and provides a rare platform for ordinary people to tell their lives in relation to the nation's turbulent recent past, hence the enthusiastic public response to *Old Photos*. Nearly all the photos come from ordinary Chinese family albums that otherwise would not be available to the public. Most of the essays come from people who have never published anything before. These authors often use plain, candid, and sometimes witty language to describe their photos, telling why and how they were taken, including stories about the people or scenes depicted, and setting forth relevant family history. Clearly, many of the contributors see the book series as a means to record their personal history and a valuable outlet for their feelings about the past. Thus, one cumulative achievement of *Old Photos* has been to create a wide-ranging and multifaceted archive of Chinese social history during this era of photography.

The present volume is based on materials from the *Old Photos* series. It consists of selected photos and their accompanying essays, taken mostly from the first eighty issues of the series; they have been translated into English and provided with annotations and introductions. Chronologically, these essays cover the period from the end of the Qing dynasty to the late twentieth century, or roughly from the 1900s to the 1980s. Most of the major episodes of the twentieth century are covered, including the 1911 Revolution, the war with Japan, the Communist revolution, and various political campaigns in Mao's era—in particular, the Anti-Rightist Movement, the Great Leap Forward, and the Cultural Revolution. Since virtually all the contributors were born in the mid-twentieth century, the majority of the photos and associated essays are about their lives in those turbulent and eventful times.

Except for a section on profiles of prominent leaders such as Mao Zedong (1893–1976) and Hu Yaobang (1915–1989), a particular focus of the book is on how ordinary people coped with the extraordinary changes brought about by revolutions and wars and on their interactions with great political affairs in the nation. Readers will learn vivid details of these lives, as told by those submitting the photographs. These are valuable stories, as they often uncover the plights and the pleasures—and everything in between—of everyday people who have been omitted from most histories. In order to better record, comprehend, and analyze China's tumultuous modern history, we need these previously untold and largely uncensored stories of the "little people" whose voices have seldom entered the historical record. This volume is a small but genuine effort toward that goal.

Transliteration

A transliteration system offers a way to present the sounds of one language in a form that makes it possible for users of another language to approximate those sounds. In this book we use the *pinyin* system (full term: *Hanyu pinyin fang'an*) originally developed by Chinese scholars during the 1950s as they set out to teach illiterate Chinese their own language. As some readers might not be familiar with *pinyin*, a brief introduction follows.[13]

In *pinyin* the initial letter you see provides the initial sound that you utter. Initial consonant sounds *b, p, d, t, g, k, j,* and *ch* are pronounced the same as in English. The initial *zh-* is pronounced "j". *zh-* and *ch-* signal that the vowel *-i* following them is pronounced "-r" (*zhi* is "jr" and *chi* is "chr"); a *u* that follows *zh-* or *ch-* takes the sound of "-w."

The most difficult initial consonant sounds used in *pinyin* are *c, q,* and *x*. None of these is ever pronounced as in English: *c* = "ts"; *q* = "ch"; *x* = an initial sound not used in English, "sy"; *qi* and *xi* are "chee" and "syee." Both of these are often followed by endings that begin with *i* or *u*: the single letter *u* = "chew" and "syew." Examples of longer endings with *i* and *u*: *qian* = "chyaan," *xian* = "syaan"; *quan* = "chwaan," *xuan* = "sywaan" (this "aa" is the nasalized *a* that comes readily to English speakers, as in "man"). The initial sound *z-* gets the sound of the *dz* in the English word "adze."

For vowel (or final) sounds, let's continue by showing the possible pronunciations of the five vowels and giving some examples of each:

13. Two articles in the present book address the evolution of systems used to foster correct pronunciation in China. See "The Sound Syllabary Displayed on a Wall at Tian'anmen" and "Eliminating Illiteracy in Villages."

The *a* sounds are either "ah" or "aa," this latter repeated from above as the nasalized sound we regularly use in American English. You'll do well if you just "say ah," as in the doctor's office. Here are examples of commonly occurring sounds: in *fan, fang, ban, bang, hang, mang*—and many others, say "ah." The nasalized "aa" sound is correct for the *–a* sound in *qian, xian, quan, xuan,* and other usages, as already mentioned.

The "e" has two sounds, "-uh" and "-eh." In Mao *Ze*dong's name, the *e* has this "-uh" sound, and also in *de, le, te, che, zhe,* and many other examples. "e" is "-eh" in these usages: *ye, jie, qie,* and—again—many others. (Of course, the last two examples combine *e* with *I,* but this principle still applies.

"i" is a bit more varied in its usages: it can be "-ee," "-uh," and "-r." The "ee" sound is most common, as in *bi, di, ji, li, mi, pi, qi,* and *xi.* It is "uh" (or very close to this sound) in *ci* and *zi*. Examples were given earlier for usages as "r": *chi, zhi,* and others.

The *u* is pronounced as "-w-" or as something close to "yooh," an umlaut-*u* as in German.

There are some helpful internet sites to provide further guidance. In choosing a site, it will help to see *pinyin* side by side with the earlier standard Wade-Giles transliteration forms and the "Yale system" developed by teachers of Chinese at Yale University during the early 1940s to train American military people as quickly as possible. The last presents the sounds of Chinese in a way that connects more directly with those who use American English. One such site is Chinese Outpost, www.chinese-outpost.com/language/pronunciation/mandarin-romanization-comparing-pinyinto-yale-and-wade-giles.asp [14]

One more basic principle should be mentioned: units in Chinese are pronounced as a single syllable. Examples: *duo* is pronounced "dwo," not "du-o" as in English; *chuang* is "chwang," not "chu-ang"; *miao* is like English "cat's meow," not me-ow.

As stated at the beginning, *pinyin* is used throughout this book. There is an argument for using long-established forms for various

---

14. Jonathan Spence's widely used textbook, *In Search of Modern China* (New York: W. W. Norton, 1990) presents *pinyin* and Wade-Giles alongside each other. This is useful, but the Yale system is a better bridge to *pinyin.*

personal and institutional names (examples: Chiang Kaishek, Tsinghua University, Tsingtao beer, etc.). Our purpose in this book is to present all Chinese terms in a single system as an aid to readers. Thus we use Beijing, not Peking, for China's capital city, Qinghua for the name of the university mentioned above, and Qingdao for the city in Shandong where that beer is brewed. And "Jiang Kaishek" is used for the leader of the Republic of China.

To conclude, one watchword: Be not afraid! One term for "dictionary" in Chinese is *cihai*, an "ocean of words." But there are only 400-plus *sounds* used to utter these countless words—so look on the bright side! Chinese has stimulated a variety of transliteration systems. Besides the three emphasized here, there is the "postal system" developed well over a century ago to establish uniform spellings of place names, and responsible for "Peking," for example. There are also transliteration systems for those who speak French, German, Russian, and other languages, each trying to get it right in their own language. So while *pinyin* presents problems, the Chinese linguists who worked it out as a system proceeded carefully, using some sounds from other alphabetic languages besides English (the source of some of "our" concerns). While no such system can be perfect, this one is manageable.

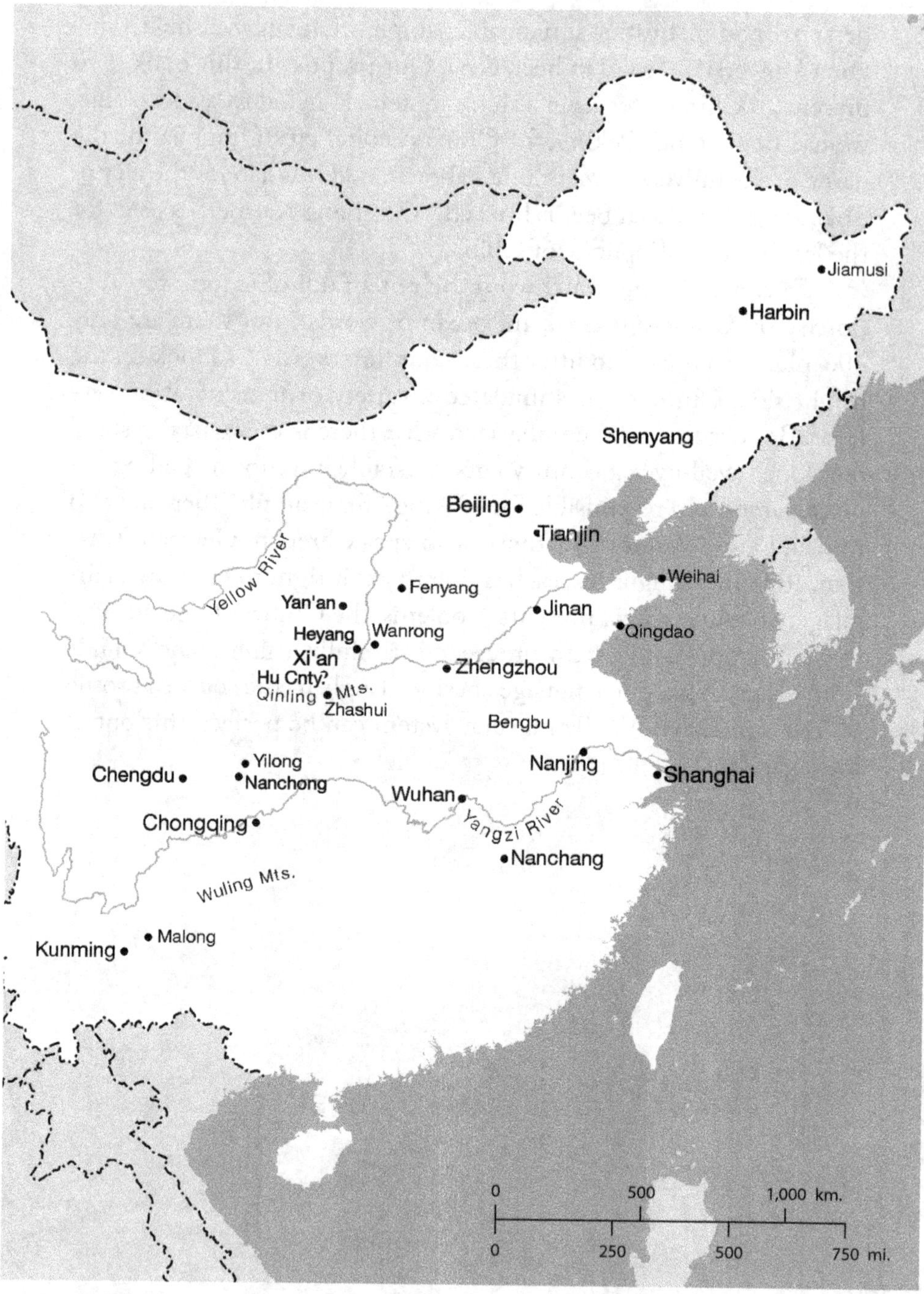

PLACES OF INTEREST

China in Family Photos

CHAPTER ONE

The Age of Scarcity

Hard Times, But Good Times

Introduction

Seldom do we see a reminiscence so appealing. Opening with stories about what the writer and her classmates had to eat (double meaning intended), this memoir moves on to youthful good times despite hardship. She looks back with pride on difficult times and is not surprised that things turned out quite well for her generation, even though they didn't have the advantages young people enjoy today. This was at the end of one of the worst periods in the People's Republic, the "Three Bad Years" (1959–1962), when millions starved to death from poor harvests and failures in distribution of scarce food resources. As the writer notes near the end of her memoir, soon after the outing to Dragon Gate, the Cultural Revolution began, and the opportunity to go to university was lost to all these classmates.

Related pieces here are "Chairman Mao Studies Economics During the 'Three Bad Years'" and "A Shaanxi Peasant's Critique of the Economy." These were also bad years for Chairman Mao, and these photos provide background to the Cultural Revolution a few years later, Mao's counterattack against those who disagreed with him.

Stories About Eating

These are all things that happened in the 1960s.

"People regard food as heaven" is an old saying passed down from our ancestors. Those who live in a time of material well-being perhaps can't identify with this, but people who grow up in a time of starvation and poor nutrition cannot separate themselves from such a feeling when they look back on their past.

I started lower middle school in the fall of 1960 (Photo 1). That was exactly the time of the "Three Bad Years," and especially during my first and second years, everything was hard to get.[1] Textbooks from the school's supply had tattered, yellowed pages. The electricity was off many nights, and if we wanted to study on our own at school, we had to take our own small kerosene lantern. If you didn't pay attention you would spill kerosene, and then you'd have to share a classmate's light. Sometimes if several of us worked beside an oil lamp, breathing became difficult because of the fumes. But by comparison, nothing was worse than the hunger, so what we wanted most was to eat. Among our grain rations there was very little rice; most of the grain was corn husks and wormy beans. Vegetables also were very rare, and even wild greens or edible tree leaves would get picked clean. In a month we would only get to eat meat once, and then only three or four pieces. So a piece of candy, a piece of meat, a few crackers with husks or bran, a bowl of plain rice, a piece of candied or salted fruit seed we might be lucky enough to buy, or some vegetable that glimmered with a spot of oil—any such simple thing as these would make us very happy. Young people today have a very hard time relating to such a desire for food, which comes from living through a long period of food shortages that made you think only of eating from morning to night. We never had to worry about making choices or even the possibility of having choices. Just to be able to eat was good in itself, and we were really fortunate if we got a full meal. Of course, what we longed for most of all was to eat meat.

Finally, the day to have meat would come. From very early on, everyone would become very excited, and the students would begin to ask each other, "Will it be 'red-cooked' meat, salted meat, or

---

1. The "Three Bad Years" were 1959 to 1962. This period resulted from the agricultural policies of the "Great Leap Forward" and, as much as from those policies themselves, from false reporting of production figures that misled authorities to believe food supplies were forthcoming. Thousands upon thousands of people starved to death; estimates of those who starved to death run as high as forty-five million by some accounts. This disaster led to challenges to Mao Zedong's leadership; the Cultural Revolution beginning in 1966 represented Mao's effort to regain full control of government and policy.

stir-fried meat with pepper?" What we were really thinking about all day was the flavor of the meat, and not at all about our classes. My desk-mate happened to be a Chinese girl who had lived in Indonesia, who was the only one in the class with a watch. So when we were in class, many notes passed through my hands, asking, "What time is it?" The whole class had only one thought: can we get out a little early today?

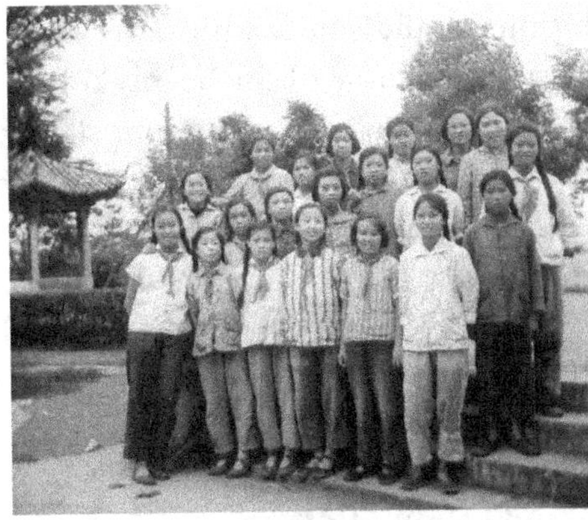

Photo 1: Our class in 1960, starting lower middle school.

At noon the Daily Life Committee would distribute the food. The usual procedure was for the students to serve themselves, but on the day when we had meat, the head teacher had to get into the action. All the students placed their bowls on the table, in a whirlwind of activity, and the head teacher would serve a piece of meat into each bowl, one at a time. If there were still a few pieces of meat left, the teacher would cut them into smaller pieces and serve them. If there were just a couple of pieces left, the teacher would toss them, and they went to the person whose bowl it fell into. If a piece fell onto the table or the floor, everyone just grabbed for it. I remember once when we had some dried fish that had been deep-fried. As it was being served, one piece fell on the floor. The teacher distributing the food said, "If it falls on the floor, it's not dirty—pick it up and eat it anyway."

Back then, every time we got something we could put in our mouths, it would make us extremely happy. The mother of one of our classmates worked in a shop that sold pickled vegetables. Once she brought us some fermented bean curd, and we picked it up with paper. A few of the girl students used matches to pick it up and eat it. Another student, whose father worked in a Chinese medicine shop, brought some dried herbs one day and gave them to his friends. We sucked on them in class after hiding them in our desks.

As compared with the boy students, the girls with their smaller rations would snack a bit more. This usually consisted of live worms saved from some of the beans. You could take them out of your pocket one by one and enjoy them slowly. When you were really hungry, you could put them in your mouth with a throat lozenge and moisten them. In the classroom next door to ours, they discovered something new: castor oil plants cooked in salted water were good to eat. Everyone clamored to try them, and the result was that half of the class had to go to the hospital because of the poison in these plants.

The Era of Patches

Now let's talk about clothes. First, this group photo of our class (Photo 2) was taken when we were in the second year of higher middle school. You can see me (in front and on the right) as a second-year student, a girl of seventeen or eighteen with two big patches on my pants and still standing right in front. I didn't even keep myself out of sight. If I had stood in the back row, no one would have seen those patches! This is the way we think and talk now, but at that time, no one thought or spoke that way.

That was a time when such tough simplicity was beautiful and even glorious. It was a national slogan that a piece of clothing was "new three years, old three years, and then patched for another three years." The kind of education we received at school taught us, "Although the working people have dirty faces and wear old clothes, and their hands are full of mud, they are most beautiful and clean. And as for the young women and the wives of the capitalist class, although they might look bright and smart, they are ugly and dirty."

Of course, people everywhere and in all times love beauty. Especially among women, who doesn't love beauty? I remember that our math teacher in lower middle school was very beautiful. Although her clothes were very simple, her skin was fair and pretty, and she wore her hair in braids that hung down to her waist. The overall air about her and her gracefulness made everyone think she was someone special. The girl students would often make comments behind her back and shoot admiring glances at her, but at school no one would dare to talk about wearing better clothes. They would be afraid this would be seen as an "unhealthy thought" or "seriously bourgeois thinking."

Among your classmates you would almost never see anyone

wearing new clothes, because most of our clothes were hand-me-downs from our brothers or sisters. I was the oldest in my family, so by comparison I got more opportunities to wear new clothes, but I certainly would not dare to wear anything new. First because I was concerned it would be too obvious, and second because it would be considered a sign of "bourgeois thought."

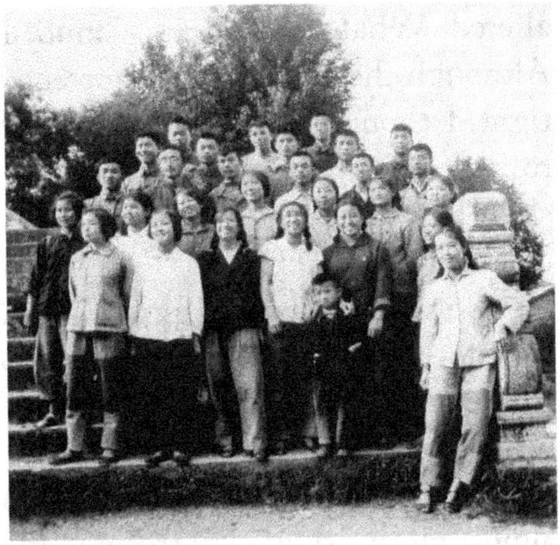

Photo 2: Our class in second-year upper middle school, 1964.

I remember once Mother gave me a new blouse and pants. I was adamant that I must wear an old shirt, and only then would I put on the new pants. Not only that, I wanted those two perfectly straight patches sewed onto the legs of the new pants, and only then would I dare put them on. There was a girl in our class who used fire tongs to curl her bangs. For this the Youth League Branch Committee called her in for a talk: the love of beauty became a risk, a luxury, and a burden.

I remember yet another thing: whenever there was some kind of activity at school, you were supposed to wear a white shirt. I had to borrow one each time. If our chorus was performing or there was a singing competition, it was relatively easy to borrow a white shirt. But if it was a school outing on National Day (October 1) or some other mass activity, it was difficult to borrow one. So one of the things I most wanted to own in my early years was a white shirt.

After I had asked many times, Mother got a white shirt for me. When I saw it I knew it was her own old white shirt that she had made over. But she told me, "This shirt is one that your father wore to attend a meeting in Shanghai!" Mother was telling the truth. Just about the time when Father was forced to put on that thorny cap of "rightist," he got an official invitation to an academic conference in Shanghai. So Father rushed back from the labor reform farm and hurried off to Shanghai, wearing this white shirt that Mother had altered. Whatever its background, I was happy to have this shirt. Although the sleeves were shortened and the shirt didn't really look right, I thought it was wonderful that I no longer needed to borrow someone else's shirt. After this, every time someone noticed this shirt, I always said—in the style of Ah Q—"We went to the big city of Shanghai to get it!"[2]

So when I started upper middle school, Mother made a white shirt for me out of poplin, with "a monogram on the collar." I treated it like a baby and only wore it when I wanted to dress up. Now I was no longer borrowing shirts from others, and instead others were coming to borrow mine! And when the shirt was returned it was almost always stained with something that was hard to wash out. I would use up the whole family's allowance of soap and still couldn't quite get it clean.[3] I had never imagined that having a new shirt might also bring new problems.

So with all this, you could think that our life was somewhat gloomy. Yes, our clothes were full of patches piled on patches, and there was almost no one who wore leather shoes. On days when it rained, rather than wear our shoes and get them wet, we would take them off and walk home barefoot. Yes, eating a piece of bread or fruit-flavored candy would bring us great happiness, and tying a shiny red ribbon to the end of your braid or having an unusual hair clasp was something most desirable. Still, the beauty of our lives was that we were vigorous and indomitable, and our youthful days were full of great, colorful dreams.

Our Great Adventure

You can look again at these two photos of us five girls: Zhao Nan, Ma Jing, Huang Sujin, Qiu Jiayi, and me. At that time, we were in our third year of higher middle school.

---

2. Ah Q was the main character in Lu Xun's (1881–1936) well-known novella, *The True Story of Ah Q*, published in 1921–1922. In the story, Ah Q is a village day laborer who habitually turns defeat into victory by reversing the logic of events in this way. Such "spiritual victory" or "psychological triumph" is now known as the "Ah Q mentality" in Chinese language and culture.

3. As this statement suggests, even everyday items such as laundry soap were rationed during that time.

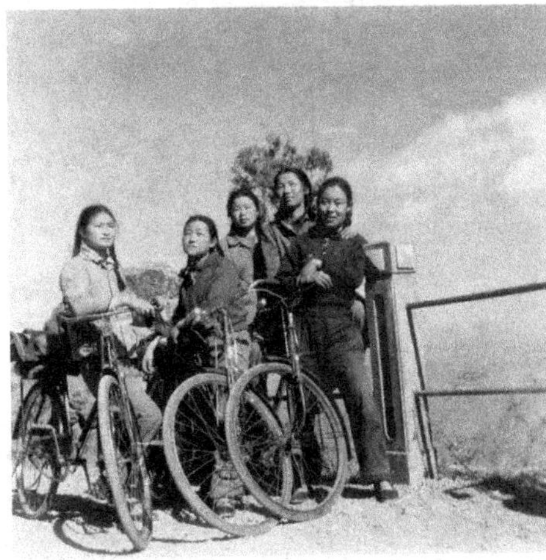

Photo 3: Nearing Dragon Gate on our outing—Dian Lake in the background.

One day we decided that on Sunday we would go up to West Mountain and Dragon Gate.[4] West Mountain is in the western suburbs of Kunming, on the banks of Dian Lake, a little more than ten kilometers outside the city. Trees that reach to the sky set off Huating Temple, Taihua Temple, Sanqing Pavilion, and other beautiful places. Of all of these, the greatest sight is Dragon Gate, laboriously carved out of the mountainside to hang suspended in air. To climb to Dragon Gate and look down at the vast Dian Lake and see the glittering water—this whole scene makes you feel relaxed and carefree.

We started early and made preparations so we could go and enjoy this sight. Because only two of the five of us could ride a bicycle, we borrowed two old bikes and took one from my home, and on Saturday evening we went to the sports field to practice riding our bikes. After a long struggle, the three of us who hadn't been able to ride could ride without falling down (there was still one who could ride but wasn't able to mount or dismount from the bike). We only got the job half done—we fell short in developing our confidence in bike-riding! So now the only question was how to leave our dorms while it was still dark and get out of the school gate, which was still locked at the time we needed to leave.

That evening, my parents as teachers left their door open and all of us slept in our family's apartment.[5] Then in the middle of the night we got up, climbed over the school wall, and went to another classmate's

4. The places listed here are key tourist sites outside Kunming. West Mountain towers above the lake, offering a dramatic view, is a natural site for temples and pavilions arrayed around the higher levels of the mountain.

5. Mention of this restriction suggests that even everyday life was closely regimented during that period.

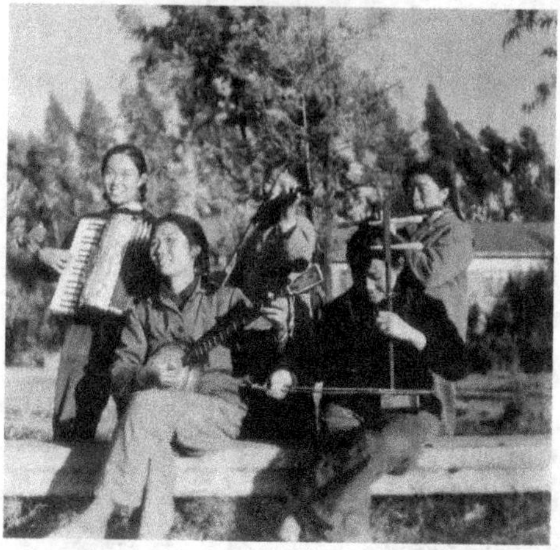

Photo 4: Making music after coming down from the mountain.

home to get the bicycles (he had let us put them there). Then we set out on our adventure, with two people riding on the bikes' carriers. We went along in our own way, stumbling and staggering.

The photo of the five of us gathered around our bikes (Photo 3) was taken when we had almost gotten to Dragon Gate. In the background you can see Dian Lake in the mist below. That day we set out under the stars and rode through daybreak until daylight. One of the bikes got a flat, so then we rode the rims. When we couldn't pedal anymore, we got off and pushed the bike, just struggling along as best we could. Finally you couldn't tell if we were riding the bike or the bike was riding us. We got so involved with simply getting along the road for Dragon Gate that we forgot to check where the sun was so we could enjoy the reflections off the lake. When we finally took a break and looked down at the lake, the sun was already high in the sky.

Photo 4 of us playing and singing was also taken that day. When we got back from West Mountain, everybody decided we still hadn't been crazy enough, so we had another round of fun back at school. At that time, we were facing the university entrance exams. Although it was practically time for the exams, as things happened the "Cultural Revolution" dropped on us out of the sky. Our rosy dream of university vanished into thin air, and our lives took a sudden turn! But we did indeed have a great third year of higher middle school![6]

6. The institution of university entrance examinations has been the most critical one for Chinese students as they near the end of higher middle school. That the author and her friends set aside their concern about the upcoming exam for a weekend adventure was regarded as daring if not reckless. The exam the author referred to was the last university entrance examination before the Cultural Revolution, and the examinations were not

Now I sometimes wonder how parents at that time could have been so relaxed and open. We were all in our third year of middle school, and we could be so bold as to go out and fool around like that, not worrying about the exam or even about our safety. If you look at kids today, they are carrying such heavy book bags, wearing such thick glasses, and constantly sighing as they repeatedly check their watches. People have advocated "basic education" for so long, and ultimately we haven't seen any improvement—maybe actually have slipped backward somewhat—as we try ever more earnestly to be "responsive." At that time our alma mater, Kunming First Middle School, was one of the best schools in Yunnan, and it turned out a lot of people with good morality, wisdom, and physical strength, good in every respect, who were able to help build a good society.

Thinking back on all this, in our childhood and youth we didn't have Kentucky Fried Chicken, McDonald's, chocolate, or ice cream, and we didn't have singers or movie stars as idols, or hand-held video cameras—and certainly our parents didn't have any conveniences. But unlike the children of today's forest of steel and concrete buildings, we had an expanse of blue sky and a boundless green Earth. Who could deny that? Although we dealt with some hardship and were made to exercise self-discipline, we grew up with good spirits and sound health.

This is by Zhu Xindi. It was published in Old Photos *no. 51, pp. 93–100.*

Epitaph to the Era of Scarcity

Introduction

This appealing memoir, presented from the writer's viewpoint as he began middle school, captures several aspects of his family life. The initial photo, taken the day before Mao Zedong's death, provides the starting point. The narrative concludes about two years later as the Cultural Revolution officially ended and life began to take on a more

held again until 1977.

rational footing than in the previous decade. The writer identifies somewhat more with his mother than his father as he comments on her work situation, how she "commuted" on foot, and how the purchase of a bicycle made her life a bit easier. While these conditions seem difficult to American readers today, they were common for Chinese women of that generation and for another couple of decades thereafter.

The parents here were university-educated, and the writer's comments on their lives as students disclose their high ideals, the books they read as university students, and the low-grade paper on which the books were printed. As was common for their generation of students, they also used a sickle to work in the fields. In the last photo of this set, the father appears relaxed as he anticipates better times ahead. Finally, the writer compares his younger brother's outlook to his own, noting the difference even a few years might make.

The Text

My mother remembers clearly that this photograph (Photo 1) was taken on September 8, 1976. This is easy to remember because it was the day before a great event happened in China: the death of Chairman Mao.

At that time, it was really extravagant for an ordinary family to have a camera. This photo was taken by our Second Aunt who had brought a camera when she came to visit Grandmother. In this photo, Mother had just gotten off from work, and she looks a bit worn out. She hasn't even had a minute to take off her work-cap, although she had already put on the dacron jacket Father bought her when he made a trip to Shanghai. My little brother was a year and a half old, and wearing a sun hat and sitting on the wooden seat at the front of the bike. It seems strange that he's looking right into the lens. I'm sitting across the back carrier of the bike wearing a worried expression, like I'm not happy about what's going on. Surely my expression was like that because I had just been forced to stop playing and go home. The bicycle in this photo is something that Mother really loved, and it was an important resource in our home at that time. Thirty years have passed, and this old, out-of-date "East Is Red" bike is still tucked carefully away at

Mother's house. It is called "the Grandpa Bike."

In 1974, Mother worked at a chemical fertilizer plant in the county seat where she did laboratory testing. To get to work every day she had to walk more than ten *li* (about three miles). Then in the evenings she often worked the night shift as well. At the time she was pregnant with my younger brother, so all this walking was not easy. So Father set out to buy a bicycle for Mother. In those days everyone thought buying a bicycle was a "big deal." The famous brands like Phoenix, Forever, or Flying Pigeon were not easy to get. You had to have a coupon from your unit. Mother couldn't even think about buying one of those famous brands, so she simply bought an "East Is Red" bike that was made in Shenyang.[7] Father took Mother some thirty *li* (about ten miles) by bicycle and it was after eight in the evening when they got home. To make it easier for Father to use, Mother was careful to buy a "man's 28" model. The one hundred *yuan* necessary to buy a bicycle took almost all our family's savings, and they also had to "borrow" thirty from my grandmother.

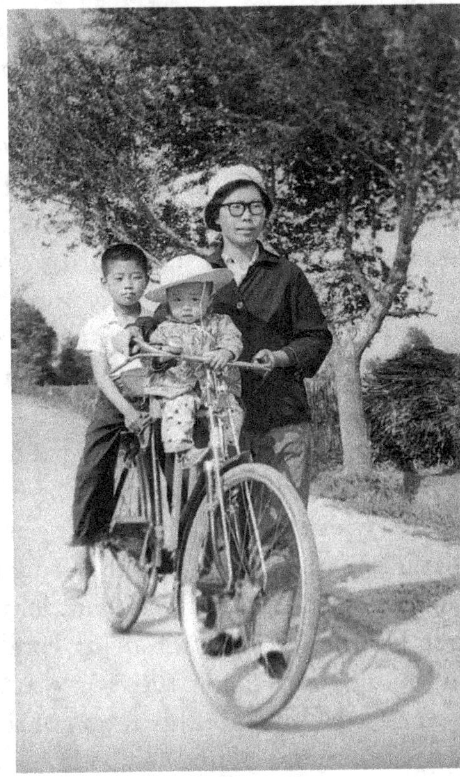

Photo 1: Mother with her new bicycle.

The first few days, Father very solemnly decked out the bike: the handlebars and the frame all had a colorful plastic ribbon wrapped around them, the axle sleeves were painted in alternating red and

7. During the heyday of bicycle use in China, the three brands mentioned did indeed have high reputations, which they maintained for decades. It is clear that the local "East Is Red" brand served the family well, as the writer mentions earlier that it has a place at his parents' home after many years.

green circles, and the bolt of the bike's lock was wrapped with a little plastic goldfish. Mother then began to learn how to ride the new bicycle. Every evening she looked forward to the chance to ride her bike after dinner, and Father went with her every time for this "teaching and practice." After Mother got it down pat, she wanted to have a photo taken of her riding the bike. By the time she was able to have such a photo taken, two years had passed, and the "East Is Red" bicycle no longer had the nice shiny appearance it had at first.

By the mid-1970s, the economy of ration coupons had become somewhat relaxed, but grains, non-staple food items, and many everyday items were still rationed with a certain amount for each person. The variety of ration coupons seemed endless. In Liaoning Province, the local system allowed three *liang* (a few ounces) of cooking oil per person, which was well below the national average. Mother told me that in the month she was recovering after I was born, the wheat allowance she used was the amount for the whole family for a year. My grandmother recalled the living conditions during her active years and jokingly said that in those times ordinary people even had to "save their hearts." The monthly ration of rice was just a few *jin* (one *jin* is very close to one pound) and a few *jin* of vinegar. Yearly allowances were just a few feet of cloth, a few *jin* of oil, and you could only get a few pairs of rubber shoes or eat only a few pieces of *doufu* (bean curd).[8] The entire plan might be much the same as a family would decide for themselves, but the government had it all figured out for you.

But then, even though you had coupons to buy something, you might not be able to find the items you wanted to buy! In 1972, when my maternal grandfather was nearing death, he kept asking for a *doufu* dish that he loved to eat. At that time, you could only buy *doufu* at the staple foods store operated by the local government. You could only go once every few days—and not go at all if it was really cold. If you got there when it was available, everyone who wanted some would snatch it up so quickly that things became chaotic. Mother

8. The Chinese term for bean curd, *doufu*, is used here in order to accustom readers to using it. This food item was introduced into the American diet as *tofu*, which means the same thing in Japanese. Japan is often identified as the source of this food even though Americans have consumed *doufu* dishes in Chinese restaurants for many years. China and Japan have shared bean curd for many centuries.

and Second Aunt tried desperately to get this for their father and made four trips to that staple foods store every day during that sad time. Grandmother held the irrelevant *doufu* coupons, crying bitterly that she couldn't fulfill her husband's last wish as he was about to die.

Every year at the Qingming Festival, my aunt would angrily exclaim that she wanted to place a serving of "boiled *doufu*" on her father's grave.[9] To this day at Mother's house, she keeps the "Certificate for Resident's Allotment of Grains and Oil." Your eyes meet the impressive words, "Highest Directives" printed on the envelope: ". . . on the issue of economizing on grain and foods. Everyone must manage things carefully; quantities are given per person. During busy times eat more, and eat less when things are more relaxed. During busy times eat dry foods, and half-dry, half-moist when things are not busy. For variety you may choose things like sweet potatoes, green vegetables, carrots or turnips, melons or beans, and taro. These matters must be grasped very carefully."

I remember when we wanted to buy vegetables we had to go to the state-operated vegetable market, which was a wooden-canopied open-air store with four sides. There were about ten large "rooms" within this space. The order of business for buying vegetables there was that you first went to a simple metal-sided building and got in line, and there you exchanged the cash in your hand for "vegetable tickets" for the vegetables you wanted to buy (these tickets were made up by the market itself). Then you took the numbered "staple items certificates" to the big canopied building and lined up to wait again. When the vegetables hadn't yet arrived, everyone noted their place in line and then took care that no one jumped the line on them. Once the vegetables arrived, the line at the counter became chaotic. Often the vegetables didn't get there on time. The big truck carrying them would get stuck in the mud, and the driver simply didn't contact the market. It was common to have to wait for several hours.

One year just as winter was coming on, I went with Mother to buy potatoes. We stood in line the whole morning and the potatoes

9. Qingming Festival is the day when families clean their ancestors' graves. The two characters in this term can be translated as "pure brightness." This festival was traditionally celebrated in the third solar month and is now observed on April 5.

hadn't arrived. Everyone was freezing, and with worried faces they stamped their feet to try to stay warm. Later we heard that because it was so cold and the road was slick, the freezing road surface caused the mules that were pulling the trucks to injure their backs so they couldn't do the job.

After you got the items you wanted, officials at the market wanted to put a red stamp on your "staple items certificates" so you couldn't buy those items again that day. Radishes or cabbages weren't really staple items and could be returned for cash. They were weighed again when sales closed out and could be thrown away after a day. Those who had the coupons—or just knew the right people—could buy such things and resell them. Who would have thought that fresh vegetables could be hoarded for sale on the black market! Just wait a bit and you could buy these vegetables without standing in line, or you could use ordinary meat coupons in exchange for a better grade of animal fat. It was enough to make you incredibly angry.

As I recall, our family was relatively well-off in supplementary coupons. If a family was small they couldn't have new clothes very often. When I started middle school, I often wore hand-me-down clothes from the adults, which usually had been repaired several times. It's worth mentioning that Grandmother still had several bundles of "foreign clothing" (these were all old items our family had accumulated when they were in business). During the difficult years these old clothes came in handy any number of times. A family's cotton-padded jackets, cotton pants, quilts, mattresses—almost all these "antiques" were used. Mother used some blue denim (*yanglanbu*) to make a pair of pants for me. When I put them on, they felt a little strange: the feel of them, the color, what others were wearing—all were clearly not the same as things of that period. It was as if they had a whiff of "the old society," and I was afraid my classmates would make fun of me. Still, when I happened to get a new piece of clothing, Mother always bought something that was a little too long. She said my brother and I were nine years apart, and it was a pity if you had used clothes that no one would wear!

During 1977, the dark haze of the "Cultural Revolution" gradually dispersed, the mood of society began to brighten, and all kinds of good news kept coming in. There was the slogan "respect knowledge, respect talent," and after the groups who had volunteered to go to rural areas came back, Father, who taught at the middle school in town, was always sighing and happy. Another good result was that both my parents got an increase in their salaries: Father's had been 43 *yuan*, and it went up to 47.5. After Mother left the factory and went to the Cultural Office in town, her salary went from 38.6 to 43 *yuan*.

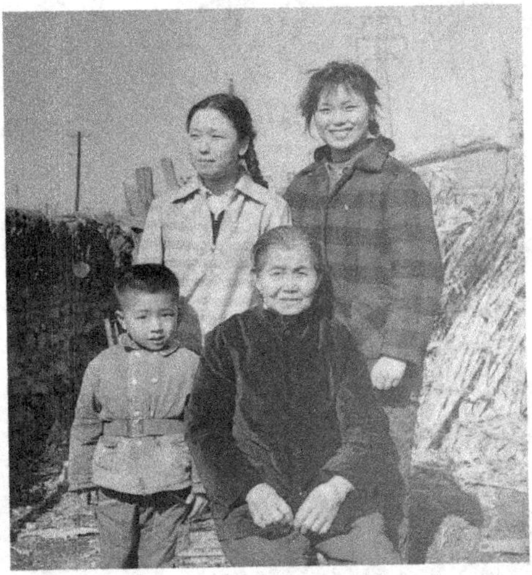

Photo 2: Mother and our aunt and our grandmother, with my brother.

When she had some leisure time, Mother began to straighten up the bookshelf that had gotten dusty from being neglected so long. The books were from the years 1960 to 1964, when Father and Mother went to Liaoning University and majored in Chinese literature. There was *The Literary Mind and the Carving of Dragons*, the *Book of Poetry*, the *Book of History*, and novels such as *Song of Youth* and *And Quiet Flows the Don*. Then there were books by Guo Moruo from the Great Leap Forward period, *The Camel Collection* and *Ballads of the Red Flag*.[10] The paper was gray and coarse, and you could clearly see little bits of rice straw in it. On the lowest shelf I could see the sickle that Mother and Father had used when they were in the volunteer corps, and although it showed signs of use, with rust and stain, it was

10. Two of the first three items mentioned in this list of books, the *Book of Poetry* and the *Book of History*, are among China's traditional classics. *The Literary Mind and the Carving of Dragons* is a major work of literary theory from the fourth century CE. *Song of Youth*, by Yang Mo, a woman writer, was published during the 1950s and showed how patriotic students of the 1930s fought against the oppressive Jiang Kaishek Nationalist government. *And Quiet Flows the Don* presents both internal and foreign struggles involving Cossacks over the wartime generations of the twentieth century. Read widely in many countries, this novel won the 1958 Nobel Prize for its author, M. A. Sholokhov. Guo Moruo, a major literary historian and theorist, produced the two collections of modern Chinese poetry mentioned last.

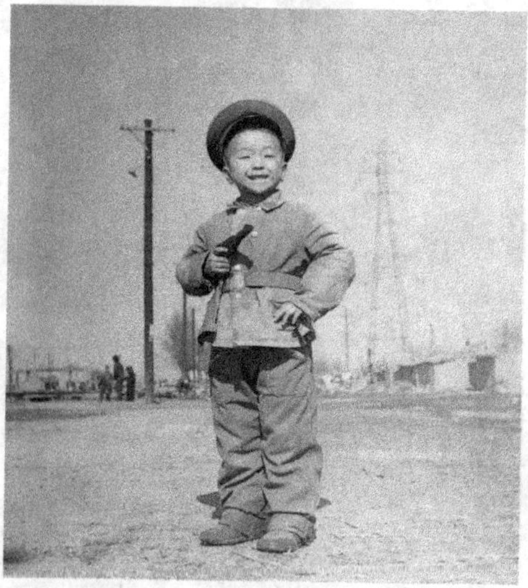

Photo 3: My little brother proudly showing off his soldier style.

solid visual evidence that they had been among China's first student farmers who wanted to follow the examples of Dong Jiageng and Xing Yanzi by going to the village "to struggle for heaven and earth" with "great ambition and lofty ideals" instead of taking up their work.[11] Having gone through so many stressful days and political movements (especially the Lin Biao affair), they had determined hearts and could not be surprised by anything.

Soon after the Spring Festival had passed in 1978, another news item of great interest came out. There would be a "National Conference on Science," and the frozen rationing system would begin to thaw out to a certain degree. In this town in the northern suburbs of Shenyang, the first watchword was that the black market had come back, and items such as matches, whitewash, light bulbs, and soap would not require coupons. Long lines formed outside of restaurants to buy *youtiao*, fried dough sticks. Grain coupons were still the most important kind of certificate, though food had begun to be somewhat more plentiful. Some people had begun—more or less semi-publicly—to trade eggs for food coupons. A little earlier this had been somewhat risky.

---

11. Dong Jiageng and Xing Yanzi were model youth of the writer's parents' generation. Dong left his home in Jiangsu to study at Beijing University, only to abandon that ambition and return home to contribute to building a better life in his village. Decades later, Dong was recognized as one of China's fifty exemplary model agricultural workers. Xing Yanzi was a young woman from a village near Tianjin who did much the same kind of thing as Dong Jiageng. She organized a brigade of other young women who worked the fields, and then as a crowning achievement she built a canal for their area, a very demanding project completed in difficult conditions. Xing served as both a local and national leader over the following decades.

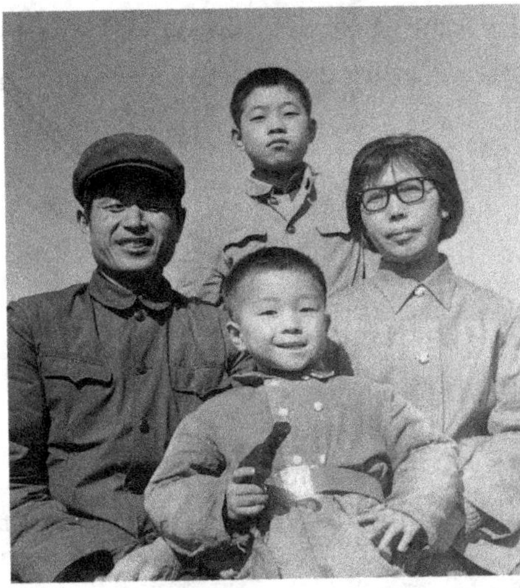

Photo 4: Our family in 1978, as life began to return to normal.

One Sunday morning when two cousins who had just returned from the volunteer corps came to visit with Grandmother, Mother suggested that we take a picture (Photo 2). With the improvement in living conditions, our family had just bought a "Seagull" brand 120-millimeter camera. The weather outside was not yet really spring, but the worst part of winter had definitely passed. My little brother was wearing new long pants with the cuffs rolled up several times and an old jacket that was both too small and too short for him by that time. On its collar the jacket had a "red neck badge" with a quotation from "the little red book" of Chairman Mao's sayings. He was also wearing a child's military dress hat and an adult-sized wide belt, also military style. In his hand he held a wooden pistol like some soldiers carry (Photo 3).

Photo 4 is our family at that same time, a snapshot that Grandmother took. Mother, who had just turned thirty-six, is clearly sad, with a glint of tears visible behind her glasses. She had just received a letter with news that a classmate, a friend of many years, had suddenly died, and she couldn't hold back the tears. Father looks very satisfied, with a smile on his face knowing that he won't again feel desolate and depressed. He squints his eyes as they bathe in the bright light of 1978.

That year I started middle school. The schools had resumed the entrance examinations, and our teachers wanted all of us to grow up and be scientists. The essay topics they gave us were often on subjects like "life" or "ideals."

This is the end of the story of my youth. I sometimes envied my little brother, who was always happy and carefree, with some new dream every day, and always full of hope. A new era had begun!

This was written by Li Fan. It appeared in Old Photos *no. 51, pp. 101–108.*

A Photo of My Father

Introduction

The brief description of this photo is melancholic, almost fatalistic. The young writer seems to know little about its background. Still, there is an appreciation of the father's simple, hard-working life and a realization that the father had managed to make a step into city life.

This photo has a history of its own. The editors of *Old Photos* always returned photos and other materials sent in by readers. But when they sent this photo back to the writer, the post office was unable to deliver their letter because the writer had moved. They continued their efforts over the years, always to no avail. This story has a happy ending—for which see the last section of this book!

The Text

My father is a common, ordinary peasant, and his life has been like that of the yellow earth, always remaining the same.

Father is indeed a peasant, but he has not been tied to the land and his village. Throughout his life he has moved around in order to make a living. When he was twelve, he started working in a construction group and learning the skills of construction work. This was a hard life, so he spent much of his life away from home and would return only once in a long time. Throughout his life he poured cement, laid brick, did carpentry work—all these manual skills. He did this kind of work as long as he was physically able.

It's sad that there are so few photos of Father, and most of these are simply views of his face and upper body used for various kinds of documents. This is the only photo of him that I have ever seen showing him in a scenic background. When I pulled this photo out of an old file, I was amazed that my father, now so stiff and grizzled,

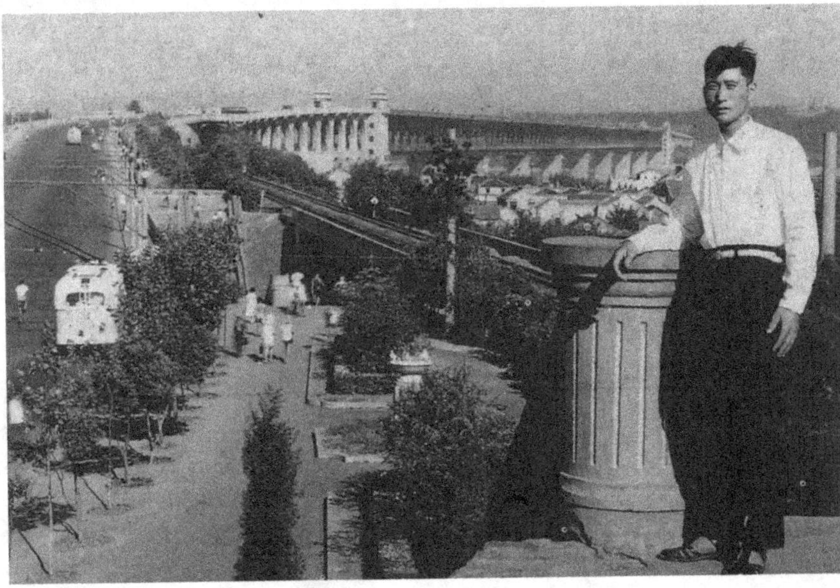

was once a young man who looked so strong and handsome. Father would have been twenty-eight when this photo was taken, but he appears to be about twenty years old.

Father was wearing a white shirt, blue trousers, and a pair of leather sandals. His shirt was tucked into his trousers, which might have been a style for city youth during the 1960s. From his appearance, Father looks only a little like a peasant. Because he had already begun making his living in the city, he had taken on the look of a city person.

As Father stood beside the bridge back then, he gazed downheartedly into the distance, perhaps because he was tired of his unsettled life, or perhaps he was under such pressure in life that he could hardly ever relax. The Wuhan bridge over the Yangzi River looks quiet and almost desolate, with few vehicles or pedestrians making their way across.

Now, when I stand beside this busy bridge here in Wuhan, the scene has changed from those earlier days. After thirty-three years, almost everything has changed so much, and the young man who stood beside the bridge looking so handsome has become an old person.

Father has indeed grown old, and he might have forgotten many things from earlier years—including this photo taken at the bridge—so I haven't asked him much about this photo and how he got it. Father can never be young again, so I try to recover his earlier life. Everyone who is old was young once, and all young people will become old and, finally, will be removed from this earth.

This is by Qiu Sanbao. It appeared in Old Photos, *no. 10, pp. 92–93.*

Sewing Machines in China

Introduction

This brief piece suggests how an everyday foot-powered machine helped transform the Chinese economy, providing thousands upon thousands of jobs and becoming a fixture in as many homes as could afford to buy a sewing machine. The article also relates the simple but ingenious methods by which foreign manufacturers built their infrastructure in China over several decades. Only toward the end of the twentieth century did the sewing machine begin to lose its role both in industry and as a fixture in households ranging from the well-off to the relatively poor. Even today as the urban economy becomes ever more consumer-oriented, one can find many sewing shops that repair clothing in residential and industrial areas and around university campuses.

The Text

Sewing machines were invented by Germans in the early nineteenth century, at about the same time as the bicycle.[12] On the page of text for this advertisement for the Likang Sewing Machine Factory, there is an illustration of the factory. From the captions you can see that this factory was located in the German city of Durlach, and at the time of this ad the factory was about eighty years old, dating from

12. Efforts to develop a workable sewing machine occupied many in Europe and the United States from the late eighteenth through the middle decades of the nineteenth centuries. Some workable devices could perform only to meet the needs of a particular job. Elias Howe received the earliest US patent for a sewing machine in 1845. Howe traveled to England in hopes of greater success than he was having in the years after receiving his patent. These ongoing activities suggest close parallel developments in Europe and the United States.

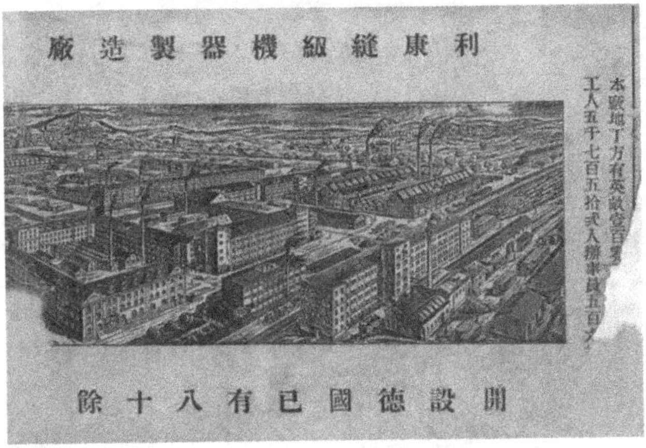

Photo 1: An overall view of the Likang Sewing Machine Manufacturing complex.

about 1830. When this ad appeared in 1914, the factory occupied more than one hundred acres, employed more than 5,700 workers, and had 560 managerial personnel. It was almost certainly the world's largest sewing machine factory at the time.

While it is difficult to determine when sewing machines first reached China, this advertisement appeared in a magazine published in 1914. According to the text of the ad, there were Likang Sewing Machine distributors in thirteen cities and towns in China, including Shanghai, Hangzhou, Hankou, Qingdao, Jiaozhou, Jinan, Nanchang, Ningbo, Shaoxing, Wei County, Zhou Village, Shahe, and Jimo. With Likang sewing machines so broadly established in China by 1914, sewing machines must have been introduced well before that year. Of course the spread of sewing machines was rather slow and depended on changes in Chinese styles of dress. With the overthrow of the Qing government and the establishment of the Republic of China, the "Zhongshan suit" and Western clothing were generally introduced, and this increased the use of sewing machines among the Chinese.[13]

The text of the ad includes this statement:

---

13. The "Zhongshan suit" is named for Sun Yatsen (1866–1925), known universally in China as Sun Zhongshan. This style is actually the term used for what Americans and other Westerners have called a "Mao suit." Mao Zedong and the other Chinese Communist leaders regularly wore this style, in part to show respect for Sun Zhongshan.

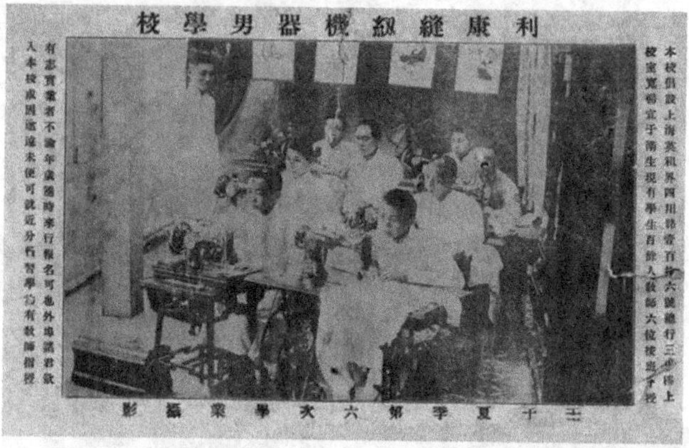

Photo 2: An advertisement for the foreign-made Likang sewing machine .

> Now that the Chinese Nation has established a republic, everything is new, and Western-style clothing shops have become common throughout the country. There are none who do not love the great convenience of these styles. It isn't convenient to make the clothes we like, so we have to buy our Western-style clothes from a store. To avoid such inconvenience, we need to invent various kinds of sewing machines and make all kinds of Chinese and Western clothes for ourselves. You can get what you'd like to have, and you can make it very quickly. Things can be made with detail and very sturdily, and it's so much better than making things by hand.

The presence of sewing machines in China began with family-run tailor shops. Later there were large and medium-sized sewing factories that could sell in larger quantities.

The Chinese compradors for the Likang Company advocated the strategy of "progress through industrialization" for enriching the country and its people.[14] In their ads they urged, "We want our brothers and sisters of the Republic of China to rouse their spirits and quickly learn the true arts of industry. If you will learn how to

14. A comprador was a native Chinese manager who worked for a foreign business firm in China, a term used until 1949. A classic study of this topic is Yen-p'ing Hao, *The Comprador in Nineteenth Century China: Bridge between East and West* (Cambridge, MA: Harvard University Press, 1971).

operate a sewing machine, you'll soon see unlimited benefits. Adopting this simple machine can help us get back our rights and create a rich country and rich family." To the present day, this manager's idea about using industry to save the country is still encouraging people to establish clothing factories in some regions in China.

The Chinese comprador of the Likang Company was an intelligent businessman, and in the text of his ad he included a cleverly worded "Three-Character Classic" on sewing machines that was easy to read and remember:[15]

> It's a machine to make clothes, invented in Germany. These machines are of good material and are made well. They're found throughout the world and have a good reputation. Easy to use, and easy to learn. It is top-grade, this Likang brand. The thread won't jump, and the needle won't break. Clothing, shoes, and socks all can be made well. It also works well with leather products. Scenic sights can be embroidered as if they were real. And you can also do flowers and grass.
>
> Come and see for yourself, and decide if you like it. Buy at a guarantee or on contract for a period of your own choosing. You've got seven years, and if something goes wrong it can be made as good as new. We have both male and female teachers who will clearly show you our methods. You can pay all at once or buy over time. The period can be limited to ten months, or you can buy one machine after another and then settle up. We treat you well; please come in and make a deal!

For the Model 28 Likang Home Treadle Sewing Machine, the cash price was fifty-four silver dollars, or sixty dollars if bought in installments—a modest price.[16] Classes were offered everywhere the

15. The "Three-Character Classic," or "Trimetric Classic," is a thirteenth-century Chinese text. It held a secure place in Chinese tradition as a text for inculcating moral and language instruction. Basic ideas were presented in sets of three characters with simple, easy-to-memorize rhymes.

16. For much of the Qing period, international trade in China was conducted according to an international standard that used silver dollars as its basis. This began when Spain

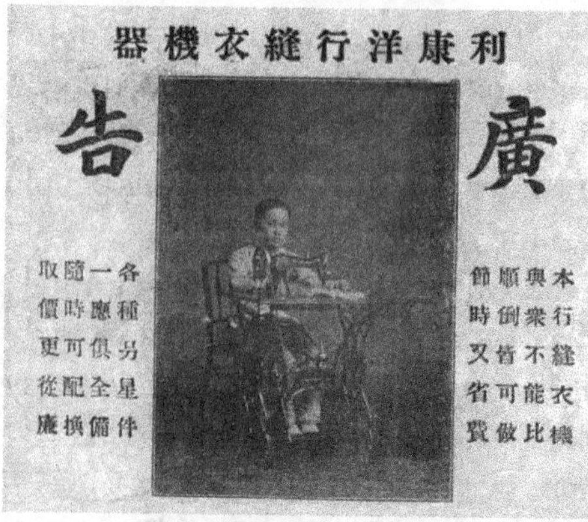

Photo 3: "Graduation photo" from the Likang Sewing School.

machines were sold, and buyers could sign up to join these classes. The sewing class for men and boys shown in the photo met on the third floor of the Likang business office at 116 Sichuan Road in Shanghai. There were more than one hundred students in the class, with six male teachers working with smaller sections. You can imagine the almost explosive sales of Likang sewing machines in China at that time. There were many Chinese children who lacked the money to go to school. Therefore, they took the sewing class as vocational training to have a way to make a living.

Many years later, in the 1970s and '80s, the sewing machine was always listed as one of the "three wheels" that people regarded as essential for marriage—a bicycle, a sewing machine, and an electric fan. Added to these was "one sound," a radio, to complete the "four main items" for marriage. For many decades the Shanghai brands Honeybee and Butterfly (earlier called Invincible) were things one could not purchase without ration vouchers. Generally speaking, beginning in the early 1990s, buying ready-made clothes became the custom in Chinese life, and sewing machines gradually faded out. Many small tailor shops also were absorbed or put out of business by large-scale clothing factories. Workers in the industry began to use electric sewing machines, and the once-flourishing sewing machine factories were converted to other kinds of production one after another.

The great era of home sewing machines became part of the past.

This is by Tan Jintu. It appeared in Old Photos *no. 37, pp. 150–153.*

---

was a powerful trading nation and the coins were "Spanish silver dollars." By about 1800, the silver dollars came from Mexico (of all places) and were called "Mexican silver dollars."

CHAPTER TWO

Currents in Revolution

Zhang Shenfu: A Forgotten Founder of the Chinese Communist Party

Introduction

This article offers much information and insight about an individual who is not well known outside of China. Zhang Shenfu was an early leader of the developing Chinese Communist organizations in Europe. He was the agent through whom two major figures in the history of the Chinese Communist Party, Zhou Enlai and Zhu De, joined the early Party. Yet Zhang himself left it soon after introducing these two individuals. As detailed here, Zhang was an intellectual, and this was how others perceived him. He cared about radical socialism but also had a deep interest in Bertrand Russell and other European thinkers of his time. As noted in the article, this characteristic kept him from being a good revolutionary.

This article is also about Zhang's relationship with Liu Qingyang, with whom he traveled by ship to France in 1921. Theirs could be called an activists' or revolutionaries' romance. Their relationship resembled marriage, but as Zhang says here, they were never formally married. Zhang himself sometimes put their relationship in jeopardy, usually because of a new romantic interest that he had developed. So he is known as a serious advocate of women's liberation, but also as a libertine or womanizer.

We have not referred to other resources for learning about our subjects in any other of these introductions. For Zhang Shenfu, such mention is appropriate. Vera Schwarcz, known for her work on the intellectuals of the May Fourth Movement before and during 1919, met Zhang Shenfu and became interested in his career. The result is *Time for Telling Truth Is Running Out* (New Haven: Yale University Press, 1992). In this very personal study of Zhang, Schwarcz pays great attention to Zhang's reflections on his life and career. Such reflection is a major theme in these translations and in *Old Photos.*

The Text

The complicated love and family relationship between Zhang Shenfu and Liu Qingyang began when they traveled to Europe to study. At the beginning of 1921, Zhang and Liu shared a voyage of thirty-five days to Marseille and then traveled on to Paris. In Paris they lived in a stylish residential building that became the meeting place for early Chinese Communist groups in Paris. Zhang Shenfu could afford this because Cai Yuanpei had arranged for him to be a teacher in the Sino-French Education Society at a salary of eight hundred francs per month. This was more than twice the stipend for a Chinese student in the Diligent Work–Frugal Study program.[1] Thus Zhang was free to study, give lectures, participate in leisurely discussions, or write.

Zhang Shenfu was born on June 15, 1893, in Xian County, Hebei. In 1911, as a young man inspired by the revolution of that year, he returned home to get married. His new bride was named Zhu Huiyi, and on her behalf Zhang changed his own name to Zhu Deyi. Although this was a traditional arranged marriage, the two of them were well-matched. In 1914, his wife died painfully a few days after giving birth. A comment he made soon afterward showed Zhang's love for her: "If she had not died so young, I wouldn't have thought of trying to find another woman."

In the spring of 1915, Zhang married a second time, again

1. The Diligent Work–Frugal Study program had been launched by Wu Zhihui, Li Shizeng, and others who had studied in France in the early 1900s. These individuals became anarchists through their experiences in Europe, and this program is usually seen as an offshoot of the early anarchist movement in China. After World War I, the program was resumed but lacked financial resources. Hence the kinds of funding problems indicated here and the student protests against the program's leaders mentioned below. For further detail see Marilyn A. Levine, *The Found Generation: Chinese Communists in Europe during the Twenties* (University of Washington Press, 1993).

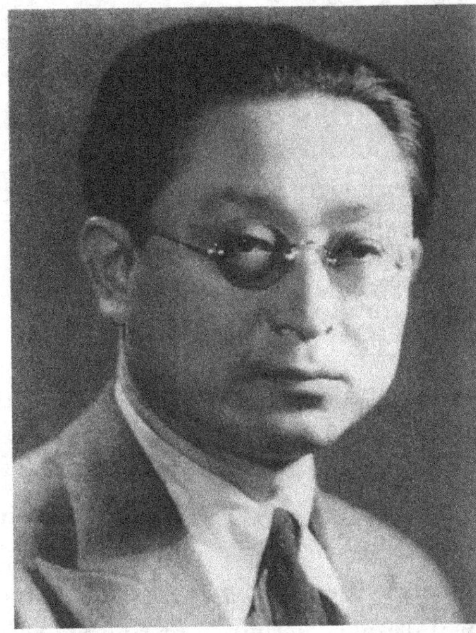

Photo 1: Zhang Shenfu as a professor at Qinghua University during the 1930s.

arranged by his family. He said that he was "seriously deceived." Although as a young man he was progressive-minded about marriage and family, Zhang was reserved in his actions. Liu Qingyang was stubborn about marriage. In 1916 when a young man wanted to marry her, she suggested an unusual way to decide. She said she would suggest a topic for an essay that they both would write, and if the young man's essay was better than hers, they would be married. While this is probably only a story to suggest her independence, the young man didn't have such ability and he was embarrassed. So after Liu's older brother intervened, this relationship ended. All of this was in the background when Zhang Shenfu met Liu Qingyang during May Fourth activities in 1919.

By the spring of 1917, Zhang Shenfu was invited by Cai Yuanpei, president of Beijing University (Beida), to become a professor of logic.

At Beida, Zhang heard about Liu Qingyang from Chen Duxiu and Li Dazhao. He knew this young woman from Tianjin was a very capable leader of the Consciousness Society, and she was an excellent speaker who could rouse a crowd at a street demonstration. This was an important reason why Zhang and Li Dazhao introduced Liu Qingyang to the Beijing Communist cell. As Zhang recalled,

> We were always looking for people to develop as Party members. Liu Qingyang was a person we most wanted to bring along as a leader. She had performed very well during the May Fourth activities; she was a student leader among

> the women. The National Student Alliance was established in 1919, and when they set out in July of 1920 to raise funds in Southeast Asia, they decided to send Liu Qingyang as one of the group for this task. . . . When she came back to Beijing in September, Li Dazhao and I asked her to come to the office of the library director to talk, but she wouldn't come. At that point she didn't wish to make a greater commitment. . . .

From this you can see Liu Qingyang was very independent and self-centered at that time. She was "especially independent."

Liu Qingyang was born in Tianjin in 1894. The obituary published by *People's Daily* upon her death in August 1979 outlined her career. She accepted progressive thought as a youth, joined Tongmenghui activities during the Revolution of 1911, and later joined in the revolutionary activities of the May Fourth Movement. She worked with Zhou Enlai and Deng Yingchao to organize the Consciousness Society in Tianjin and was a leader in the student movement there.[2] In November 1920, she joined the Diligent Work–Frugal Study program in France. After returning home, she organized the Patriotic Women's Association and remained active, especially in women's organizations, over many years.

In 1920, Chen Duxiu, Li Dazhao, and Zhang Shenfu often talked about how to build the party.[3] Before going to Europe, Zhang visited Chen Duxiu in Shanghai, partly to bid farewell and partly to develop a consensus in their views. Chen Duxiu delegated Zhang Shenfu to build the party and recruit new members in Europe.

---

2. Like so many of China's twentieth-century leaders, Zhou Enlai began his career as a revolutionary while a university student and in the atmosphere of nationalism against Japan. Zhou graduated from Nankai University in Tianjin, where he met Deng Yingchao, whom he later married. Zhou and Deng's careers intersected with Liu Qingyang's repeatedly over several decades, first during their sojourns in Europe.

3. Chen Duxiu and Li Dazhao are regarded as the founders of the Chinese Communist Party. Chen was on the Beijing University faculty as a professor of philosophy; Li was the university librarian and a scholar of history. The Party was not formally established until the summer of 1921, and in this early period was made up of a number of "study societies" or "cells." [I have changed "small group organizations" to "cells" a few times in this article for brevity.] As described in this article, the early leadership was largely personal and party discipline had not yet developed.

Photo 2: During their period in Europe, this photo taken in Berlin shows Zhu De (first on the right in the front row) and others with Zhang Shenfu (first on the left in the front row).

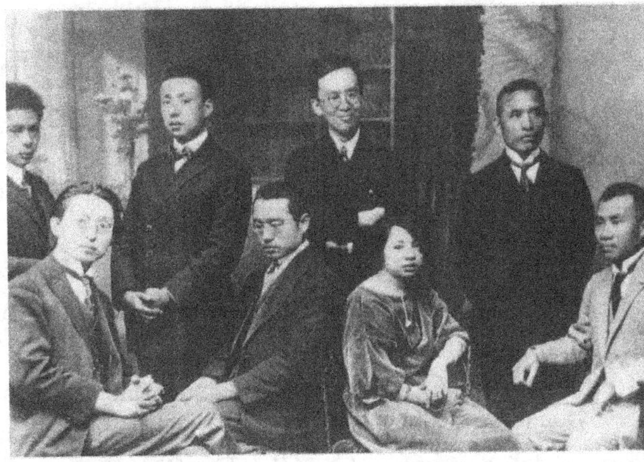

So it was that Zhang and Liu traveled to France together. On the ship bound for Marseille, Liu Qingyang straightforwardly declared her love for Zhang Shenfu. Zhang said, "We began living together in 1920. It was on our voyage to France that she first told me she loved me. She accepted my views and followed my lead. I persuaded her to join the Party. . . . When we returned from Europe, she lived with me—at first it looked like she was a concubine, but we didn't care what other people thought." The feelings between them developed especially quickly. Before she boarded the ship on November 23, 1920, Liu Qingyang was still a virgin, while Zhang Shenfu had been married twice. After their ship docked at Marseille, they went on to Paris and in January 1921, the two of them started living together. All the way until their final parting in 1948, they never went through the formal procedure of having a marriage ceremony.

In spring 1921, Zhang Shenfu established the first Chinese Communist cell in Europe. Through his introduction, Liu Qingyang, Zhou Enlai, and Zhu De all joined the Chinese Communist Party. As to how they joined the Party, Zhang Shenfu recalled,

> I simply wrote a letter to Chen Duxiu in Shanghai—only this. When I came to Europe, Chen Duxiu gave me complete authority and responsibility to induct Party members. He knew me and trusted me. If I thought someone was suitable, I just wrote him a letter. So when Zhou Enlai

> and Liu Qingyang joined the Party, all that was required was that I write a letter, and this was also the way that the Paris cell was started. This way of entering the Party can't be compared to the way things were done later on. It obviously depended on the character of like-minded people.

In that same year, Liu Qingyang gave birth to a baby boy, who later was given to a French orphanage because they wanted to join Zhou Enlai in Berlin. From Berlin they went to Moscow in 1924 to attend the fifth Communist International Convention.[4]

The Berlin cell was doing very well, expanding its activities and increasing its membership. Zhu De, Sun Bingwen, and others joined the Party in the second half of 1922.

In late autumn of that year, in a lounge at Gottingen University, eight Chinese students had their picture taken (see Photo 2). On the far left is Zhang Shenfu, with his hair combed very neatly. At the far right is Zhu De, looking a bit uncomfortable in his western-style suit.[5] Before going to Europe, Zhu De had told Chen Duxiu that he wanted to join the Communist Party, but he had never gotten Chen's agreement. After he arrived in Berlin, he told people, "I'd still like to make progress." This time around, he and Sun Bingwen were accepted into the Party organization. Facing Zhu De here is his beautiful wife at that time, Chen Yuzhen. Standing beside Zhu De is his good friend Sun Bingwen, who was also a friend of Zhang

4. Schwarcz notes that the infant son was left with a French family, and that Zhang and Liu took the son with them when they went back to China. But he died soon after they returned. See *Time for Telling Truth Is Running Out*, pp. 1–2, 78–79.

5. Zhu De became a partner with Mao Zedong in building the Communist movement and eventually headed the Red Army. Born to a poor peasant family of Hakka descent in northern Sichuan in 1886, Zhu was able to receive an education and then enter a military academy in Kunming, Yunnan. He joined the anti-Qing revolutionary movement headed by Sun Yatsen and became a subordinate of Cai E, a regional military leader. As the revolutionary cause ebbed, Zhu located in Kunming and gradually fell into habits of opium use and visits to prostitutes (his wife had been killed during the course of his revolutionary activity). Realizing that a new idealism was developing in China's eastern cities, Zhu went to Shanghai. From there he was able to arrange for this sojourn in Germany. As noted in this article, Zhu had been refused membership in the Communist Party, so he persisted while in Germany. As things happened, Zhu, the major Party leader, was accepted into membership by Zhang Shenfu, who within just a few years would leave the Party.

Shenfu. During Jiang Kaishek's effort to remove the Communists from the Guomindang in 1927, Sun Bingwen was seriously injured. Standing in the middle of the back row, arms folded at his chest and with a self-confident smile, is Zhang Bojun. He had joined the Party in China before making the trip to Berlin. He and Zhang Shenfu were also good friends, and they cooperated in a number of political activities—for example, Third Party activity in the late 1920s and the Democratic League in the 1940s. During the Anti-Rightist Movement in 1957, Zhang became a well-known rightist.

At the time of this photo, Zhang Shenfu, Liu Qingyang, Zhou Enlai, and Zhu De were auditing some courses at Gottingen. Zhang Shenfu was following the latest developments in mathematics and physics. Zhu De was auditing in the social sciences. Liu Qingyang and Zhou Enlai were also involved in social activism, so neither of them was present for this photo. Besides their activity in Party organization, Zhang Shenfu and Zhang Bojun were active in scholarly work. Zhang Shenfu read widely in such fields as German philosophy, Bertrand Russell's theories, and contemporary sciences, even including Freud's theories. He tried to stay abreast of these kinds of developments. He also corresponded with Bertrand Russell and even became an expert on Russell.

While he believed in communism, Zhang Shenfu did not reject other aspects of Western thought, and he advocated that a revolution in thought should be carried forward alongside social revolution. Social revolution should be the same as change in human nature. It could not be limited to political revolution. On China's internal politics, Zhang Shenfu believed that the Communist Party should not ally with the Guomindang: "Our goal is not a united line, but the dictatorship of the proletariat." In his articles for *Youth* (*Shaonian*, the publication of the Youth League in Europe), he used the pen name "R." He explained his choice this way: "In using this pen name, I was showing what I was excited about at that time: Russell, Red, and Revolution."

With Zhang Shenfu at the center of the small group in Berlin, when they met on Fridays for study and discussion they discussed the theories of Marx and Engels. Zhang served as an expert on logic and talked about the philosophical background of their thought. At times he also brought in Hegel's philosophy.

In late February 1923, Zhang Shenfu encountered opposition from his comrades in Europe. The members of the Youth League voted to expel him from the Party. There were a number of reasons for this, but primarily they were dissatisfied with his conceitedness. Besides being the de facto leader of the European branch, he had also made important academic contributions. But this success made him intellectually proud, and his pride became part of his personality. He had not changed despite efforts by other members of the Youth League to get him to do so. In the meeting some pointed out that when someone challenged his directives to the organization, he would threaten to resign. This had happened more than once, and it was contrary to Party discipline, which called for mutual respect. Zhang believed the Youth League was a mass organization under the leadership of the European Branch leaders, and Chen Duxiu had acknowledged this in the fall of 1922. But the Youth League leaders wanted to make the League the basis for the Party's general organization in Europe. The Youth League membership resented Zhang's arrogant approach to leadership and at a February 1923 meeting at Billancourt (in France) they voted to expel Zhang, who nonetheless continued in his role as correspondent between the European Branch and the Chinese Communist organization.

Photo 3 is from 1923. From left to right here are Zhang Shenfu, Liu Qingyang, Zhou Enlai, and Zhao Guangchen (Liu and Zhou were old friends from Tianjin). Zhang Shenfu stares into the lens, a smile at the corner of his mouth and his right hand resting in his pocket as if he were about to start a lecture. He is more relaxed than the others. Liu Qingyang is stylishly dressed wearing a fur wrap. Her outfit is half-Chinese and half-Western, perhaps reflecting her outlook as well.[6]

Back in China by January 1925, Zhang attended the Fourth National Conference of the Chinese Communist Party in Shanghai. At this meeting, the question of alliance with the Guomindang

6. On the same day, these four people had a photo taken as they boated on Berlin's West Lake. This photo came to be called *Photo in a Small Boat*, and the version most often used shows only Zhou Enlai seated in the boat, with the other three out of the picture. This photo—with the four people in the boat—is the cover photo for Vera Schwarcz' *Time for Telling Truth Is Running Out*.

became a major issue and the subject of heated debate. Zhang Shenfu made a speech opposing this alliance and wanted to change the Party's line. He believed that the working class had enough strength to accomplish their mission. Zhang later stated, "Cai Hesen said I was laughably naïve! I got angry and walked out of the meeting. Zhou Enlai (one of the five chairs at the meeting) stopped me as I was leaving and told me he agreed with me. He asked me not to leave the Party." This episode drove Zhang Shenfu from the Party organization, but this was not the first time. Previously at a meeting in Guangzhou, where he couldn't stand the hot, humid weather, he left over some small disagreement. When he thought about this later, he had no regrets: "Finally, I left. Zhou Enlai stayed on, but I think I'm a person who prefers to break and not bend." Zhang used a simple phrase to characterize Zhou: "He will bend and not break."

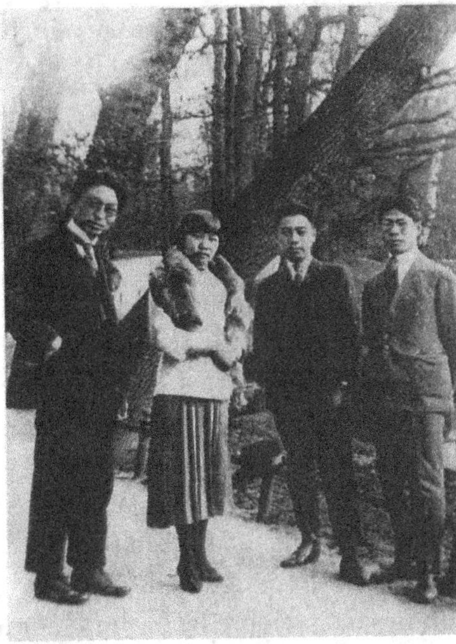

Photo 3: Berlin, 1923. From left, Zhang Shenfu, Liu Qingyang, Zhou Enlai, and Zhao Guangchen.

As to why Zhang Shenfu left the Party organization, Zheng Chaolin assessed him thus: "He is a theorist, an intellectual of integrity, but he is not someone who can make revolution."[7] To put

7. Zheng Chaolin also joined the Chinese Communist Party in Paris in 1922. He eventually joined the Chinese Trotskyist movement, who splintered from the Communists and centered on Chen Duxiu after Chen was removed from the Party's leadership following Jiang Kaishek's purges in Shanghai and Beijing in spring 1927. Zheng Chaolin spent seven years in prison before 1949 and was imprisoned for an even longer period (1952–1979) during the Communist era. Zheng's memoirs were published in 1986 and are often referred to in accounts of early Party history such as this one. Gregor Benton has translated Zheng's memoirs in *An Oppositionist for Life: Memoirs of the Chinese Revolutionary Zheng Chaolin* (Prometheus Books, 1997).

this another way, his obituary in June 1986 said that Zhang "was removed from the Party." His family had a different idea and said that he "retired from the Party" and that Zhang was an old friend of the Party.

When Zhang Shenfu left the Party, he became absorbed in studying philosophy, while Liu Qingyang continued as a Party member. They lived together despite their differing political involvement. They had to be alert to the activities of the warlords and the Guomindang forces. Their life was so unsettled that Liu had to have two abortions.

In Beijing in 1927, they were working with Li Dazhao. In March, in order to evade the forces of the Northern Expedition, they and Li Dazhao took refuge in the Soviet embassy.[8] After Li's arrest on April 6, Zhang and Liu secretly escaped to Wuhan. That summer Liu began an assignment as head of the Women's Section. Under continuous pressure from the Guomindang, the two of them moved again in August, this time to Shanghai. At the end of that year, Zhang Shenfu took a position as professor at Shanghai Continental University. To make things easier after the birth of their daughter, he translated several articles on Western philosophy to supplement his income. Liu Qingyang gradually slipped further away from her underground work for the Party and pursued a more mundane family life as wife and mother.

Photo 4 was probably taken in summer 1935, by which time Zhang Shenfu was professor of philosophy at Qinghua University in Beijing. The location is outside their apartment on the campus at Qinghua. Zhang Shenfu looks relaxed and confident, and Liu has a thoughtful expression. Liu's left hand rests on the shoulder of her older daughter, Liu Fangming, a gesture of motherly affection. The younger daughter, Liu Fangqing, sits on the step below holding her oversized Western-style doll. The view here is of a peaceful household, and again Liu Qingyang enjoys the roles of wife and mother. But events in the larger world would disrupt the placid life of this "professor's home." The Japanese were now advancing day by day and

8. Li Dazhao, one of the founders of the Chinese Communist Party, was killed soon after this in a raid on the Soviet embassy in Beiping (Beijing).

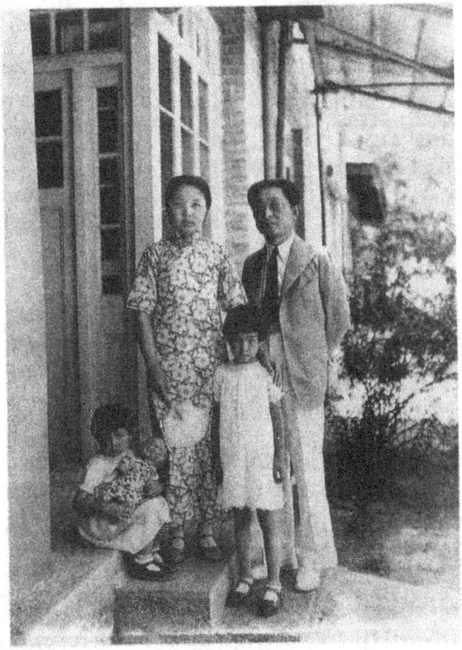

Photo 4: A family photo outside their home at Qinghua University.

the relationship between Zhang and Liu would not remain placid much longer.

Zhang Shenfu had changed gradually in his outlook toward women, and this became an important factor in his relationship with Liu Qingyang. Zhang had published his first book, *What Has Been Written* (*Suoshu*), in the autumn of 1930. This was a collection of his writings on freedom (including sexual freedom), women's liberation, and social revolution. The book and the thinking it grew out of marked Zhang's transition from political activist to advocate of social revolution, from soldier to scholar, and a modification of outlook from spirited idealist to cool materialist. In "Women and Revolution," he wrote, "It seems that the great majority of women intellectuals today still have the ambition to find someone and get married. They are one way before marriage, but become another way after marriage. . . . This point—dependence—is indeed the fatal flaw of women."

Zhang had seen Liu undergo such a transition. As their relationship began, Liu Qingyang absorbed his views on strength of character and independence in thought, and the couple worked together in the political struggle. But Liu Qingyang, who at that time was a radical revolutionary youth, gradually became a peaceful housewife. Zhang Shenfu became a well-known academic and a self-styled advocate of women's rights. So he was invited to serve as an editor of a supplement of the *Mighty Bulletin* (*Da Gongbao*), called *Knowledge of the World* (*Shijie Zhishi*). In this role he entered into another romantic story. Zhang Shenfu had always believed that men and women have a difficult time understanding each other, and he got a letter from a reader who signed her name "How can it be?" "If you were a woman," this writer asked, "what would you do? Could you honor and respect, and *marry*, someone whose mind and knowledge were less than yours?" Zhang Shenfu's reply was not scholarly but personal and very direct: "Really, I am not able to respond to your question because I am not a woman. I am not able and do not claim to be able to understand a woman's feelings. All I can say is that I have great admiration for such a bold and intelligent woman."

A few months later, Zhang met the woman who wrote that letter. She was Sun Sunquan, the principal of Beijing's No. 1 Girls' School. The relationship between Zhang and Sun became intimate, and Zhang lived with Sun for some time. They and Liu Qingyang—and many others—joined the December Ninth Movement demonstrations in 1935. Zhang and Liu were among those arrested, and Zhang lost his position at Qinghua University. Liu was released from jail first. She returned to Wuhan and again threw herself into the movement. Zhang also went to Wuhan after getting out of jail, showing that at least they still agreed about much having to do with politics. Yet Zhang did not break off his relationship with Sun Sunquan. This led Liu Qingyang to confront them and vent her hatred and distrust. So Zhang and Liu separated at this point.

With the war against Japan escalating, Zhang Shenfu went to Chongqing, where he edited a magazine called *Wartime Culture.* A young woman student from earlier years, Dong Guisheng, served as his secretary. With Zhang and Dong working closely together, seeing each other up close day after day, gradually reverence turned into adoration, and the two began living together. When Sun Sunquan learned of this new situation, it was her turn to become uncontrollably angry. She sent Zhang an emotional letter: "When a person can't devote himself completely to his work and blindly gets everything confused . . . what a loss to us all!" Zhang claimed his relation with Dong was platonic. Soon afterward, Zhang ended his relationship with Dong and arranged for her to marry another of his students. Zhang Shenfu had various relationships with political movements, and presumably for different reasons. He had multiple relationships with women as well.

Finally, during the first part of 1944 Zhang Shenfu became like a "roaming crow coming back to its nest" or "the prodigal returning." In Chongqing, he and Liu Qingyang lived fairly close to each other, and both were active in the work of the Democratic League, so they saw each other often. Zhang set out to restore his earlier good relationship with Liu, and in the first draft of his book *Women and the Constitution*, he said clearly that he was abandoning his pursuit of sexual freedom and wished to establish "a long-term relationship of faithfulness and mutuality between a man and a woman." Thus as they rebuilt their love for each other, they found they could work together for their livelihood and their home life.

The year 1946 was the peak of Zhang Shenfu's political career. Photo 5 records this period of history. This was one of Zhang's favorite photos, but not all of its meaning is obvious. Zhou Enlai and Deng Yingchao, husband and wife, are at the far left. As compared with their time in Europe, their roles and positions had changed greatly and their friendship had lasted for a long time. Not only is this a lively record of the friendly feelings from before, it is a moving image of the ferment of that time. How could Zhang Shenfu still be in a spot in the middle of this photo, a situation that outsiders would find hard to understand? No wonder that in the 1980s Zhang Shenfu could exclaim that he still couldn't believe what he wrote:

> Those were the days when I was most important. There can be no doubt that this was the happiest time in my life. After eight years away, I had just returned to Beijing. In March 1946 I had just returned from Chongqing with my family to our home in Beijing, and then I was flying back and forth between Beijing and Nanjing to prepare for the first National People's Assembly. At that time, I was one of the major leaders of the Democratic League, a really important position. Whenever I opened my mouth on the subject of China's future and constitutional government, the reporters swarmed around to interview me and report my views!

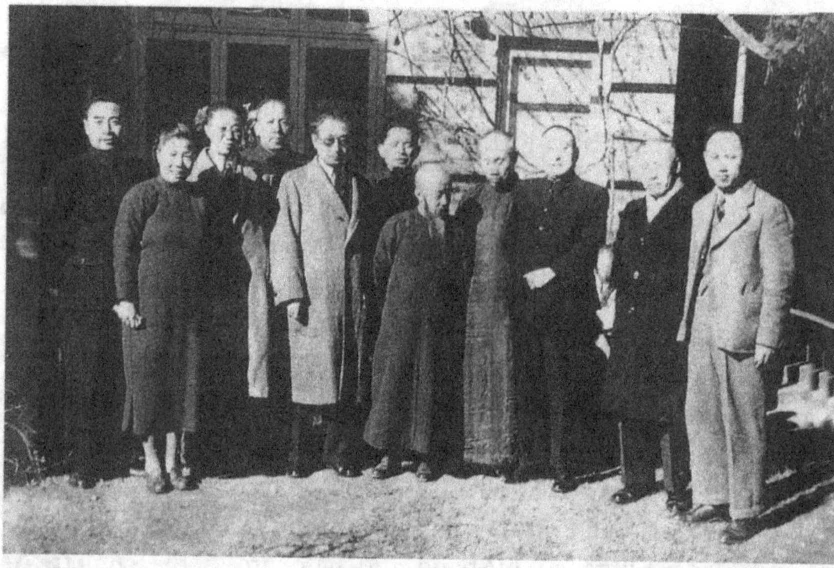

Photo 5: Taken on November 17, 1946, following talks between Nationalists and Communists, this group of Democratic League members were seeing off Zhou Enlai from Plum Garden New Village in Nanjing. From left: Zhou Enlai, Deng Yingchao (Zhou's wife), Luo Longji, Li Weihan, Zhang Shenfu, Zhang Bojun, Shen Junru, Dong Biwu, Huang Yanpei, Zhang Junmai, and Wang Bingnan.

A closer look at this photograph reveals more about the people in it. From the left they are Zhou Enlai, Deng Yingchao, Luo Longji, Li Weihan, Zhang Shenfu, Zhang Bojun, Shen Junru, Dong Biwu, Huang Yanpei, Zhang Junmai, and Wang Bingnan.[9] The photo was taken when these leaders of the Democratic League were seeing off Zhou Enlai and other Communist Party representatives from the place in Nanjing where they met for talks between the Nationalists

9. The representatives of the Democratic League shown in this photo were major figures in what is often called the "third force," that is, neither Nationalist nor Communist, during the 1930s and 1940s. These individuals were liberals, almost all influenced by Western thinkers and movements. Respected by both major parties, they were relegated to serving in advisory roles at most. Several of these individuals became government ministers in the early PRC years. They and many of their followers served in the People's Political Consultative Conference as representatives of "democratic parties and groups" (*minzhu dangpai*). For further information on "third force" figures, see Roger B. Jeans, ed., *Roads Not Taken: The Struggle of Opposition Parties in Twentieth-Century China* (Westview, 1992); also see Edmund S. K. Fung, *In Search of Chinese Democracy: Civil Opposition in Nationalist China, 1929–1949* (Cambridge University Press, 2000).

and Communists. The date was November 17, 1946, when the Communist Party and Democratic League withdrew together from the Nationalist-controlled first National Conference. This development indicated that talks between Nationalists and Communists had broken down and civil war would begin soon afterward. Deng Yingchao shows a relaxed smile while Zhou Enlai's is a bit less obvious, but this is enough to indicate that the Communist side was pleased. The expressions of the two men wearing traditional-style robes is not the same as the others in this group. Shen Junru, the legal scholar, short in stature, is looking to one side and seems to be wagging his head and heaving a sigh. Dong Biwu wears a pained expression and seems to be full of concern. Conditions at the time, like the weather, were cold, and relations between the Nationalists and Communists had again cooled. This photo, as part of the historical record, hangs in the Museum of the Chinese Revolution.

After this time Zhang Shenfu busied himself with democracy, peace, and unity, and publicly advocating through his writings on behalf of these causes. Then his article "An Appeal for Peace," which appeared in *The Observer* (*Guancha*) on October 23, 1948, created a disaster for Zhang. In the article he said,

> The most important thing for us now is peace; speaking negatively, we must break the present conditions, and speaking positively, we may return to peace. If we cannot work out a way to end the fighting there will be no way to have peace soon, and people throughout the country, or at least in the Northeast and in North China, certainly those who live in cities in those two parts of the country, will have no way to live, and our nation will lose its vitality and will greatly lose face.

When the naïvely outspoken Zhang Shenfu wrote the following, he provoked disapproval among the upper levels of the Democratic League and the Communist Party:

> Perhaps some people think that there is now one side that is indeed moving along smoothly and vigorously, and wants

> to win once and for all—that they can succeed in one go. To appeal for peace at this time might change their will to fight and cause them not to succeed so quickly or lose their focus. But such a view shows that these people do not recognize the present conditions of most of our people's life.

Some said Zhang expressed these views because he didn't understand the situation at the time and didn't see the import of the Communists' slogan "Cross the Yangzi and Liberate All of China." I think the opposite, that this was simply a manifestation of Zhang's outspoken nature, consistent with his earlier behavior. Evidence of this is that he didn't have any particular explanation for what he had said, and he later admitted it was a mistake.

Neither side thought well of this article: the Guomindang said that Zhang had "confused the people's minds," and the Democratic League said he "had opposed the Democratic League's political views." When the League met in Hong Kong on November 15, 1948, they went even further and voted to strike Zhang's name from their rolls.

Liu Qingyang sent a letter to Shen Junru and Zhang Bojun of the Democratic League, saying that she was outraged at Zhang Shenfu for his corrupted thinking and betrayal of the people. Her letter also said that she was severing all relations, public and private, with Zhang. One month earlier, in November, Liu Qingyang had moved with her children to Shijiazhuang (Hebei), which was already liberated.

Although Zhang was deeply saddened, he understood Liu's decision: "She had no choice but to break with me. . . . she had to protect herself and the children . . . after Liberation, although we never talked directly, she often asked mutual friends how I was doing."

Zhang Shenfu underwent many changes over the years, but there was one thing about him that didn't change—his self-confidence. He thought he still held a place in Liu Qingyang's heart, even that they might rebuild their family once again.

The on-again, off-again relationship between Zhang Shenfu and Liu Qingyang was not such that it could ultimately be called either a tragedy or a romance. Changes over time and the mental or spiritual changes in intellectuals—or the maturing of individuals and organizations—are complications worth noting and remembering.

This is by Meng Dong. It was published Old Photos *no. 27, pp. 42–59.*

Workers' Uprising in Shanghai, April 1927

Introduction

Photographs of the popular militia groups involved in the actions described below are rare. Labor organizing had become effective during the early 1920s, especially in such cities as Shanghai and Guangzhou. A "First Guomindang-Communist Alliance" had been launched in 1923 at Sun Yatsen's behest and involved the Communist International in Moscow, who sent agents to China. The Chinese Communists were effective in organizing workers, thus preparing for the events of spring 1925. The "May Thirtieth Movement" of that year became a vast popular rebellion provoked when British garrison forces retaliated against a workers' strike in Shanghai by firing into a crowd, killing eleven people and injuring dozens more. There were popular demonstrations in Beijing and other cities over the next several weeks, with students joining workers. Further violence came on June 23 in Guangzhou when foreign troops killed fifty-two people and injured more than a hundred. Violence subsided and the movement wound down, but those weeks of turmoil were essential to the breakup of the GMD-CCP alliance.

Sun Yatsen's death in March 1925 led to a succession struggle that lasted most of two years and was ultimately won by Jiang Kaishek. Sun had appreciated Jiang's loyalty and his strength as a military leader, but he was not necessarily Sun's preferred successor. Jiang used his military prowess in leading the "Northern Expedition" to carry out Sun's goal of uniting north China with the more southerly areas of the country where the Nationalist movement had its major support. As the brief article here points out, these worker militias first rose in support of the Northern Expedition. Thus

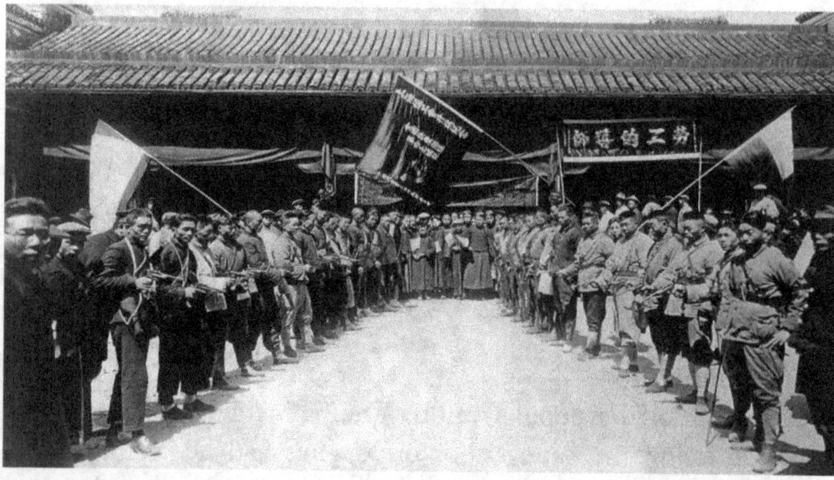

Photo 1: A group photo taken after the Shanghai General Union's flag-raising ceremony, March 1927.

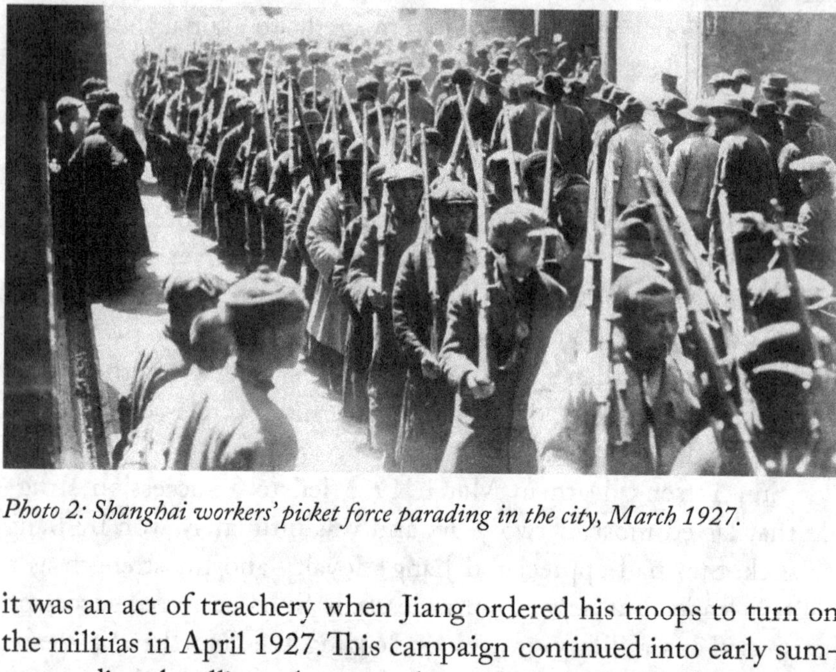

Photo 2: Shanghai workers' picket force parading in the city, March 1927.

it was an act of treachery when Jiang ordered his troops to turn on the militias in April 1927. This campaign continued into early summer, ending the alliance between the two movements and forcing the CCP remnant under Mao Zedong and Zhu De to their rural base in the Jinggang Mountains straddling Jiangxi and Guangdong. Note that these events in Shanghai in April 1927 were directly related to Jiang Kaishek's raid on the Soviet embassy in Beijing that brought the death of Li Dazhao, a founder of the Chinese Communist Party and forced Zhang Shenfu and Liu Qingyang to flee.

The Text

During the time of the Northern Expedition, to support the advancing armies, the working class in Shanghai brought down the reactionary government of the Northern warlords. Under the leadership of the Chinese Communist Party, they carried out three armed uprisings.[10]

On October 23, 1926, and February 22, 1927, two armed uprisings failed because preparations were inadequate, the strength of the effort was weak, and the leadership was not unified.

In mid-March 1927, when the First Army of the Northern Expedition entered the Longhua district of Shanghai, the revolutionary spirit of the workers and the masses reached a high tide, so the Shanghai District Committee of the Chinese Communist Party decided to launch a third armed rising of Shanghai workers. The organizer and leader of this rising was Zhou Enlai, Central Military Committee Secretary and also Secretary of the Jiangsu-Zhejiang Region Military Committee.

On March 20, the Northern Expedition army occupied Longhua, and conditions were ripe for an uprising. On March 21, beginning at noon, a general strike began throughout the entire city. Students boycotted their classes and store clerks joined in. The strike echoed throughout the city. Beginning at one o'clock in the afternoon, the workers' militia units in South City, Hongkou, Pudong, Wusong, Hudong, Huxi, Zhabei, and seven other districts began their attacks on the enemy at the same time. Very quickly, Shanghai was enveloped in the earth-shaking sound of gunfire.

In this intense military action, police offices and the army base came under attack, both the telephone and telegraph offices were occupied, and electricity and water were cut off. The people of the

---

10. The "Northern Sea" (*Beiyang* in Chinese) warlords were dominant leaders in China's northern provinces. Some of them controlled Beijing, and by 1927 these provincial and local overlords formed the major opposition to Jiang Kaishek's Northern Expedition, a mission to bring those northern areas under the control of the Nationalist Party (Guomindang).

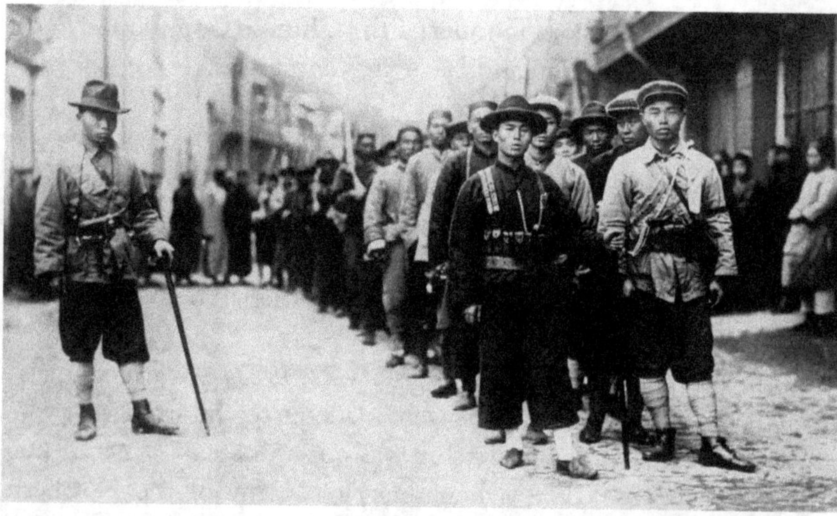

Photo 3: Shanghai workers' picket force on patrol in the city, March 1927.

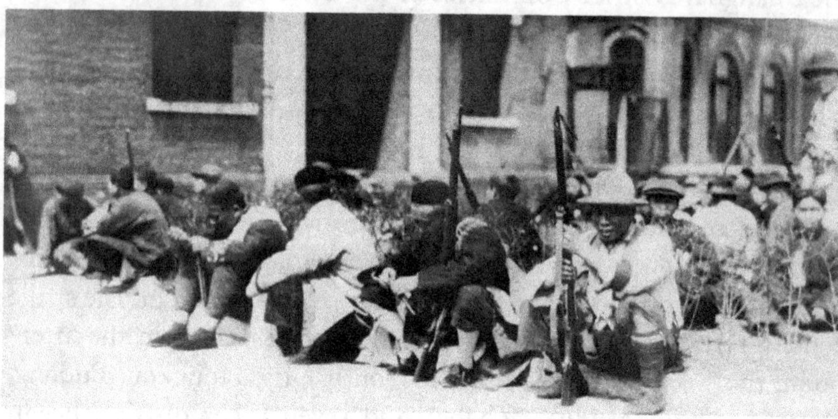

Photo 4: Worker picket linking up in the industrial area of Shanghai, March 1927.

city also joined in, either in support from the rear or at the front in the fighting, with gunfire all around them. The workers in the uprising fought hard without thinking of their own safety. They fought heroically, and by six in the evening of March 22 they had defeated the enemy. The people shouted and jumped for joy, celebrating the liberation of Shanghai, China's largest city and its industrial center.

After this victory, representatives of all groups in the city convened a citizens' assembly and established a Shanghai Citizens' Government.

But not long afterward Jiang Kaishek launched the April 12 anti-revolutionary coup, carrying out a "party rectification" that broke totally from the Communist Party, forcefully broke up the picket force, and carried out a vast massacre of the Communists.[11]

The set of photos printed here recorded the conditions of the popular joint actions organized and carried out by the Shanghai Workers' General Union. These photographs show the workers holding their steel weapons, their lines straight and orderly, their heads held high, with a "one for all and all for one" spirit. These materials that record the third armed uprising of the Shanghai workers are a rich resource that is seldom seen.

This is by Wu Yan. It appeared in Old Photos *no. 69, pp. 26–29.*

The "House of Democracy" in Wartime Chongqing

Introduction

Movements seeking genuine democracy have been an ongoing phenomenon in China's twentieth-century political history. These movements have produced many capable, courageous leaders whose fortunes have borne out the simple truth that political power requires a kind of military power that liberal leaders have disdained. Whatever their frustrations and limitations, China's aspiring democrats are admirable figures.

This tribute by Xian Jigen to his father and associates covers quite a range of subjects. It provides background on much of a century in the history of Sichuan, one of China's largest provinces and "the country of natural riches" (*Tianfu zhi guo*). Sichuan was deeply involved in the Republican Revolution of 1911–1912, and Xian Ying and many of his associates had their beginnings as cadets in a variety

---

11. This purge of the Communists was a critical episode in the history of the Party, who went on to retreat to the mountainous Jingangshan border area between Jiangxi and Guangdong.

of military training schools fostered by both the Qing government and its opponents. Many progressive young military men persisted in their early hope that China could build a democratic society.

Sichuan and its largest city, Chongqing, developed special importance when the Republic of China's president and military leader, Jiang Kaishek (Jieshi), moved the capital to the city in 1938 to gain a measure of safety from the invading Japanese. This strategic retreat began with impressive patriotic fervor, but over several years Chongqing became a focus of infighting and more than a little corruption. Jiang and his associates grew increasingly aggressive against those who failed to show enthusiastic support. The Chinese Communists, headquartered in Yan'an (Shaanxi) during these wartime years, regularly had representatives in Chongqing. Zhou Enlai often resided at "Red Crag," a site on the Gele Mountain ridge that forms Chongqing's backbone. Zhou and other Communist emissaries interacted with leaders of the democratic movement at the "House of Democracy," and the liberals hoped the Communists would support their goals.

As the Anti-Japanese War began to wind down, Xian Ying's House of Democracy became a major meeting place for leaders of the Democratic League, so the full range of members in this group called it Xian's institution. These heady days would fade, and the fate of the House of Democracy and of Xian Ying during the difficult decades of hectic revolutionary activity conclude this story.

The Text

In "Reminiscences of the Democratic Movement," Guo Moruo[12] wrote as follows:

> At Shangqing Temple stood the institution started by Xian Tesheng, called "Te Garden." Members of the democratic movement regularly gathered there. By late 1945 it had become the major center of the Democratic League, and

12. Guo Moruo (1892–1978) was a writer deeply involved in the movement toward modern literature in China. He was one of the founders of the Creation Society, an important writers' group during the 1920s. Guo gradually became attracted to Marxism and eventually joined the Chinese Communist Party, serving in various official capacities.

> Zhang Lan, chairman of the League, lived there. Te Garden was expansively large, located on the south bank of the Jialing River, and the view was excellent. The signboard with the words "The House of Democracy," a gift for all in the organization, is in my calligraphy, and I also wrote a poem alongside this. When you talk about the democratic movement in Chongqing, this place deserves many volumes. It is sad that no one has recorded more about what we experienced there in Chongqing.

Xian Ying was my father, also called Tesheng. He was born in 1885 in a small village in Xichong County named Taipingchang, a poor area where people worked hard all day and still could not make a good living. Our grandfather was a poor farmer with two sons and a daughter. Only at fall harvest every year—and only after the sweet potatoes were harvested—could they get enough to eat. When the leaves of sweet potatoes and some minor cereals were boiled together, they called it "duck's-foot porridge" because the leaves of sweet potatoes look like the web of a duck's foot.

As Father got a little older, the family saw he was intelligent and provided the money for him to go to a private school. He took the exam for military training in 1908 and entered the Bianmu Academy in Chengdu, where he joined the Revolutionary Alliance (*Tongmeng hui*). This school changed its name to "Army Short-Term Training Academy," and here Father became classmates and friends with Liu Xiang, Yang Sen, He Guoguang, and other Sichuan military men.[13] Father and Yang Sen were upper and lower bunkmates in the school dormitory.

When the Republic of China was established in early 1912, Zhang Lan was appointed Public Relations Officer for Northern Sichuan, and he appointed Father head of his security guards. Yang Sen and Liu Xiang were made members of the guard. Zhang was thirteen years older than Father, but they were like-minded men

13. Liu Xiang later became involved in warlord struggles in Sichuan and eventually became governor of Sichuan. Yang Sen's career paralleled Liu's, but Yang gave way to Liu in Sichuan and ruled Guizhou for many years during the 1930s. His return to Chongqing shortly before 1949 is discussed below.

who had a relationship as teacher and student, and as friends, despite the difference in their ages. Thus, after "Xian House" was built in 1931, whenever Zhang Lan came to Chongqing he stayed at Xian House, where a room was set aside for him. We children called it "Mr. Zhang's Room," and our parents called it "Biaofang's room" (his style-name, *zi* in Chinese, was Biaofang).

In 1913, Father was twenty-eight years old and he got a place in the Beijing Army University's program for advanced studies. There he met several others who became prominent military leaders. When Father saw that Yuan Shikai was set on restoring the imperial system and had signed the sell-out treaty of the "Twenty-One Demands" with Japan, he went to Guangzhou to join the "Campaign to Save the Nation."[14] After Yuan's defeat, Father became angry at the muddled effort of the Northern Expedition of that period and returned home to Sichuan, where he joined the staff of the Sichuan Army. In 1920, Liu Xiang became commander of the 16th Division, and Father served as his chief of staff. The following year the Sichuan area defense system was set up, and Liu Xiang became a general in charge of defending Chongqing. Again he appointed Father as the overall field chief of staff.

In 1925, Liu Xiang appointed Father commander of the 10th Army, with some related political duties. When the Sichuan armies were reorganized in 1928, Father left that position and became head of the popular army's arsenal. Not long after the May Fourth Movement, Father along with Luo Wei, Yuan Chengwu, and others established *New Sichuan News* (*Xin Shu bao*) to bring the new thought and new culture to Sichuan.

---

14. Yuan Shikai succeeded Sun Yatsen as president of the Republic of China in early 1912. Yuan's role as a premier military leader developed through his service in the Qing Dynasty's modernized army during the first decade of the twentieth century. He headed the Beiyang military clique, the dominant group during the last decade of Manchu rule. A clever political manipulator, Yuan held himself out of action as the revolution unfolded in late 1911 before agreeing to try to rescue the dynastic government. Yuan's power made him the choice to succeed Sun Yatsen, who idealistically yielded to Yuan in recognition of the latter's military predominance—and his pledge to respect the Republic. Yuan bided his time before the attempt to restore the monarchy in early 1913. Xian Ying was among the many political and military figures throughout China who successfully resisted Yuan's attempt.

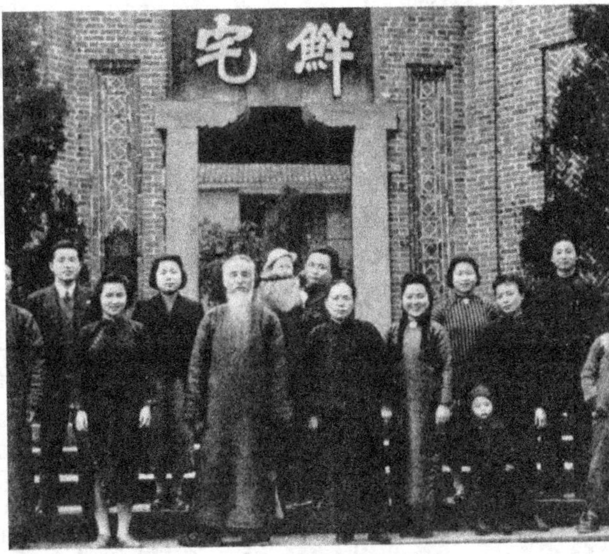

Photo 1: A family photo from 1944.

In the 1920s, Father bought a parcel of land from the Jiangxi Provincial Association in Chongqing.[15] This was about seventy *mou* (twenty-five acres). There was a small hilltop in the middle of this piece of land, and Father built his own home on that spot in 1931. He called the place "Te Garden," and he went on to add more than ten other buildings of various kinds. He also added a paved driveway, which connected the road to Shangqing Temple. This entire estate was also called "Te Garden." Finally, he added a pair of signboards at the entrance inscribed with this name.

Before joining the army, Father was married to a woman named He. When he returned after a few years of service in the military, he found she was pregnant. According to lineage rule, an unfaithful married woman would be punished by being driven out of the family. In an isolated village like his, such punishment often made women commit suicide. Father figured out a way to keep things manageable. She stayed with him at home, and after the baby was born he let her leave to find her lover.

Our mother, Jin Shushen, was born to a poor family of city

15. Provincial associations were organized by people from the same home province who were working in another city. They usually built meeting halls that were also social centers. These associations played several roles and were important institutions in traditional Chinese society.

people. She wasn't educated, but she was good at managing money and able to deal with all sorts of things. She had an orderly approach to managing the family and she was very severe in bringing up her children. We children said that our family had a loving father and a stern mother. Mother never saw herself as the mistress of Te Garden. She still went to the fields to work with the servants after the garden was built. Besides the buildings, the remaining space in the garden was turned into paddy-fields or grassy areas, a greenhouse, and a tennis court. Sichuan people love to eat pickled vegetables, and Mother was good at making all kinds of pickled vegetables, sweet jams, and broad-bean sauce. People called her broad-bean sauce "the lady's broad-bean sauce." She would make seven or eight crocks of this sauce each year, each half as tall as a person.

Father never showed any traditional male chauvinism in Mother's presence. (Indeed, Mother ran things.) We all recall that on each New Year's Eve, Father would hold a meeting of the entire family in the guest hall. The general program would be to remind us of some basic principles: don't forget where you came from; behave according to moral principles; be prepared to struggle throughout life and do not make fame your purpose; and always remember the two words "work" and "humility." Finally, he would summarize the year's events within the family and plans for the future. He would then ask all family members to express their views.

As 1935 began, Liu Xiang contacted the Chinese Communists through Zhang Lan and started cooperating with the Communist-led front to resist the Japanese. So Father began participating in the activities of the Communist-led United National Military Front. Soon Father, Zhang Lan, and one or two others began representing Liu Xiang in talks at Chengdu with representatives from the Chinese Communists to arrange for secret assistance against Jiang Kaishek in the anti-Japanese effort. They delivered 100,000 silver dollars for food, salt, cloth, and medical supplies to meet the urgent needs of the Red Army in northern Shaanxi.

In February 1935, Liu Xiang became chairman of the Sichuan provincial government and set up eighteen administrative districts. Each district was to have a supervisory official who would also lead the local militia. Over the next few years, Father served as supervisor

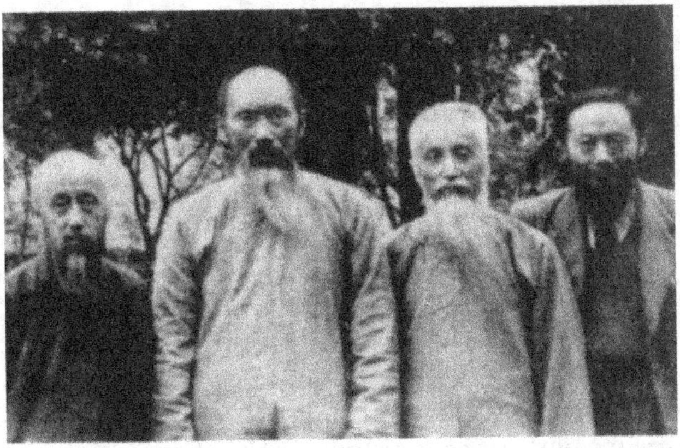

Photo 2: from left, Shen Junru, Zhang Lan, Xian Ying, and Li Gongpu. In the garden during the mid-1940s

for two of these districts. On the eve of the Anti-Japanese War, the Chinese Communists sent Yu Jiangzhen, Jia Ziqun, and others back to northern Sichuan to commence their duties. Father became further involved in this under Zhang Lan's supervision. As supervisor in Nanchong, Father started the Jianhua Middle School with Zhang Lan as principal and Father himself as administrator so they could support progressive young people. One goal was to support these youths by covering expenses for travel to Yan'an.

By the late 1930s, this cooperation continued with Communist leaders who came to Chongqing. Zhou Enlai and Dong Biwu arrived in Chongqing in 1938.[16] When Dong told Father that Zhou Enlai would like to meet the head of Te Garden, Father extended a warm welcome. In their talks the next day, Zhou inquired whether

16. Zhou Enlai appears in several of these articles. He was born in 1898 and died in 1976. An activist as early as the May Fourth Movement in 1919, Zhou eventually became premier under Mao Zedong and, as indicated in this article, he was almost universally respected. Zhou is often referred to as "Premier," as in this article. Dong Biwu was an important Communist leader but is not so well-known as Zhou. Dong (1899–1975) was born in Hubei to a declining gentry family. He passed the first of the traditional civil service examinations but then began a modern education at Wuchang just as the Revolution of 1911 broke out in that city. Dong quickly joined the Tongmenghui and would later tell people that he was a revolutionary from that time on. He became a founding member of the Chinese Communist Party in 1921. (Along with Mao Zedong, Dong was the only person present at the Party's founding meeting at Shanghai who later held a major position in party and government.) Dong's arrival in Chongqing here in 1938 marked the beginning of his role in CCP-GMD liaisons throughout the Anti-Japanese War until 1945.

Father might offer his home as a place for the Communists to discuss national affairs with people from various parts of society. Zhou noted that if Father agreed, our home might assume some risk. When he asked if Father would be afraid, Father told him, "From early in my life I have wished to serve our country, and I have not changed my mind. To sum it up, yes, I want to do this, and no, I am not afraid." From that time on the Chinese Communists had another place to meet in Chongqing besides their headquarters at Red Crag Village.[17]

In December 1938, Wang Jingwei, who already had a long-time rivalry with Jiang Kaishek within the Guomindang, left Chongqing and announced his plan to form a pro-Japanese government as a means to manage China's national crisis.[18] Wang's action brought out another aspect of the democratic movement's views. Zhang Lan, Huang Yanpei, and several others in the movement wrote an article criticizing Wang for his plan. This article was published in *New Sichuan News* (*Xin Shu bao*), the newspaper that Father managed.

---

17. Red Crag Village was the Communists' headquarters in suburban Chongqing. It was the center of many important activities during the period when Chongqing served as China's temporary capital and remains a revolutionary monument especially devoted to Zhou Enlai and his delegation from Yan'an.

18. Wang Jingwei (1883–1944) became a revolutionary activist during the late Qing period. An early and trusted associate of Sun Yatsen in the Tongmenghui, Wang rose alongside Jiang Kaishek in the early years of the Guomindang. Wang contended with Jiang for leadership of the party after Sun's death in 1925. Wang headed the left-wing element in the party and emphasized the socialist aspect of Sun Yatsen's principles. Jiang came from a military background and emphasized the theme of nationalism, which centered on party and individual leadership. At the time of Sun's death, the Nationalists were allied with the younger Chinese Communist Party. Jiang was suspicious of Marxism and of the influence of the Soviet Communist Party. Jiang prevailed in his struggle with Wang by attacking Chinese Communists and Soviet representatives in Shanghai and Beijing in April 1927. Jiang continued by launching a Northern Expedition intended to complete China's unification. Wang Jingwei went to Wuhan where he and the Guomindang left wing continued efforts to challenge Jiang. Gradually the two wings came together again, with Wang a ranking leader in the Guomindang. By the time war with Japan broke out in 1937, Wang gradually came to believe that resistance to Japan could not succeed. Many in the Guomindang had long-standing contacts with Japanese progressives, and Wang hoped that cooperation would reduce death and destruction in China. Wang is almost universally regarded simply as a traitor in China, and a number of his associates in the government he headed in Nanjing were tried for treason after the war ended. Wang himself died of illness on a visit to Japan in 1944. Those sympathetic to Wang tend to see his effort as self-sacrificing, if misguided.

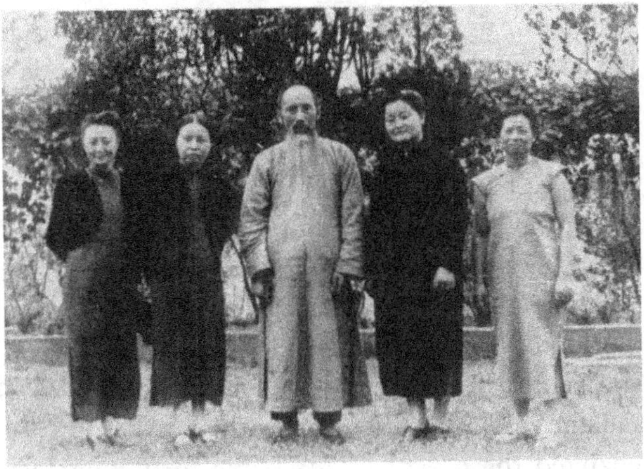

Photo 3: Zhang Lan with some of the women in the democracy movement, from left: Liu Wang Liming, Liu Qingyang, Zhang, Shi Liang, and Zhang Manyun. In the garden, from the mid-1940s

In January 1939, Jiang Kaishek ordered an attempt on the life of Wang Jingwei at his temporary residence in Hanoi. It failed. Jiang's action in turn prompted those in the democratic movement to disclose the truth about this incident. Then in fall 1939, Zhang Lan, Liang Shuming, Huang Yanpei and several others in the "National People's Senate" put forward a call for Jiang's Guomindang to end "party government," implement "constitutional government," and concentrate all of the nation's strength on the task of defeating Japan. This proposal was discussed and agreed to at Te Garden before their announcement was made. In November 1939, Liang Shuming proposed the establishment of a "Society of Comrades to Unify and Build the Nation."

When the National Government moved from Nanjing, Chongqing's role became much more important than before. Important people from every field—government, military matters, the economy, and culture—came to Chongqing, and there was a great deal of social activity. Tranquil Teyuan became the stronghold of the democracy movement. Patriotic people from all kinds of groups came to Teyuan to meet. Sometimes as many as forty or fifty people came almost every day. It was no small matter just to have enough tea and cigarettes. Mother went along with this without complaining. She hated Jiang Kaishek's political power and knew Father was doing the right thing. Shi Liang once told Father, "Mr. Te, you have provided a place for the democratic movement and hosted progressive people from everywhere, and your wife's contribution is half of this."[19]

After the "New Fourth Army Incident" aroused the anger of people throughout the country in January 1941, Zhang Lan, Luo Longji, Zhang Bojun, Xian Ying, and seventeen other people secretly established the "Alliance of Chinese Democratic Political Groups."[20] Father was made a member of the Alliance's central committee. On "Double Ten" that year this "Democratic Alliance" put out a proclamation of their formal establishment, and Teyuan really did become the center of activities. On September 19, 1944, the Democratic Alliance held a national representative conference at Teyuan and changed the organization's name to "China Democratic Alliance." This became a third major political party and an alternative to both the Nationalists and Communists.

On February 16, 1943, Zhou Enlai and Dong Biwu came to Teyuan with Lin Biao, who had just arrived from Yan'an. Lin was then serving as commander of the 115th Division of the Red Army. They met with a large group of people in the democratic movement, including Shen Junru, Huang Yanpei, Zuo Shunsheng, Zhang Bojun, Luo Longji, Yang Gengguang, Deng Chumin, Zhang Shenfu, Liu Qingyang, Tao Xingzhi, Leng Yuqiu, Han Zhao'e, and Xian Ying.[21]

---

19. Shi Liang (1900–1985) was the most prominent among women activists in the democracy movement. She appears in two of the photos for this selection. She became a lawyer and was known for defending the poor and those with difficult cases. Shi Liang participated in the democratic movement and thus was a participant in activities at the House of Democracy. After 1949 she was appointed Minister of Justice.

20. The "New Fourth Army Incident" occurred in January 1941 at a time when civil war activity was supposed to be suspended. The New Fourth Army was a Communist unit operating in southern Anhui Province when Jiang Kaishek ordered military forces of both parties to move to the north of the Yangzi River. As the New Fourth Army started this movement, its nine thousand troops were attacked by a much larger Nationalist force—as many as eighty thousand. Approximately two thousand of the New Fourth Army escaped alive.

21. This list of "people in the democratic movement" includes most of the prominent figures in this movement. Many of these are discussed in Roger B. Jeans, ed., *Roads Not Taken: The Struggle of Opposition Parties in Twentieth-Century China* (Boulder, CO: Westview Press, 1992). Jeans' study of Zhang Junmai, a major figure in the democratic movement, offers great insight into the broader movement. See *Democracy and Socialism in Republican China* (Lanham, MD: Rowman & Littlefield, 1997). Also see Edmund S. K. Fung, *In Search of Chinese Democracy: Civil Opposition in Nationalist China, 1929–1949* (Cambridge, UK: Cambridge University Press, 2000).

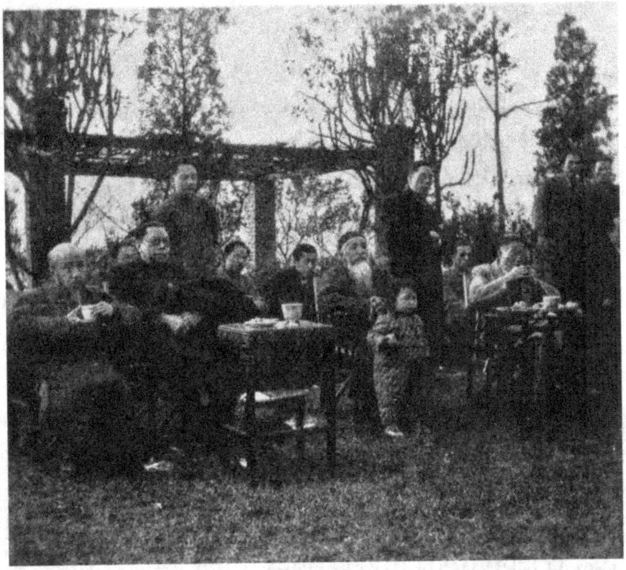

Photo 4: New Year celebration in the garden, 1946

Zhou Enlai led the meeting and one of the important items was Lin Biao's report on the situation of the Northern Army in resisting Japan under the leadership of the Communists. After the meeting, Father had prepared a dinner for all those who attended. Besides this, several people involved in the Guomindang left wing attended including Li Jishen, Chen Mingshu, and General Feng Yuxiang. Some other progressive younger people from all groups came to Teyuan to discuss national affairs.[22]

22. Li Jishen, like several others in this selection, launched his career as a military leader in one of the Qing government's military academies. He came out of a Guangdong Province grouping of army leaders and joined the Republican revolutionary cause. He often disagreed with Jiang Kaishek as leader of this movement and from time to time organized movements intended to resist Japanese aggression. Chen Mingshu also was a progressive military leader from Guangdong. Feng Yuxiang (1882–1948) was born in present-day Hebei Province, the son of a low-ranking military officer. He had little schooling but showed talent as a military leader and began to associate with reform activity. He also began a lifelong effort at self-education and in 1914 became a Christian. His work to propagate Christianity—including mass baptisms with fire hoses—led to his appellation as "the Christian General." Feng's military movements and his political associations were complicated. By the late 1920s he had developed enough influence to be a factor in Guomindang squabbles. He was in and out of formal positions in Nanjing and other provinces. When Japanese aggression began in earnest, Feng opposed Jiang Kaishek's lackadaisical efforts at resistance so that Feng came down on the side of the democracy movement as described here.

As the situation developed both internationally and within China, Teyuan became even busier in 1945. On the evening of February 14, Zhou Enlai was relaxing at Teyuan and held a dinner for Sun Ke, Yu Youren, Tan Pingshan, Chen Junqu, Yang Gengguang, Wang Kunlun, Qu Wu, Shen Junru, Zhang Bojun, Zhang Shenfu, Tao Xingzhi, Li Huang, Zuo Shunsheng, Leng Yuqiu, Deng Chumin, Xian Ying and about twenty other people. That evening Zhou Enlai updated everyone on the harmonious talks with the Guomindang and asked everyone for their opinions in order to advance the current discussions.

Zhou Enlai would also come to Teyuan from time to time to hold receptions for foreign journalists. Besides such political meetings, Deng Yingchao (wife of Zhou Enlai) sometimes held meetings of the Women's Alliance. Deng Chumin's "Social University," New Year celebrations, and various other kinds of teas and receptions were held at Teyuan. Tao Xingzhi also held gatherings of school leaders and students who performed folk music or *yang'ge* dances. Guo Moruo also brought the famous performer Zhang Ruifang for poetry recitations and singing performances.

Zhou Enlai was very thoughtful and solicitous of people and always took an interest in every individual. One evening he saw that our household was especially busy and happy, and he let his bodyguards know it was my mother's birthday. Zhou then had the bodyguards go out and buy a piece of cloth as a present for her. Every time Mother recalled this, she spoke with great respect for the Premier.

The Premier often said to Father that people would never forget those who had done good things for them. After 1949, when the Premier met father at various kinds of meetings, he would recall these earlier times and once again express thanks to Father. Father's contributions were recognized later, as in 1949 he was made a member of the Southwest Administrative and Military Committee and a delegate to the second session of the National People's Political Consultative Conference. He also was elected to the first National People's Congress.

Because Teyuan was important to the Democratic movement, some Guomindang leaders regarded Father and his home as a thorn

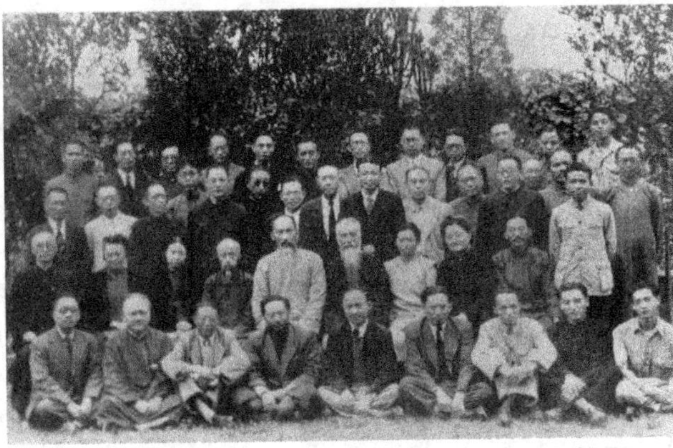

Photo 5: Group photo of the members of the Democratic League in Chongqing in the mid-1940s, again in the garden

in their side. Guomindang special agents hounded Father and his associates in various ways. Of all their activities at Teyuan, the most annoying was that the chief special action agent, Dai Li, himself set up shop at Teyuan.[23] He rented one of the outbuildings, Number Four, and for a time Teyuan was taken up with all kinds of petty goings-on. Even some of the serving staff were bought out by these agents. One evening when a meeting was underway, the spies planned to set off a bomb that would have destroyed Teyuan. One of them suddenly thought to check who was inside. He stealthily climbed to a spot outside a window and then leapt down after seeing that Chen Cheng (a high-ranking figure in this organization) was also at the meeting.[24] Teyuan thus avoided a disaster.

As this threat came to nothing, Jiang Kaishek sent Chen Lifu and others to Teyuan to talk with Father. They suggested fame and fortune as enticement in trying to get Father to cease making Teyuan available to the democratic movement, but Father steadfastly refused them. Thus Teyuan really did live up to its name as the House of Democracy. To the Guomindang, another problem was that non-party

23. Dai Li was chief of Guomindang "special services," or master spy, and was directly or ultimately responsible for all forms of harassment mentioned here and in all other situations. Dai Li (a pseudonym) was Jiang Kaishek's most trusted subordinate. His career and its multitudes of effects are the subject of a masterly study by Frederic Wakeman, *Spymaster: Dai Li and the Chinese Secret Service* (Berkeley: University of California Press, 2003).

24. Chen Cheng was a major Nationalist military leader in the Anti-Japanese War and the Civil War between the Nationalists and the Communists.

newspapers came directly to Teyuan for news about democracy movement activities instead of believing Guomindang news sources.

At the beginning of 1946, Guomindang special agents perpetrated the "Cangbai Tang Incident" and the "Jiaoyangkou Incident," in which Guo Moruo, Li Gongpu, Father, and more than sixty others were injured.[25] This was followed by attacks on February 22 on the *Xinhua Daily* (*Xinhua ribao*) and the *Democracy Tribune* (*Minzhu bao*) offices. The latter was the voice of the Democratic Alliance. Zhou Enlai expressed his concern and sent Li Weihan to Teyuan to express his opinion about the leaders of the Democratic Alliance.

After victory in the Anti-Japanese War, the Guomindang moved the capital back to Nanjing. Teyuan continued as the southern headquarters of the Democratic Alliance and made important contributions. In July 1946, when Guomindang special agents assassinated Li Gongpu and Wen Yiduo as part of the white terror, groups of many kinds convened there for memorial gatherings. Along with Shi Liang and Deng Chumin, Father successfully urged the Guomindang provincial governor of Sichuan, Zhang Qun, to arrange a memorial service.

After the liberation of Chongqing, a "black list" of Guomindang enemies came to light, and special agents planned their assassination. Zhang Lan's name was first on this list and Father's was second. Fortunately, the Liberation Army arrived in Chongqing just in time. The Party underground also helped, and none of these plans were carried out.

Dong Biwu told Father personally that after he and Mother returned to their old home in western Sichuan, "Xian House" would be made into a memorial hall to remember activities on behalf of democracy during that period. But during the Cultural Revolution, a Red Guard group in Chongqing liked the garden and made it their general command center. Unfortunately, Daguan Pavilion was burned out and reduced to ashes in the midst of the fighting. All that

25. These incidents occurred in January 1946 at places where members of the Democratic Alliance gave public talks on current questions and political issues. Actions by special service agents to break up these gatherings resulted in the injuries to Alliance members mentioned here.

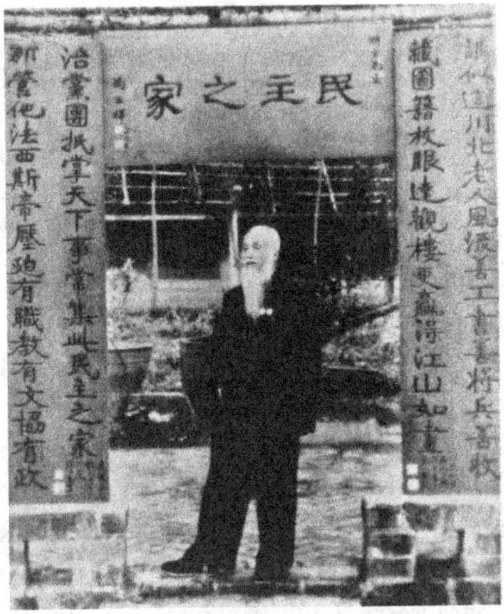

Photo 6: Xian Ying in the early 1950s, also in the garden behind the house

is left now is a small part of a side building and a small courtyard with a room where we as small children studied and practiced martial arts. In April 1992, the Chongqing municipal government designated Teyuan one of the city's important cultural relics.

The time before and after the victory in the Anti-Japanese War was the most glorious time for Teyuan. In August 1945, when Mao Zedong came to Chongqing for talks with Jiang Kaishek, he visited Teyuan three times.

On August 30, Mao Zedong came for the first time with Zhou Enlai. When he came in the main entrance to the Xian Mansion, he saw the placard that said "House of Democracy," done in the Official Script style. (This placard is now hung in the Red Crag Revolutionary Monument in Chongqing.) The name on the recipient's space is Father's name, "Tesheng," and in the space where the presenter's name appears, it says "Feng Yuxiang." Mao Zedong said, "This is the home of democracy and I too have returned home." Father said that because his old friend Zhang Lan had lived there for a long time, the headquarters of the Democratic Alliance were there and people from all fields in the democratic movement had used this place to come together and talk about the concerns of our country. So Dong Biwu bestowed the title on us. General Feng, who did this calligraphy, went there as a guest from time to time.

The couplets at the entryway were done by Zhang Lan himself. In his Hunan accent, Mao Zedong read aloud, "Who is as great as this old gentleman from northern Sichuan, good in calligraphy, good at leading troops, good at collecting art and history? Fix your eyes on the Daguan Pavilion, where we have a political body and seek to control the affairs of our nation. We often gather here at the House of Democracy." Mao Zedong also said, "Dong Biwu has left his name, General Feng Yuxiang has presented some calligraphy, and Zhang Lan has done these couplets. We can call this the place of 'Three Great Men.'" Finally, he said, "Today we are meeting at the House of Democracy. Hereafter we must work hard together to live in a democratic nation."

On September 2, Zhang Lan and Father held a dinner on behalf of the Democratic Alliance and invited Mao Zedong, Zhou Enlai, and Wang Ruofei. Accompanying Alliance members were Huang Yanpei, Shen Junru, Leng Yuqiu, Zhang Shenfu, Zuo Shunsheng, and others.[26] After dinner the hostess invited Mao Zedong to write in our guest book, and with his fountain pen Mao wrote four large characters, "Our prospects are bright." He also said that the way ahead would not be easy, but the future was bright. As they continued to talk, Mao recalled the time that Father contributed 100,000 silver dollars to the Red Army on behalf of Liu Xiang.

On September 15, Mao visited Teyuan a third time and talked with Zhang Lan and Father for about three hours.

In 1950 Mao Zedong invited Father as a guest at a dinner at Huairen Tang, and as he shook Father's hand, he said, "When Chongqing was liberated, you had a bit of a scare!" Father replied, "Thank you, Chairman. There was a scare but no real danger. We were fine."

Father and Yang Sen had been classmates in earlier years, but during the 1920s their careers took different courses and they had no interaction for many years. In May 1948, however, Yang Sen was made mayor of Chongqing with a parallel appointment as garrison commander, combining great political and military authority in one individual. With his political needs at this time, he sought to arrange a political marriage with Father and our family between one of his marriageable daughters and our third brother, Jiheng. Both Father

---

26. Mao's visit to the "House of Democracy" presumably came soon after the visit Mao arranged in Yan'an, where Huang Yanpei and other Democratic League leaders visited upon the end of the war against Japan. See the later article "Mao Zedong Charms a Liberal Educator-Entrepreneur."

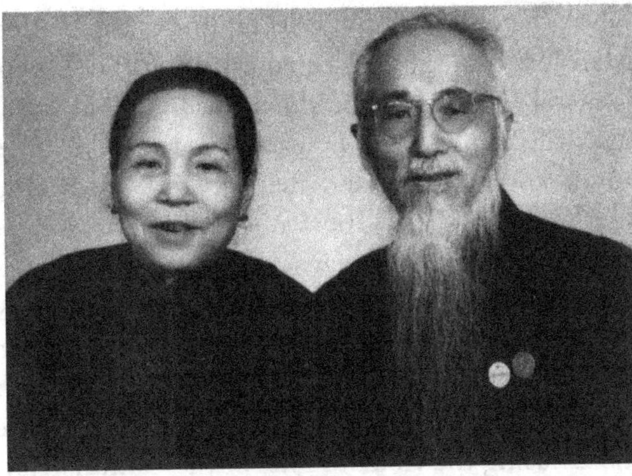

Photo 7: Xian Ying and his wife, Chongqing 1954

and Yang Sen could be assertive and willful men, but now they decided to work together. First they discussed this marriage possibility with our brother Jiheng, who accepted. Thus the Xian and Yang families entered into a relationship.

With their families now linked, Xian Ying and Yang Sen began talking about how to prepare for the expected arrival of the Liberation Army in Sichuan and Chongqing. Yang had been a long-time enemy of the Chinese Communists, and he feared reprisals when they reached Chongqing. Xian and his son suggested a military action against remaining Nationalist Army elements as a way to demonstrate Yang's changed allegiance. Yang decided to leave some of his family members behind, and Xian pledged to look after them. Yang decided not to go to Taiwan (as Jiang Jingguo had personally suggested to him), but only to go to Hong Kong and then return to China, perhaps within a year. The approaching Liberation Army leader let it be known that Yang Sen would not be hurt if there was no resistance as his forces approached Chongqing. Ultimately Sichuan gave in with almost no resistance to the Communists. On the eve of the liberation of Chongqing, Father made a great effort to protect and save Communist Party comrades at Zhazedong Prison.[27]

27. Zhazedong Prison was located in Shapingba in suburban Chongqing. Communist Party members imprisoned there suffered greatly. Conditions in this prison along with other CCP activities in Chongqing were important parts of the popular 1950 novel *Red Crag* (*Hong Yan*), which is titled after the location of Zhou Enlai's office in the

When the Cultural Revolution began, the Red Guard saw "Xian Mansion" as a good place and made it the "Chongqing Red Guard General Command Headquarters." At that time, our fifth brother was still living in the house, with a total of nine people including adults and children, and they all were ordered out immediately by the "Commandant." There was nothing he could do, but he did ask to take the signboard that says "House of Democracy" with him, as it could be used for a bed-board. Thus the old sign was salvaged. After the Cultural Revolution came to an end, this sign became a valuable artifact, and it is now exhibited at the Red Crag Revolutionary Commemorative Hall. But "Xian Mansion" was reduced to ashes in a fire during the violent fights of the Cultural Revolution.

In 1957, Father was mistakenly adjudged a "Rightist." He was pardoned in 1961. After that, he was made a specially invited delegate to the People's Political Consultative Conference. As the Cultural Revolution began, he inevitably suffered. In 1968, Father died in anger. In a reversal of policy launched in 1970 when more moderate policies were begun, Father was recognized as a patriotic figure in the democratic movement. In August 1977, Father's ashes were placed in a room at Babaoshan Cemetery in Beijing.[28] In 1997 we children decided to move Father's remains to a private cemetery in the Gele Mountains in Chongqing so that Father and Mother could rest together forever among the green pines and cypresses.

This article is by Xian Jigen. It appeared in Old Photos *no. 22, pp. 80–93.*

Chongqing area. A museum devoted to the Zhazedong prisoners' experiences was later built nearby, and its location is near the gate of the Sichuan Foreign Languages Institute, now Southwestern Culture University.

28. Babaoshan Cemetery lies in Beijing's western suburbs and has long been a venerable burial place. After 1949 a new section was opened to honor those who died in the cause of the Communist revolution—a Chinese equivalent of Arlington National Cemetery. The graves of many well-known activists are found there.

CHAPTER THREE

Building a New Nation

First Lessons in Manual Labor

Introduction

This memoir tells a great deal about railroad construction and is of interest simply for this reason. Similar railroad-building efforts were made during the late 1960s and 1970s in many parts of the country, and were done using these same techniques. During these years the Chinese were creating one of the world's great railroad systems. In the relatively early stages described in this article, the project was a triumph of planning, engineering, and sheer human effort. As the writer observes, this project continued during a decade of upheaval. At a more basic level, the writer recalls specific experiences that describe what everyday life was like: daily tasks, going to movies, relationships between young men and women, and social status concerns.

The Text

In November 1968, just when the Cultural Revolution was at its peak, there was a central directive calling those who would graduate from universities, middle schools, and vocational schools in 1968 to report for work assignments. Although this was delayed at least half a year beyond the usual time for assignments, we were lucky in comparison with the classes of 1966 and 1967, who were delayed for at least a year. So in December 1968 I graduated from the Beijing Railroad College (today known as Beijing Transportation University) and was assigned to work in the Zhengzhou Railway Bureau.

The Railway Bureau's orders told us to report on December 25. On December 23, Mao Zedong's latest directive came via telegraph: "Educated youth should go to the villages and receive further education from the poor and lower-middle peasants; this is a necessity." Although we were not among those whom Mao Zedong referred to as "educated youth," those in all the trades and enterprises wanted to carry through on "the latest and highest directive." So the Railway Bureau revised its plan and allowed us to go to a basic unit and do the kind of work we had been trained for. Those who were assigned to the Zhengzhou Railway Bureau included about 180 people from the three kinds of schools, and we were then divided into a large construction brigade, a labor brigade, and a basic construction brigade—all of which offered an intense experience of discipline through labor. I was assigned to the large construction brigade.

The full name of the construction brigade was Zhengzhou Railway Bureau No. 1 Line Construction Brigade. Within the bureau it was called the "First-Line Brigade," with its center at Xinxiang. After a few days of study at the beginning of 1969, I along with about a dozen others from different schools arrived at Xiaoshangqiao Station as the Fifth Sub-Team of the Construction Brigade to begin our life of discipline through labor.

Xiaoshangqiao was a small station near the Beijing-Guangzhou line. It was a historic place, as this was where Yang Zaixing, a general under the leadership of Yue Fei, the national hero of Song, fought the Jin army and was killed in a hail of arrows when his horse sank in the muddy bottom of the Shang River.[1] At the point where the Xiaoshangqiao line branched off, there was a line of old railroad cars that served as the dormitory for the Fifth Sub-Team of the Construction Brigade. These dormitory cars were made up of old-style heated cars with more than ten sets of upper and lower bunks in

1. Yue Fei lived during the transition from Northern to Southern Song period (twelfth century CE). Yue urged the Song emperors to defend all of their territory from the Liao, a frontier people and state in the northeast. For his passionate devotion to this goal, Yue has remained a hero to Chinese of later generations, especially to diehard patriots. There is a memorial hall to Yue in Hangzhou, which became the capital of Southern Song. Han Chinese rulers would not regain control of the northern territories until the Ming period and the early years of the fifteenth century.

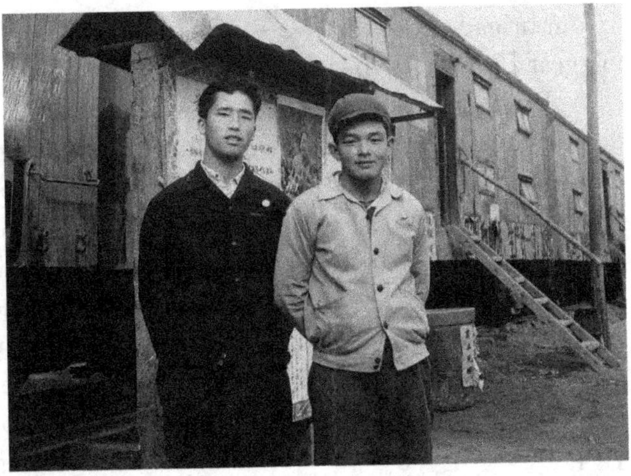

Photo 1: Two of the writer's friends outside the dormitory cars.

each car. There was also a stove, so the place was not cold. The train included the construction group's office, a technology office, a clinic, a library, and a kitchen—the entire sub-team of more than four hundred people could carry on their lives in this string of railroad cars. Photo 1 is one I took of my friends Han and Zhang at the time, with our dormitory train in the background.

The brigade had three major jobs: replacing ties, replacing rails, and maintaining the roadbed. Replacing ties meant removing the old wooden ties and installing concrete ties. Replacing rails meant replacing rails that weighed forty-three kilograms per meter (this was called "forty-three rail") by putting in a heavier rail at fifty kilograms per meter ("fifty rail"). The third job, "sifting roadbed," is something those on the outside probably wouldn't understand very well. A railroad line is subject to the effects of wind and sun over long periods of time, plus the speed and pressure of trains, dirt, moisture, and oil that seeps into the gravel. The roadbed can harden into pieces like clods in a field. Of these three jobs, everyone agreed that sifting roadbed was the hardest work. Our Fifth Sub-Team did all these jobs, and all depended on simple, hard, physical labor with hand tools.

How great was the labor done by our brigade? During the 1960s, every city had a set grain allowance for various categories of people. For example, an ordinary city resident received twenty-six *jin* of grain, a male university student got thirty-four *jin*, and a worker in our brigade got fifty-two *jin*.[2] We always heard this was one of the biggest allowances in the country.

As the brigade carried on its work, the train had to suspend its regular operation, which in our professional jargon was called the "seal-up point." Every time we came to a seal-up point, three hundred people gathered and got aboard, and the big train danced on down the track with all the metal clanging. At these times no one made idle chatter—you don't want to spit in a place you might wish to drink from! And it was also true that even on cold days, your back got soaked with sweat. The two hours for these "seal-up points" were the high point of a day, the tail end of a process that included preparations from when work began that morning. We would work at a place for a little more than five hours a day, but because the place where we worked each day would advance a few hundred meters daily, we had to walk a little farther each day. When we got nearer to the next station, maybe twenty kilometers away, we had to move our quarters. In February, a steam locomotive came, hooked us up, and moved us from Ta'he City to Yan City station about eight or nine kilometers away, where we carried on the same kind of work we had been doing.

Winter that year was especially cold. It snowed several times just as the year began, so finally the ditches on either side of the tracks got filled. All you could see besides the expanse of white across the level land were the two black rails extending into the distance. A few times we had sleet before the snow began, and as the sleet dripped onto your body, it froze immediately so that your work coat was encased in a suit of "armor" that made this swishing sound as you walked. There was no way to get away from it; all you could do was tolerate the wind and rain. When I saw the bright lights of trains rolling by and the passengers sitting comfortably in their warm coaches, I could only think of masses of maintenance workers in the cold and snow, as if I saw behind the curtain of Tonya in *How the Steel Was Tempered*.[3]

2. 1 *jin* = 500 grams—approximately 1 English/American pound.

3. This refers to the novel by the Soviet writer Nikolai Alexeevich Ostrovsky (1904–1936), *How the Steel Was Tempered*, in which Pavel Korchagin broke up with his teenage girlfriend Tonya Tumanova in a railway station on a snowy day. For an English

In April 1969, the Ninth National Congress of the Chinese Communist Party met, but conditions remained unsettled throughout the country. The Xuzhou Branch railroad office was consumed with factional struggles and no one was going to work. Military officials of the Railroad Ministry met and sent down an urgent order for the Zhengzhou Branch to send people to help, and at the end of April our subgroup received an order to go to Xuzhou. Starting at the first small station, Taoshan, south of Xuzhou, we repaired the worst areas along the tracks as necessary to maintain the Beijing-Shanghai line. We went directly to Bengbu and the bridge over the Huai River there (once you got onto this bridge, you were into the area covered by the Bengbu Branch railway office). We worked on a stretch of 160 kilometers of track and moved our dormitory train seven or eight times. During this time our living and working conditions were really difficult. Sometimes we would be uncoupled on a side track near the river, and every night we could sleep soundly despite the whistles and clattering of passing trains. In our spare time, we walked around the area near the railroad. This was the area where the famous Battle of Huaihai took place.[4] We would identify the names of places mentioned in [Mao's article] "Military Principles of the Battle of Huaihai." We could not help but sigh deeply when we saw all the poor villages and peasants in the area.

The new Yangzi River Bridge at Nanjing was completed at the end of 1969. Newspapers and radio broadcasts carried this great

---

translation, see Nikolai Alexeevich Ostrovsky, *How the Steel Was Tempered* (Moscow: Progress Publishers, 1964).

4. The Battle of Huaihai proved to be decisive in central China as the Communist People's Liberation Army (PLA) continued to advance southward in its showdown with Jiang Kaishek and his Nationalist armies. The period of this battle was November 1948 to early January 1949. The cities of Xuzhou and Bengbu were indeed nearby, as the writer of our article notes. This battle is also called the Battle of Xu-Beng, a shortened form of the two place-names. Jiang Kaishek was by this time using military equipment and munitions from the United States, but one major result of this battle was that these assets were largely exhausted. Railway lines were often critical, as was the case in this battle. The Nationalists hoped to cut off the Communists' advance toward their capital at Nanjing. The Communists created some advantages for themselves, including the infiltration of Nationalist divisions and the information this provided. These factors and the actual fighting brought victory to the PLA, but losses were heavy on both sides. This is regarded as one of the major twentieth-century battles in China.

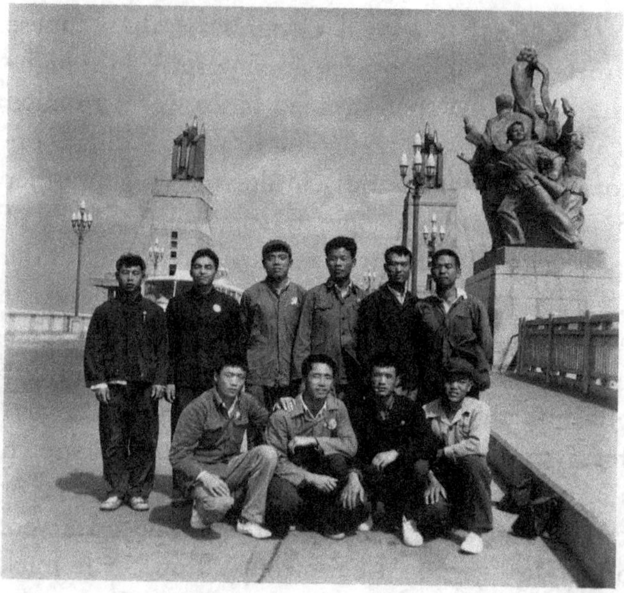

Photo 2: The young workers visit the new bridge in Nanjing in 1969. The writer is second from left in the back row.

news day after day, and it made us young people want to go see this great structure. As we were rather close to Nanjing, we went down to see the bridge. My workmates and I got a photo of this outing. (See Photo 2—I am second from left in the back row.)

When we finished our work at Xuzhou, our dormitory train was pulled to a spot on the north bank of the Yellow River at Zhongyi Station on the Beijing-Guangzhou line, where we were to continue with repairs. By summertime, the wheat growing along the tracks had ripened from green to yellow. The work was hard, but by this time I had gotten used to it. During the summer, our work had one nice advantage. You could get up quite early and get most of the work done while it was still cool, and then you had the rest of the afternoon off. We heard that a nearby village showed movies, so we could walk ten or twenty kilometers to see the movie! Not long afterward, I was transferred to the brigade's team for Propagating Mao Zedong Thought. To spread the "Spirit of the Ninth Congress,"[5] we

5. This refers to the Ninth National Congress of the Chinese Communist Party held in Beijing, April 1–24, 1969, at the height of the Cultural Revolution. The Congress officially ratified the political purge of Liu Shaoqi (1898–1969), then president of China, and formally designated the military leader Lin Biao (1907–1971) as Mao's heir.

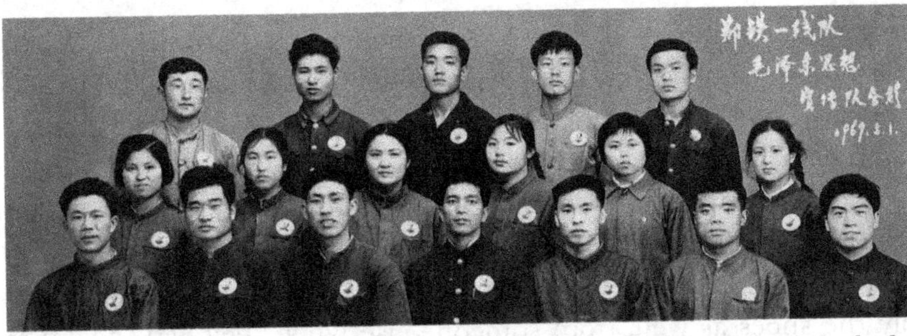

Photo 3: The work brigade's propaganda team. The writer is third from right, back row.

prepared a number of small performances that we presented to other subgroups in our brigade and in nearby villages. (See Photo 3 for all the members of our team. I am third from right in the back row.)

The members of our brigade had two kinds of status. About one-third of us were formally trained professionals working on the railroad, called "road workers." This group's work cards had black covers. The other two-thirds of the brigade were contract workers with red-covered work cards. Most of them were from Xuchang and Ta'he or nearby places. Most of them had worked in the brigade for at least two or three years, and most were young men in their early twenties. When we entered the brigade, we showed our black-covered documents, which made the other guys really envious.

These young contract workers were all simple and capable, and they interacted enthusiastically with other people. We were all about the same age, and most were Henan people, so we melded quickly as a group. After we had worked together for a while, I saw that many of them had good heads and were lively and humorous in conversation. They were smart, even wise, and a few were quite crafty! I often thought that if they had been born in a city and had the opportunity to go to school from the time they were small, they would be my equal in every respect. That I had a better position than they did was to a great degree due to the social environment and the differences between city and country.

As I spent more time there, I developed more knowledge and understanding of Henan villages and peasants. The practical problems that these young people faced were finding a mate and a place to live. For most of these young people, the difficulties of poor economic conditions at home were their biggest problems. When the young women from these villages saw that a relationship would not provide a good living, they would break it off and move on to another person. They were looking for a male who was a "formal worker," so our black work cards were often borrowed to be used as "stage props." Some of the young men who came from seriously disadvantaged backgrounds would borrow the clothes, and probably the shoes as well, from someone else when they had their first meeting with one of the girls. A few would even borrow a watch. (At that time the stainless steel "Shanghai" brand watch cost 120 *yuan*, which was a really "big deal" if your monthly wage was forty *yuan* or so.)

Building a house was another really expensive undertaking for most of the contract workers. At that time it cost eight hundred *yuan* for a three-room brick adobe house, the equivalent of twenty months' wages. To realize this "ideal" (or for some, a dream), they could only scrimp and save, eating only the most meager food. Some workers saving to build a house ate the cheapest food every meal. I myself have seen a few workers eat their *mantou* [steamed buns] dipped in salt water. Young people saving for a house or an apartment today don't have to make anywhere near such sacrifices. Since opening and reform began, what has impressed me most is the improvement of farmers' houses.

In mid-1969, we began work on the Jiao branch line. At the time we called it the "group war on the Jiao branch line." This was directly controlled by the Wuhan military district and the Hubei and Henan provincial military districts. Every county along the route had organized a militia unit. Because a militia made up of peasants had a great capacity for work, the excavation and the basic fill work were done quickly, but building bridges, cutting tunnels, laying rail and other such jobs had to be done by the trained railway crews. We of the Fifth Sub-Team were transferred to help with the "Jiao branch line group war," and on September 28, 1969, our dormitory train stopped at Loyang East station.

The first project for us was to build a bridge connecting the Jiao branch line to the Longhai line, the Loyang-Zhengzhou Highway, and the Zhongzhou [Henan] Canal. This was a six-pier two-level

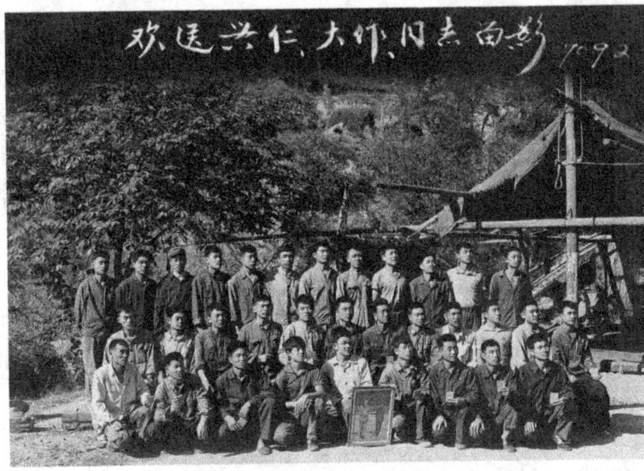

Photo 4: The team shortly before receiving new work assignments, summer 1970.

bridge. Of course, winter was very cold that year and I remember that in making concrete I was handling the mixer. On the night shift when we stopped to eat, I had to leave the water running all the time or the gummy skin would freeze in the delivery pipe. In a season like that, you can imagine how hard it was to come crawling out of your warm bed in the middle of the night to go join your work shift!

At that time, it was the fashion to write slogans in big characters on architectural structures. I could do artistic characters, so before we finished the bridge I was chosen to carry out this task. Even today when you cross this bridge, you can still faintly see the slogans I wrote back then: "Long Live the Invincible Mao Zedong Thought," "Long Live Chairman Mao's Revolutionary Line," and other such phrases.

When we finished building that bridge in the spring of 1970, our work shifted to laying track. We slowly advanced to the north a few hundred meters every day, and finally we reached Baihe [White River] Station. We lived in a village near the station, called Chenjiagou. We lived in cave dwellings, a kind of housing we had not used before. Inside it was especially cool but also quite damp.

At the time it was already hot outside. Our group chief asked me to deliver the hot water. At first I would carry a load two or three hundred meters and then needed to rest a while. My endurance got better. In less than a month, I could carry a load and walk more than ten *li* [three miles] without batting an eye. We had worked hard at our jobs for those two years, but deep down we hoped we might get some professional post before long. Finally, in July 1970, we received notice that we were to go to the brigade headquarters for a period of study so we could receive our next assignments. The brigade arranged a farewell gathering for us and brought in a photographer who took a group photo for us in front of the cave where we had lived. They gave me what was a popular memento in those days—a ceramic statue of Chairman Mao, which stands on my desk today.

This article is by Liu Dazuo. It was published in Old Photos *no. 45, pp. 121–27.*

Managing Exhibits for Propaganda

Introduction

This reminiscence recalls the years when socialism flourished in China, through its successes in a small rural county in Shaanxi Province. This is Zhashui County, south of the provincial capital of Xi'an and south of the Qinling Mountains that divide north from south in Shaanxi. It was truly an "out-of-the-way place." Yet Gao Xin, the writer here, and his colleagues brought an awareness of larger events in China to that county and contributed to some impressive economic achievements.

While revolutionary art is described mostly in passing, it is an important subject here. Many people around the world have some impression of revolutionary posters and other forms of popular art from the years when the Chinese were self-consciously building socialism. Much of the effort involved rural areas, where literacy was still limited. Thus visual experience remained powerful, and simple popular art communicated socialist goals. "Mao posters" were abundant, but peasant art in many local forms featured peasants working together, the work of "barefoot doctors," or villagers welcoming a young person "sent down" to teach in the local school. As noted in the article, Beijing University had famous folk artists on its faculty at this time. Peasant art retains great appeal today, but now it generally depicts the colors of everyday rural life.

The writer's sensitive portrayals of his colleagues offer insight into the lives of individuals during those times. He comments on the personalities and talents of each person, their strengths and their flaws, during a challenging period. The sketches of their later years, after an era of further change, offer pathos as well as insight. Sometimes modest socialist heroes did not age gracefully.

The Text

Beginning in 1963, the movement to "Build Socialist Education" began in Shaanxi. Zhashui is a small county in southern Shaanxi, tucked away deep in the Qinling Mountains. Its economy and culture are backward. Even among such remote areas of Shaanxi, Zhashui County was simply "out of it," almost forgotten by other parts of the province. But once this movement started, everything was done well and people didn't come across as being backward. In the Socialist Education Movement (also known as "the Four Cleanups") they made some impressive achievements. In Shiping Commune, they confiscated 137.5 *mu* (*mu*=one-third of an acre) of reclaimed wastelands and 6,740 kilograms of grain. By the end of that year, forty-eight households had no food at home. In Xialiang Commune, most of the leaders were criticized, removed from their positions, or otherwise pressured to change their ways, and probably even more of the commune members were corrected.

Such achievements led to publicity and propaganda. Near the end of 1964, the decision was made to establish the Zhashui Class Struggle Exhibit Hall. Led by the propaganda section and supported by the agricultural and industrial sections of the county government, with the help of other local groups such as the Youth League and the Women's Federation, the project was started in January 1965. Two months later, the county leadership convened a meeting of poor and middle-class peasants and began work on the exhibit hall. While it is difficult to explain completely how this happened, Zhashui led the way in establishing an exhibit hall! At the time the exhibit hall was begun, Comrade Hu Yaobang was sent to Shaanxi as the Third Secretary of the Northwest Region

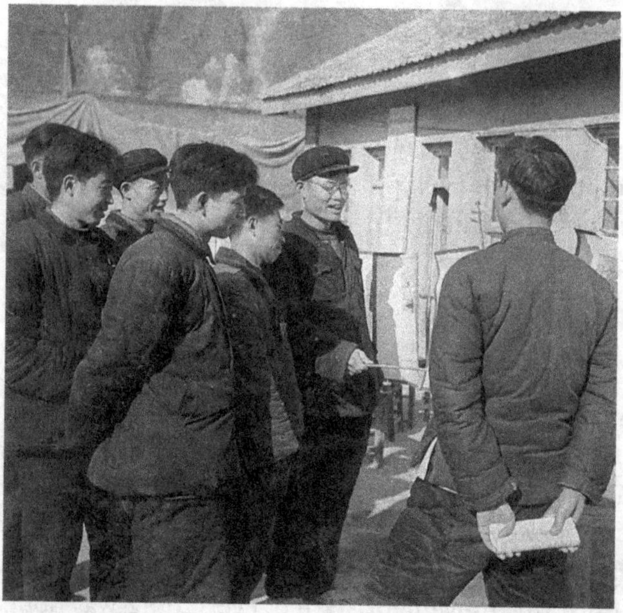

Photo 1: Members of the exhibit team listen as Yuan Zhenya describes the project. The writer is first on the right (back to camera).

and First Secretary of the Shaanxi Party Committee.[6] Soon after he arrived in Shaanxi, he made an inspection tour of rural areas. Hu Yaobang quickly downsized the scope of the Socialist Education Movement, which at the time was at its height, and sent his investigation report to the CCP central committee. Before his report reached Party Central, the Northwest Region Bureau started to criticize Hu.

In the specific aspects of managing the Zhashui Class Struggle Exhibit Hall, Xie Zilu dealt with written materials. Others of us were responsible for the artistic side. This included Lei Zaiyang and Qu Zhongxi from the Cultural Bureau, Dr. Li Xingmin from the county

6. Hu Yaobang (1915–1989) was among the most highly respected leaders in the Chinese Communist Party. He entered the Party as a teenager and joined the Long March. He always identified closely with ordinary people. His service in Shaanxi during the 1960s was one of many assignments along the way to his becoming Party Chairman in the mid-1980s. Hu was criticized from above for lax leadership, which led to popular protests calling for the greater reform that Hu sought to carry out. Hu was pressured, again from above, and resigned in 1987. His death in April 1989 made him a rallying symbol for the student and popular demonstrations that led to the tragic Tian'anmen Square episode of June 4, 1989. (Also see "A National Leader's Local Experience: Hu Yaobang in Nanchong" in this collection.)

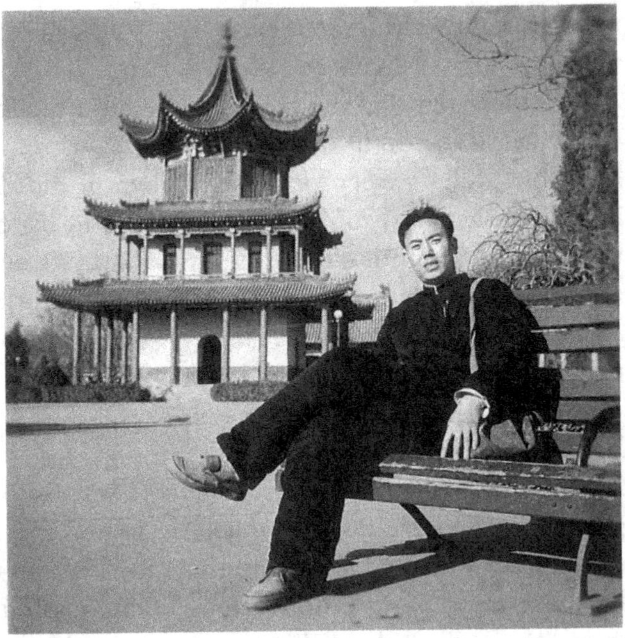

Photo 2: The author at Revolution Park in Xi'an after his day-long trek on December 19, 1964.

hospital, and me, who had been transferred from the bank. Teacher Zhang Zhenfang had the job of hand-copying the explanatory materials. The person who did oral explanations was not a beautiful young woman like today, but rather an ordinary young fellow who could speak with a desirable northeastern accent, Yuan Zhenya, the telephone operator at the post office. (These are the people shown in Photo 1.)[7]

When we were getting started with the exhibit hall, I was directed to go to Xi'an to buy materials. At that time there was no highway from Zhashui to Xi'an. You had to trudge along, step by step, cross the Qinling Mountains, a hundred-fifty *li* in a day-long hike to Xi'an.[8] Your back and your legs ached, and only after you slept for a day could you think about moving your body again. On my trip to Xi'an that time, the sports shoes I was wearing got completely ruined. That day in Revolution Park in Xi'an someone took

7. It is generally accepted in China that the best standard speech (*Putonghua*) is spoken in the northeastern provinces.

8. A *li* is one-third of a mile, so the trek described here would be about fifty miles, a challenging distance to cover in a single day in any kind of terrain.

my picture, and although it looks like I still had plenty of energy left, my shoes look really worn out! (See Photo 2.)

In those days, all the cadre in Zhashui were from outside. There was a saying that went "The mountains and rivers of the motherland are to be loved—except for those in Zhen'an and Zhashui." (Zhen'an is a neighbor county to Zhashui, known for its big mountains.) Today it only takes forty minutes to get to Zhashui from Xi'an, by taking an eighteen-kilometer-long tunnel that is the world's second longest. Zhashui is now called Xi'an's backyard garden. Forty years ago, this wouldn't have occurred to you in your wildest dreams.

Xie Zilu, Writer

Xie Zilu was older than most of us. He was from Luonan and had gotten the opportunity to study.[9] He joined the revolution on the eve of Liberation and shouldered a gun to serve in the army—except he never actually went into battle! When Zhashui was liberated, he stayed there as a cadre. Zilu was a capable writer and speaker, but he had never thought to write professionally. He gained confidence as he took on the challenge of writing and editing the texts for the exhibits. From chaotic materials he could see what should be criticized and discarded, what should be played up, and what should be left unsaid. Zilu gave the exhibit hall an organized format and presentation, and he did the originals for the picture-story books we wrote.

Not long ago when we had our reunion in Zhashui, Zilu gave me a draft copy of a play he wrote, *Fierce Fighting in the Roaring Gullies*. (See Photo 3.) He used materials from the exhibit hall, and his theme was opposition to following the capitalist road. I recall that it was performed during that year. I spent several evenings reading the play and was impressed with the way he showed what is right through a sense of the absurd. Zilu had worked hard on all the material that doesn't reflect his own character. Shortly before he left his post with the exhibit hall group, he wrote an autobiographical piece called *Spring Thunder*, which he had printed through the industrial

9. Here and later in this article, the author notes that others in the group that worked on the exhibition hall were from other counties—Luonan, Shangzhou, and Lantian. All are southeast of Xi'an, the provincial capital, and most are contiguous to each other and near Zhashui.

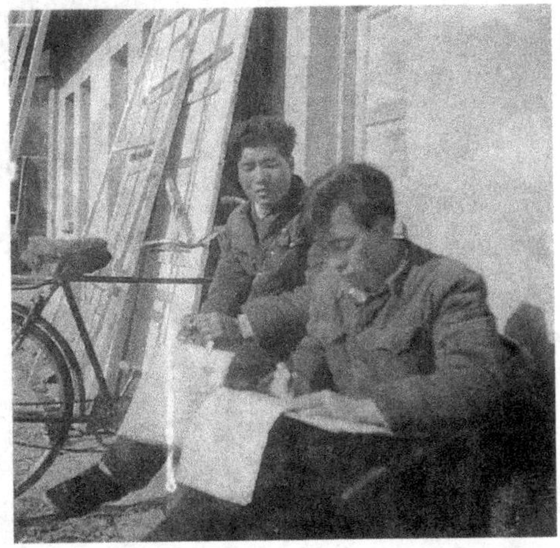

Photo 3: Xie Zilu outside the exhibit hall, making revisions on his play, Fierce Fighting in the Roaring Gullies.

management committee back in his home county of Luonan. I heard this was well-received.

From Zilu's ability as a writer as well as his intelligence and wisdom, you would think he could have had a career as a writer. But he didn't. This might be because of his own weaknesses—or was it perhaps because he lived in the wrong era?

Lei Zaiyang, Revolutionary Artist

Lei Zaiyang was from Lantian County, and at that time he had come from the Great Northern Wilderness of Manchuria to the Zhashui Culture Office. (See Photo 4.) I got very excited when I heard that he had come from the Beijing University folk arts program. At the time I was enthralled by the work of the Manchuria-based woodblock artists Chao Mei, Zhang Zhenqi, and Zhang Lu.[10] Zaiyang had also studied woodblock prints. He could act and direct, and he knew how to make people laugh. Every day he wore a grimy old army uniform, and he interacted well with everyone. The sign and slogans at the colorful main entrance to the Class Struggle Exhibit Hall were his work, and the parallel pictures on the two sides of the

10. The writer's enthusiasm for folk art reflects the common view that this form of popular art was not only excellent but also effective propaganda. An array of posters featured Chairman Mao and other leaders, or model peasants and workers, as inspiration for socialist ideals. This genre remains today in a variety of regional styles in lively peasant painting, now less didactic in approach.

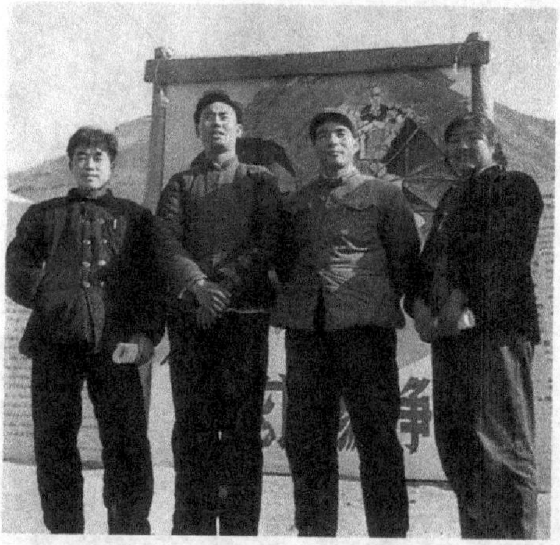

Photo 4: Qu Zhongxi stands at left; next is the writer, and third from left is Lei Zaiyang.

entrance were based on his originals. One of these showed workers, peasants, soldiers, students, and tradesmen marching forward under the red flag. The other was "Read Chairman Mao's Works, and Do What Chairman Mao Says." Another of his pieces showed poor and middle-class peasants struggling under the landlord's whip with fiery eyes blazing their readiness to resist and fight. The bold inscriptions here said "Never forget the bitterness of class. Always keep our vengeful feelings nurtured by blood and tears!" The series of picture stories in the exhibit hall were left to a few young people in the museum.

Zaiyang was the best-known culture and arts person in the county seat. During the Cultural Revolution, it was natural for such a figure to become a faction head. After this his life was not easy. It was probably in the early 1970s when, if you met him on the street, he lacked the spirit he had shown in those earlier years. He seemed to want to avoid people. Then even later on, I met him once at a performance after I had been transferred to the regional office of cultural affairs. Zaiyang brought the Zhashui group to participate, and only when they arrived did I realize that he too would be performing. We only spoke a few words before we headed off on our own ways.

A few years later when I went to Zhashui for a meeting to review the draft of the local history, a young woman comrade came over to speak with me. She was crying as she came into the room, and she called out, "Uncle Gao Xin!" She told me she was Lei Xuelin.

She was the typist for the local history project and Zaiyang's older daughter, and said that that her father was always talking about me. She remembered that I had visited their home twenty years earlier. Only then did I learn that Zaiyang had died. He had a cerebral hemorrhage just as he turned fifty. His emotions, and even more the difficulties of his life, had been the cause of his sickness. I thought I would visit his family the next day, but another local leader dissuaded me. He told me that Zaiyang's wife and small son were having mental problems, and that it really was no longer a family. I left some money and asked them to give it to the Lei family. Two years later a friend from Zhashui visited me, and when we talked about Zaiyang, I learned that Xuelin had committed suicide. How sad for Zaiyang! Who would have thought from the earlier part of his life forty years ago that he would experience such difficulties?

Qu Zhongxi, Folklorist and Social Analyst

Among those of us who painted, there was also Qu Zhongxi. We were both from Shangzhou. He told me that we had been classmates in middle school, where we were in the same grade but not the same class. His home was in a village. The family was poor, so he went to a teachers college.[11] After graduation he was assigned to teach in Zhashui. He could play the *erhu* and sing, and he could paint and write, so he was transferred to the county Cultural Office (see Photo 5). Not long ago during another visit to Zhashui, I heard that he too had died a few years ago. He was not yet fifty when he died. Of the four of us who did paintings back then, Zhongxi was one of the most active. Early in his time in Zhashui he taught in a primary school out in the country. He loved to sing, and he had learned many Zhashui folk songs—some sad and some risqué. Today these are seen as an aspect of folk art and are treasures we are encouraged to value. Back then, though, they were seen as "yellow" folk songs, lewd words and tunes that class enemies used to poison people's minds and paralyze their will for revolution. So they could only be sung on the sly. Sometimes we would be chatting amidst a pile of paintings, and Zhongxi

11. Teachers colleges or normal universities were free of tuition and fees. In addition, students were entitled to a living stipend.

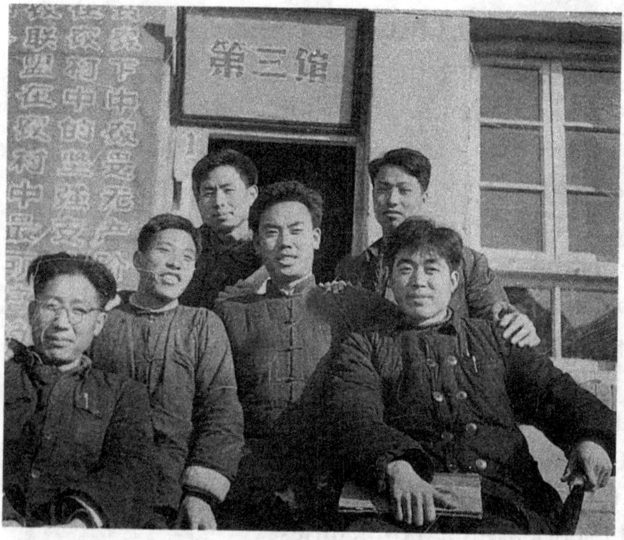

Photo 5: Zhang Zhenfang sits at the lower left. The writer is in the center here, with Qu Zhongxi at lower right. First on the left in the back row is Li Xingmin. This photo is from January 2, 1965.

couldn't resist our demands that he sing. I can still see him, singing in a light voice, eyes closed, deep into the song:

> During the first watch there was a knock,
> and my little lover came to my front door.
> Mother asked me what it was, and I said,
> "That's just the sound of the wind blowing."
>
> In the second watch there came another knock,
> My little lover came to my bedroom.
> Mother asked what it was, and I said,
> "That's just my shoes, walking on the floor. . . ."

One day Zhongxi was doing a section of a painting that showed a landlord beating a poor peasant. He made it according to our standard representation of a landlord—a short fat fellow wearing a robe and a formal hat, a tobacco pouch in his hand and a servant girl beside him. Zhongxi said:

> Are there any landlords like this in the mountains? The ones we knew there were in fact pretty poor themselves, and those here in Zhashui don't look like this at all! Before

> the land reform they did hard labor to cultivate the land like everyone else. They drank flour soup, boiled sweet potatoes, and ate cabbage at best. They too could get so tired from work that they'd spit blood. They might hire someone to help for a few days; that is what got them labeled as a landlord. A mountain landlord would not be hired as a servant of a landlord from somewhere else simply because he is too poor!

Actually, landlords are different in some ways. For example, generally they keep their houses clean. For the exhibit hall I painted a cadre, showing how he liked to visit landlords' homes to enjoy good meals, which compromised his class standpoint. I too have gone to the countryside and eaten a meal with a landlord. (I was sent by my superior.) And what did we eat at the landlord's house? Nothing more than plain noodles and boiled potatoes, but still, although the room was just four walls and nothing special, like a poor peasant's house, it was kept very clean. You didn't have to worry about what you were eating. Who likes to eat with a pig oinking and poking around in the room?

Last winter I went with the manager of the bank to Laolin Commune to check up on their credit union. It was already dark when we got there, and the accountant from that office went and got a woman to cook. With a dim light and a low fire, she produced a watery soup and took some pickled cabbage from a large crock, chopped it up several times on the pot-lid, and added some salt. The bank manager and I ate on the *kang*.[12] When people are hungry, they think soup like this is tasty, and pickled cabbage is certainly good. But there was something in it that you couldn't finish chewing. We took it out and placed it on the lid of the pot and then spat out everything. Do you know what we had been eating? It was the skin of a rat; you could dimly see the gray hair in the light of the kerosene lamp. Who knows when it got into the pickled cabbage pot? Even

12. A *kang* is the platform, usually at the rear of a rural home, that is outfitted with pipes or heating ducts from an attached stove that is usually the family's cook-stove as well. The *kang* is the one part of the house that can be kept warm during wintertime, and it serves as the place where guests are welcomed and also as the family's bed.

before I finished telling the story, several people already felt nauseated and urged me, "Don't say any more. Let's just do our painting!"

Li Xingmin, Doctor

Yet another of our artists was Li Xingmin, a doctor. He was a few years older than I, and was the best-known doctor in Zhashui. He took care of the county leaders, and he also treated the cart-pushers and hawkers. When a child from a mountain family got sick, they'd head to the county hospital and say, "Go find Dr. Li!" He said that when he got a call in the middle of the night, he could hear wolves howling as he got away from town. Sometimes he would get a telephone call from the countryside, and after walking as fast as he could for twenty or thirty *li*, he would arrive to find that the person had already died. He would just help dig a grave and bury the person. He was good at the arts. He could do the female parts of Shaanxi opera, singing shrilly and posturing coyly.[13] When we sat around talking, he was the one who was quiet and just listened.

Xingmin came from an upper-class family background, which is why he was always careful in his speech and prudent in his behavior. He went through life as a disciplined person, and such self-discipline kept him safe (until the Cultural Revolution, when he was accused of crimes—but no one was safe then). Later he taught at the Health College and after retirement moved to Xi'an. During this part of his life he studied the preparation of medicines and was able to make some money at this. When a group of old friends got together for a meal, he was always the one who paid the bill. Xingmin might not have been happy about that, but people assumed he was rich and could always do such things. Recently when I showed him the photos of our time at the exhibit hall, he said it was difficult for him to look at the photos. He felt as if the photos were from another world.

13. Shaanxi opera is one of a number of regional forms of Chinese opera. Until the past two or three generations, female parts were played by men. These roles were seen as difficult, and if a male performer did them well, he was regarded as an opera star. (Compare the article elsewhere in this collection, "'Play-Friends,' a Grass-Roots Sichuan Opera Group.")

Zhang Zhenfang, Teacher

Another in our group was Teacher Zhang Zhenfang, who was also from another area. He was the oldest of us, more than fifty. He was usually quiet but pleasant. When we sat around and talked, he almost never said anything, but he would occasionally raise his eyebrows or nod his head. His writing style was graceful and correct, without a single error, and his determined nature showed in his brushwork. When he was doing the copying, if he found an error or had a question, he would not change anything on his own but would ask us to check the original and make the change there. Then he'd continue copying. This way he would have written records if anything went wrong, and he would not be responsible for the error. We thought he was overly cautious, which sometimes made us a bit unpleasant toward him. Once when he and I were talking, I mentioned articles that I had written. He said he had read them, and they were really remarkable. He said he had studied *Water Margin* but had given it up.[14]

During the Cultural Revolution I went to Fenghuang once, where he was working then, and visited him at the middle school there.[15] He had been through rectification and his spirit was different from those few years before. In his small gray room, he had hung an enlarged photo of himself that showed him as he was young. Only then did I learn that he was originally from Henan and had lived in Yuan Shikai's old residence there.[16] After Liberation when he graduated from the university, he was assigned to a city to teach. Things kept piling up about his upper-class family background, and finally he was assigned to teach in this small town of Zhashui. Teacher Zhang had never married. What woman would want to share

14. *Water Margin* (*Shuihu zhuan*) is one of the most popular traditional novels in China. It tells the story of a band of 108 outlaws formed during the decline of the Song Dynasty. This novel is often compared to the story of Robin Hood in Western culture, although the swashbuckling stories of this band of Chinese outlaws are characterized more by internal loyalty than devotion to justice. Still, this novel has built a notion of the struggles of ordinary people against poor governance and the bond among them that can contribute to a better life for all.

15. Fenghuang is a small town near the regional city of Ankang and south of the Qinling Mountains. Geographically, this area could be a part of northern Sichuan Province.

16. Yuan Shikai's family home was in east central Henan province. Some years after his birth in 1859, his family moved to a hilly area and built a more easily defended home, which they named "Yuan Family Fortress."

the contempt of others? An elderly woman sat on the bed. He had brought his mother from Henan and the two of them lived together. Later they died and were buried there, a place far away from their hometown. Recently when I went to Fenghuang again, I asked about Teacher Zhang, but no one knew anything about him.

My Paintings About a "Historic Anti-Revolutionary"

I did a set of paintings on the "historic anti-revolutionary" Ke Changrui.[17] The accompanying text said he was from an area near Fenghuang and had been a low-level party secretary under the Guomindang. After Liberation he had slipped onto the revolutionary side and had gone to Zhen'an County as an elementary school teacher and continued his anti-revolutionary activities. The materials I had gave few details, but I had to carry out this assignment. Only later did I learn that when Ke Changrui started middle school, Jiang Kaishek's regime had already entered its last days. The Guomindang was still trying to recruit party members, and his supposed entry into the party and any activities he was supposed to have undertaken for it were handled by others. He was deceived by Guomindang operatives.

When the town went over to the revolution, false documents were turned in to accuse him. He was confronted with these and ordered to explain. Ke Changrui couldn't explain anything, and the case went nowhere. During the Anti-Rightist Movement, he attempted to clear his name, only to be officially labeled as a "historic anti-revolutionary." He was discharged from his position and sent to do labor reform. He needed to get his case cleared, but the more he demanded justice, the more difficult that became.

Finally, he decided to pretend that he had become a deaf-mute. Still later he went off to a spot in the mountains and made a living by working for others and gathering medicinal herbs. He lived in great difficulty as he waited for his case to be heard again. After many years, he got justice, and he began to learn to speak again and lead

17. The category of "historic anti-revolutionary" designated people who had belonged to other political groups and opposed the Communist revolution. Individuals so designated became victims of ongoing and sometimes vicious reprisals. As his story shows, Ke Changrui was a victim of circumstances but could do nothing about it.

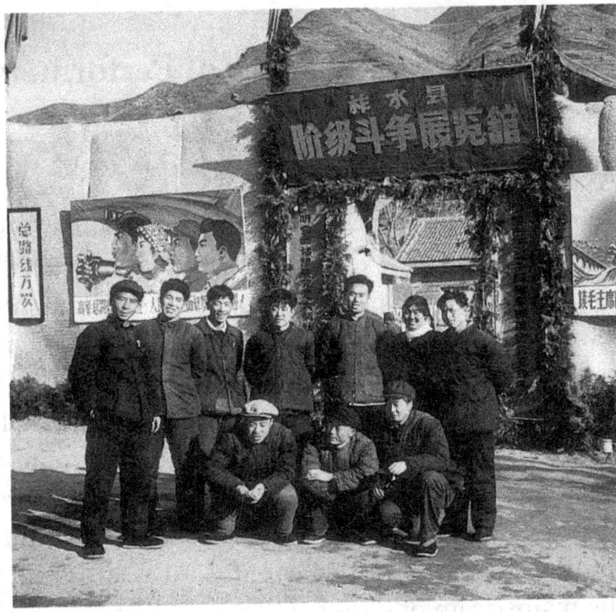

Photo 6: This photo was taken at the opening of the exhibit, when leaders came to see the exhibit. The author is third from right on the rear row.

a regular life. Ke Changrui returned to his earlier teaching position, and he always worked very diligently. He was welcomed by students and his fellow teachers, and eventually he became principal of Qingtong'guan Elementary School. But because he had been named an anti-revolutionary, he had lived half his life in an almost sub-human way. When he went back to his home village, his wife had left him and remarried, and his children had already left home. Although he was compensated with several thousand *yuan*, could any amount of money bring back the good years he had known earlier?

A Closing Thought

These were some of our activities and the later stories of some of our colleagues back then. It is worth noting that on the day the exhibit hall opened, the county leaders came. They praised us for our contribution to the revolution, and we had our picture taken with them outside that colorfully decorated gate, all to spread praise and make an impression. For dinner they spread a big banquet, and we ate a meal of Cantonese-style preserved pork and white rice—food that was hard to come by.

This is by Gao Xin. It appeared in Old Photos *no. 53, pp. 19–29.*

Not a Photo of a Group of Performers

Introduction

As the title says, this photo is not quite what it appears to be. The writer notes that the young man with the bicycle surely saw himself as stylishly modern. From another point of view, this entire group of young managers in a munitions factory all sought to present themselves in a modern form. Until some decades or merely a few years before 1948, when this photo was taken, the standard posture for a photograph was for the hands to be "empty" and placed on a person's knees. This practice tended to continue even into the 1940s in rural areas. The writer doesn't mention the potted plants, presumably also placed so as to make the photo look more natural. Jincheng is in southeastern Shanxi. The county seat was a small city but hardly in the mainstream of Chinese life at that time.

By 1948, when the writer says this photo was taken, the war with Japan was already over, but civil war had begun. This part of China was almost certainly controlled by the Communists for most of the 1940s, and the establishment of the group of weapons factories mentioned in the first paragraph supported the rear-guard and guerrilla activities of this region. That these factory leaders were so young is another indication that such operations grew out of the need to fight a war, and also that the Communist movement involved many in the younger generation of that time.

The Text

This is a photo of ten colleagues from the Number 2 Arsenal of the Provisions Division for the Shanxi-Hebei-Shandong-Henan Military District. This photo was taken in 1948 at Xinbi Village, Jincheng County, Shanxi, where this factory was located at that time. The Number 2 Arsenal and three other arsenals were responsible for weapons and ammunition for that military district.

My father is second from the left in the back row of this photo, the alert-looking one holding an *erhu*. He was a great marksman. At that time, age twenty-two, he was chief of the engineering section. First on the left in that row, the one with the moustache and

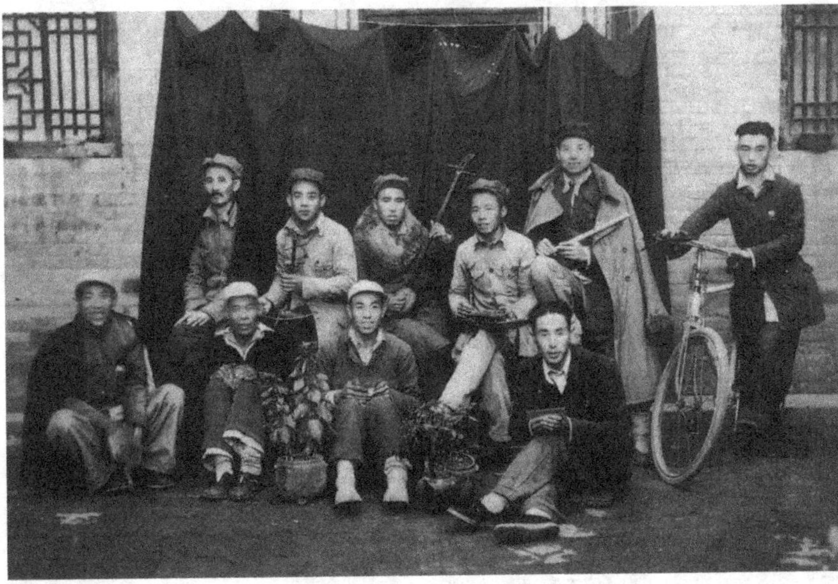

octagonal cap who looks like a Red Army man you would see in a movie, was the chief of the general affairs section and overall manager. At forty-five he was the oldest of the factory's leaders. At a glance you can see clearly that the one with the bicycle liked "the new style" and wanted to "keep up with the times"—his hairstyle and the little turned-down collar show this. For the photo, he grabbed the only bicycle the factory had. At twenty-three, he was the leader of the factory's workshop. Right next to the bicycle, holding a flute, is a square-faced fellow of twenty-seven, who was the head of the repair section and in charge of repairing firearms. At that time our munitions factories were not yet able to make firearms, so they mostly reworked firearms (captured from the Japanese). Up to that time the factories made hand grenades and land mines and only repaired firearms.

The first person on the right, front row, holding a brass harmonica, is the head of the foundry section, twenty-eight years old. The young man with the bright expression in the middle of the front row was my father's close friend. He was the head of the fulfillment section. While with the Taihang Mountain Area Army Youth Training Group, Father became very sick and was completely dependent on this friend to take care of him. Father's gratitude to him, even a half-century later, remained as strong as ever.

Up to that time, most of these men had never been photographed. The factory invited a photographer to come from a photo studio in the town of Dayang. The men thought they shouldn't be photographed without holding something in their hands, so each of them brought a musical instrument from home. In the front row, one is holding a harmonica but the other three are holding small cats. Then there's the "old Red Army" factory manager with a placid face, who is evidently more experienced, as he put his hands naturally on his lap. He looks calm, posing in the usual way when one has one's picture taken.

This item is by Xue Yanping. It appeared in Old Photos *no. 7, pp. 80–81.*

Tractors in Our Home Area

Introduction

Mechanization of agriculture in China was a halting process, complicated by the nation's changing foreign relations. William Hinton, the American known for ongoing effort in this activity, was among the first to assist, introducing tractors and other equipment in areas where they were workable through the United Nations Rural Relief Agency (UNRRA).

This kind of assistance ended after 1949, and the Soviet Union and its allies became sources for tractors and other machinery. Development was uneven, as the issue of agricultural mechanization divided Mao Zedong from Liu Shaoqi as differences between them grew during the early 1960s. Along the way, small workable "iron oxen" (*tieniu*) grew in appeal and met the needs of Chinese farmers in areas where agriculture had to be done on a relatively small scale because of the lay of the land.

There is a second term for "tractor" in Chinese that is generally applied to the larger machines used in American farming. This is *tolaji,* which directly translates as "pushing and pulling machine." This is used in the Chinese original of this article.

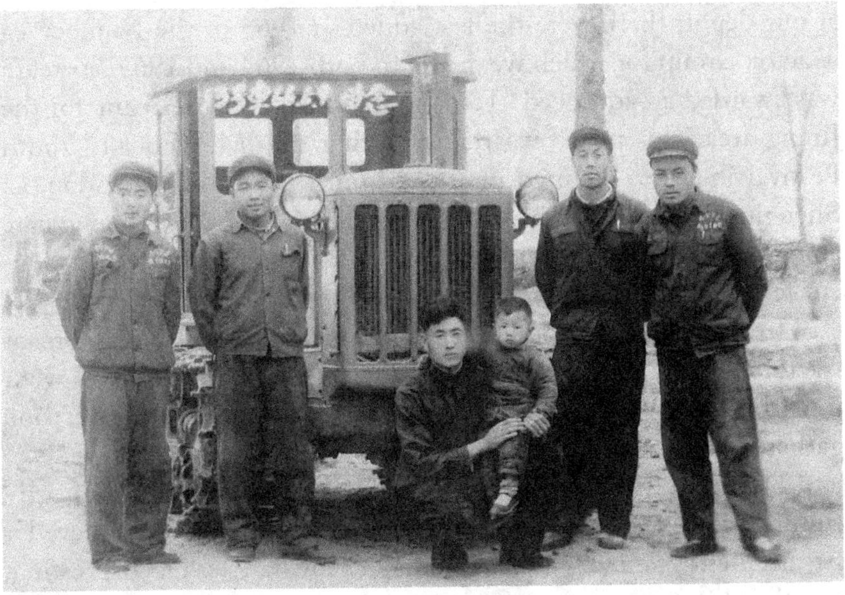

The Text

In the winter of 1959, along with more than sixty colleagues from all the communes in our county, I received a notice to gather at the Machinery Section of our county Agriculture Bureau. From there, we would go to Jining for training in driving tractors, their principles of operation, and the use and care of these machines. Then we would get practical training in Yanzhou at the China-Hungary Friendship Tractor Depot, where the training would conclude after six months.

Before the wheat harvest in 1960, Teng County received three higher-grade East Is Red-4 tractors, two imported wheeled tractors, a group of other kinds of agricultural vehicles, two T-3 tractor-harvester combines produced in Beijing, two large grain hullers made in Jiamusi, and sets of several kinds of farming tools.[18] Together, these made up the county tractor depot. All who had undergone the training became our home area's first tractor drivers, and I fulfilled my dream of becoming a tractor driver.

Before the Cultural Revolution, eight more tractors were added

18. A "wheeled tractor" describes almost all tractors, but note that the tractor in this photo is driven by a tracked mechanism.

at our depot; these were the last added in 1963 to the Number 13 Tractor Group for which we drove. In both 1964 and 1965 our team was awarded an advanced "Ten Thousand *Mu* Tractor Team" for the Jining area. This photo was taken during the "May Fourth" Youth Festival in 1966. Pictured here are, from the left, Min Fandou, Li Shengju (the writer), Shi Yingzhou and his son, Nie Guangyuan, and Qin Maoyu. (Lu Chuanzhu, one of our members, was not present for this photo.) At the time, I was chief of the tractor team.

In those days, automatic tractors were rare, and our group of tractors made a very good scene wherever we went. When we went into the local villages, we accompanied the leaders and everyone gathered around to have a look. We felt proud of ourselves!

Since the Reform and Opening Up, the mechanization of agriculture has developed quickly.[19] My hometown has truly achieved a streamlined system of production in our plowing, sowing, harvesting, and threshing.

This brief article is by Li Shengju. It appeared in Old Photos *no.47, pp. 118–119.*

19. This refers to China's post-Mao economic reform, initiated in 1978 by Deng Xiaoping.

FOUR

Professionals Serving the New Nation

An Anthropologist on Crutches Studying an Ethnic Minority

Introduction

This article introduces the effort made during the 1950s to identify the great variety of minority peoples in China and determine their populations and living areas. This was in order to work out a system of representation for them in the new political institutions called "People's Political Consultative Conferences." This discussion of Pan Guangdan's research on the Tujia people, written by his daughter Pan Naigu, deals with only one of many such projects, which ultimately concluded that there are a total of fifty-six ethnic groups. One of Pan Guangdan's accomplishments was to show that the Tujia people are indeed distinct from the Miao. Overlapping of groups in many areas, especially in the southwest, has led many to observe that the original task of identification never reached its logical conclusion and that there might well be more than fifty-six ethnic groups.

For his efforts Pan Guangdan and many of his colleagues were named "rightists" in a new wave of political campaigns. So the effort to assess the roles of minority peoples, one of several idealistic projects begun after 1949, collided with a calamitous political and social feature of the People's Republic—campaigns that turned individuals and groups against each other and brought disaster to incredible numbers of people. Exoneration of Pan and his colleagues after the landmark Eleventh Party Congress of 1978 recognized earlier errors and, in cases like the Tujia issue, correction of the policy involved.

Pan Naigu's article also shows how academics in many fields

set out to rebuild their disciplines after the Cultural Revolution. In academic life as in other activities, "reform and opening" set out to correct the extreme politicization of life. In sociology and anthropology this effort was led by Fei Xiaotong, whom Pan Naigu credits for this initiative. In those fields and many others, China once again was opened to visits by groups of scholars from many other countries and to a range of other changes that included the opening of archives to foreign scholars and opportunities to do the same kind of local research that Pan Guangdan had done in his time. Again, as his daughter observes, those earlier generations of scholars had made excellent beginnings before the ultra-leftist trends from the late 1950s to the late 1970s decimated objective scholarship.

The Text

My father, Pan Guangdan, would be one hundred years old this year. As people speak well of his academic achievements drawn from both Eastern and Western scholarship and the unavoidable dangers of his truth-seeking spirit and optimistic outlook on life, they also talk about his unerring ability to distinguish the characteristics of the Tujia people.[1] The path of his career in studying the Tujia people was difficult, but his ultimate success lay in having the respect of others in his field and advancing their efforts.

In 1991 we participated in a meeting to consider ways to advance this work in several areas in the adjacent provinces of Hubei, Hunan, Guizhou, and Sichuan. Professor Fei Xiaotong[2] spoke in this meeting and said his teacher, colleague, and neighbor Pan Guangdan

---

1. The standard Chinese term here translated as "people" is *zu*, as in "*Tujia zu.*" This term *zu* is often rendered as "nationality" by Chinese translators, in the approach also used by Soviet sociologists. Both modern nations sought to deal with a variety of concerns about the many minority groups of people within their borders. Most Western analysts would accept "ethnic group" as an accurate translation of what is meant by this Chinese term.
2. Fei Xiaotong (1910–2005) was China's leading sociologist. Having studied in England and the United States, Fei like so many intellectuals was victimized during the late 1950s and again during the Cultural Revolution. His rehabilitation beginning in the early 1980s was indeed essential in rebuilding the disciplines of sociology and anthropology. Throughout his career Fei was concerned to develop a genuinely Chinese approach in these academic fields. See the biography by David Arkush, *Fei Xiaotong and Sociology in Revolutionary China* (New York: Columbia University Press, 1981); also see Gregory Guldin, *The Saga of Anthropology in China* (Armonk,NY: M.E. Sharpe, 1994).

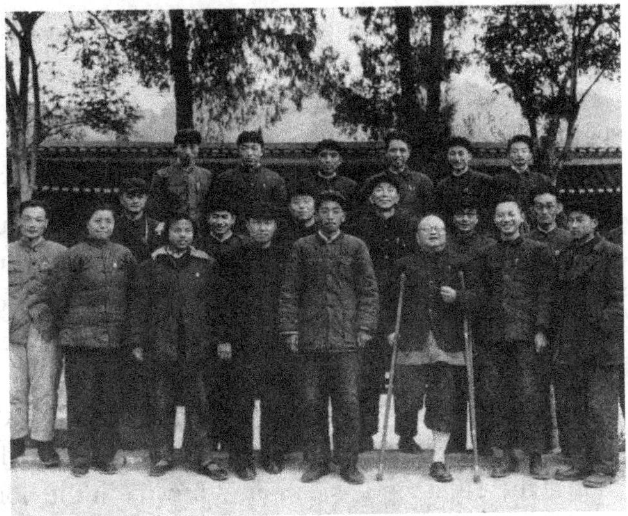

Photo 1: Pan Guangdan and leaders of Changyang County, Hubei. December 1956. (All photos by Zhang Zudao.)

had, beginning in 1956, done a great deal of research on historical change and geographic distinctions in this region, and he had done this through personal investigation.

> I always wanted to come here [said Professor Fei], but I never got the opportunity. I have meant to come for thirty-five years and now feel that I have almost made it up to him. I have come with his daughter, who is my research assistant and now works in the Sociology Research Institute at Beijing University. We also have with us Zhang Zudao, the photographer who accompanied Mr. Pan on his research trips. I hope all of us will continue the work of Pan Guangdan and not only develop a better understanding of the history of the Tujia people, but also continue to learn more about their current development in these areas.

In 1979, as Professor Fei's assistant, I began to participate in the work of reconstructing the study of sociology. After 1984, when Professor Fei completed his initial survey of small cities and county seats in Jiangsu Province, he decided the focus should be shifted to border areas and ethnic minority districts. By 1991 he was able to begin research in the internal region of the Wuling Mountain area and to bring together his studies of urban-rural development and of development in ethnic districts.

The Wuling Mountain area includes parts of Hunan, Hubei, Sichuan, and Guizhou.

It includes the Tujia-Miao autonomous prefecture and the city of Dayong in western Hunan, the Tujia-Miao autonomous prefecture in western Hubei, the Qianjiang area of Sichuan, and the Tongren area of Guizhou—about thirteen million people, 53 percent of whom are ethnic groups. This is a multi-ethnic area, and the majority Han population is also found throughout, mostly in the lower elevations. Most of the Han people live in lowlands and on transportation routes, while the ethnic people live in valleys or high in the mountains. The Tujia tend to be found mostly in the northern parts of this region and the Miao mostly in the southern parts. It is a complex pattern with relatively less concentration of population and alternating or crisscrossing pockets of population.

Professor Fei's research trip covered twenty-one days. He visited Fenghuang in western Hunan, Jishou, and went to Xiushan, Youyang, and Qianjiang in eastern Sichuan, then to Xianfeng, Enshi, and Laifeng in western Hubei, then back to Longshan, Yongshun in western Hunan, and from Dayong up into the mountains. Except for Dayong, all these were places where my father went in earlier years. Through my years of involvement in the reconstruction of the discipline and investigative research in border areas and ethnic districts, I gradually realized that it is important to review and understand the methods and thought processes of the older generation of scholars, the knowledge they developed, and the way they conducted themselves in their research. My father's research on the Tujia was one of these pioneer efforts.

After New China was established in 1949, the first question encountered in the work of establishing relations with ethnic peoples was the patterns of their distribution. Each of the groups should be masters in their own house. They should have political equality and their own representatives in people's assemblies at every level. So we needed to know how many ethnic groups there were, who they were, and how many there were in each area where they lived. All such questions needed to be included in a large-scale research effort that

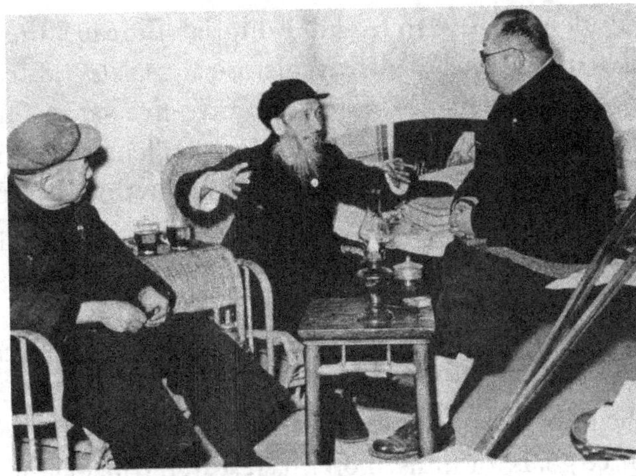

Photo 2: Pan Guangdan interviewing Tujia elders, Changyang county, Hubei, December 9, 1956.

would not be simply academic but also linked to policy and to the administrative system. Thus with the concern of our Central leaders during the years 1950 to 1957, teams of interviewers and researchers went to every known area where ethnic peoples lived to spread the word about the policies concerning them and carry out studies of these groups' societies and history.

In 1950 a woman teacher, Tian Xintao, of the Miao autonomous prefecture in western Hunan, participated in a group observing National Day festivities. This gave her a chance to talk with some of the Central leaders; she told them she was not Miao but believed there was another ethnic group with a different language and different customs. She asked that investigative research into this matter be undertaken. Her response then led Central leaders to emphasize this question and direct organizations in this field to launch investigative studies.

In 1952 Father was transferred to the Central Nationalities Academy in Beijing and was made director of the South Central Group of the Research Bureau. In 1953 he accepted the duty of studying the Tujia ethnic group. After researching historical registers and local histories, he prepared some 1,350 notecards on the Tujia and Ba peoples. He wrote a lengthy study entitled "The Tujia People of Northwestern Hunan and the Ancient Ba People," which was reprinted in *Collected Studies on China's Ethnic Issues*, number 4, November 1955, which was published by the Central Nationalities

Academy's Research and Editorial Bureau. The reference list for this article included fifty historical registers, fifty-two local histories, thirty personal notes, and fifty-five various traditional classics. By examining the Tujia's previous millennium of migratory history, geographic distribution, and cultural myths, my father carried out research to show that their origins lay in the Ba people—the Tujia are descendants of the Ba. Father's article received widespread approval from historians, and it had great influence.

In 1956 Father, as an assistant head of the Nationalities Organization under the People's Political Consultative Conference, joined their investigative team's visit to northwestern Hunan. Father set three goals for himself on this research trip: (1) for the scientific task, to compare what he had learned through his research with the conditions he and his colleagues observed and to correct any errors and supplement as necessary; (2) as a member of the Consultative Conference team, to develop an understanding of what the Tujia people themselves regarded as the distinctive characteristics of their ethnicity, the extent to which these had become general among them, and what basis those characteristics had in their own self-awareness; (3) again from a scientific point of view and as an observer, he sought to understand why the question of distinguishing the Tujia as a people had gone unresolved for so long.

In November 1956 Father, again through his position with the Nationalities Organization, joined a research trip to the Tujia area of Hubei and Sichuan. Accompanying him on that trip were Yang Zhongye, a reporter from the Shanghai newspaper *Wenhui bao*, and Zhang Zudao, a reporter for *New Outlook* (*Xin Guancha*); Zhang had been a graduate student in Qinghua University's Sociology Department in 1949, and his work in observation and as a photographer made him uniquely qualified for this assignment.

On the first research trip to northwestern Hunan, Father's group spent forty-two days, starting in Wuhan and ending up in Changsha, and they visited seven of the ten counties in the autonomous prefecture—Jishou, Fenghuang, Huayuan, Guzhang, Baojing, Yongshun, and Longshan. The second trip, to southwestern Hubei and eastern Sichuan, took fifty-seven days, starting in Wuhan and passing through Yichang several times, also going to Chongqing and

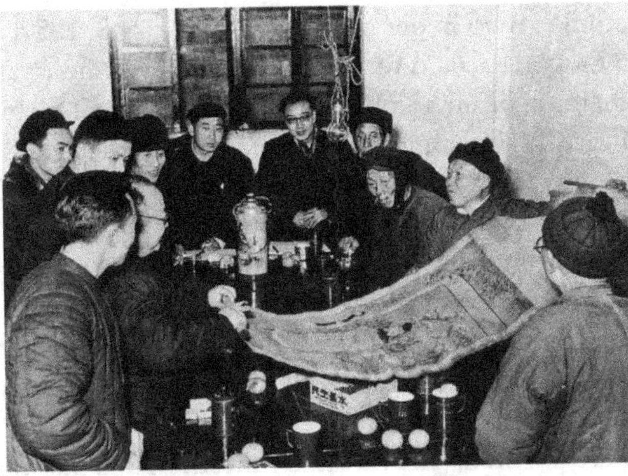

Photo 3: Pan holding a discussion with elderly Tujia people in Youyang, eastern Sichuan, January 3, 1957.

in the basic research effort visiting Tujia communities in Changyang, Fengjie, Wanxian, Zhongxian, Wulong, Pengshui, Youyang, Xiushan, Qianjiang, Enshi, Lichuan, Xuan'en, Laifeng, and Badong.

As Zhang Zudao recalled, on the second trip Father always took a direct personal interest in everything. He observed everything closely and was never quite satisfied that he had heard enough local reports. He always wanted to listen to the speech of the local elderly people, and he especially sought to hear their stories and their religious myths. The Tujia didn't have a written language, so these stories were passed on by word of mouth. Father wanted to hear the language to get everything clear and determine what was accurate and true. And this still wasn't enough. He wanted to see the actual things and visit the actual places, even though the distances were far and the way difficult. He intended to verify everything he possibly could. On Father's ability to overcome difficulties with an amazing willpower, Liu Zhenqian said in a memoir, "How can you catch tigers without going to their den?" Liu observed, "This gentleman is missing his right leg and can walk only with crutches. He gets about only with difficulty. He is very near-sighted. Daily life is very difficult for him. So why would he want to go to northwest Hunan with all its hills and valleys and such difficult terrain?"

After this research was completed, Father prepared "A Report on Our Visit to the Tujia of Northwestern Hunan," and in October 1956 the State Council recognized that the Tujia are a unique ethnic group in our country. On January 3, 1957, in accordance with a Party Central directive, the Central Administrative Office notified their subordinate offices in Hunan, Hubei, Sichuan, and Guizhou, formally directing them to recognize the Tujia people as an ethnic group. And on September 20, 1957, the Tujia-Miao autonomous prefecture was established in western Hunan.

But when the Anti-Rightist Movement began in that same year, Father's research group was charged with criminal activity. Tian Xintao, the woman in Hunan, along with Father and some others in his research group, were all declared to be rightists, and aside from them, many intellectuals and cadre in the local areas studied also were targeted.[3] At the same time, the Tujia question in eastern Sichuan and western Hubei was set aside. Not until after the reforms of the Third Plenum of the Eleventh Party Congress in 1978 following the Cultural Revolution were these things corrected. In 1983, Hubei province established the Tujia-Miao Autonomous Prefecture, and in the Yichang area set up Wufeng and Changyang Tujia Autonomous Counties. Also in 1983, Sichuan Province established the Tujia-Miao Autonomous Counties of Youyang, Xiushan, and Qianjiang, the Miao-Tujia Autonomous County of Pengshui, and the Shigui Tujia Autonomous County. Guizhou had Yinjiang Tujia-Miao Autonomous County and Yanhe Tujia Autonomous County. It had been fully thirty years since Father started working on the Tujia question when these issues of the distribution of the Tujia people and their autonomy were finally resolved.

When I went with Professor Fei in 1991, I had opportunities to talk with some elderly Tujia people in Jishou, Yongshun, Enshi, and Youyang, who had met Father or had gone out with him on his investigations. Their sincere attitude and their heartfelt respect for Father were very moving, and their comments reflected their thoughts about knowledge and intellectuals. One of them said:

---

3. Presumably this political attack was launched by people who had a stake in the political and financial value of maintaining only the Miao people's distinctive ethnicity. Such maneuvering was repeated countless times in the Anti-Rightist Movement of the late 1950s and the Cultural Revolution from 1966 to 1977. While Chairman Mao's system allowed for such power plays, local actors often caused the problems that occurred and did great harm to many thousands of these so-called "rightists."

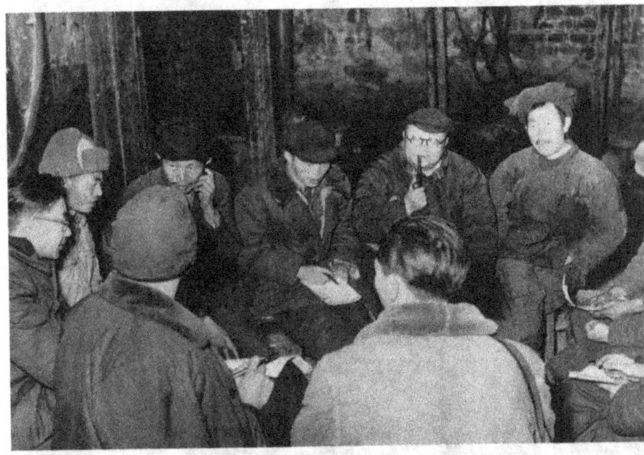

Photo 4: Pan leading a discussion with elderly Tujia people in Xiushan county, eastern Sichuan, January 6, 1957.

> We all understand that the Tujia people would not have gotten recognition without the Party's policy on ethnic groups or without the experts' work to validate our culture. Mr. Pan's article was a great contribution. Then he came and did his research, observing and asking questions, and he saw things accurately. Although back in those days there were some who disagreed with him and put great pressure on him, he stood up to their pressure and held to the conclusions of his own research. When we learned that he had been declared a rightist, we all shed tears. Elderly Tujia people revere him because he suffered injustice on their behalf.

Another person said:

> He was physically handicapped and it was not easy for him to get around, but he came to our home area where the terrain is so difficult. He went out early every day and only returned late at night, but even then he didn't rest. He would go through what he had done that day and make notes.

In Yongshun they regarded me—his daughter—as one of themselves. They had me put on a Tujia outfit that they had made especially for me and had me play the role of the new bride in their play "A Sad Marriage." All of this expressed their profound respect for Father. I hope he might know of this from his grave and be consoled.

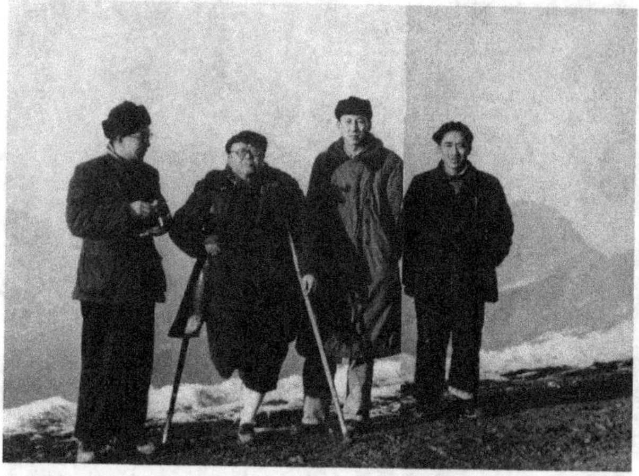

Photo 5: Atop Lucong Peak, Laifeng County, Hubei. From left, Yang Zhongye, Pan Guangdan, Zhu Jiaxuan, and an unidentified member of the group, January 16, 1957.

To recognize the existence of the Tujia people has been a task for our country, and the result of the research that Father and his colleagues carried out was the fulfillment of their duty, but the value of this work and its significance to society do not end there. As early as 1985, Professor Fei in recalling Father's point of view made an important point for future work in ethnic studies:

> It is just as Pan Guangdan said: the history of our motherland has been a process of contact, exchange, and blending of many groups of people whose ways are indeed different from each other. This process has been ongoing and is still unfolding. Beginning with this insight, from now on the task in our research should be to consider these issues from a broad perspective, as well as from detailed investigation, and bring these along together. The broad-scale perspective will enable us to advance the study of the composition of the Chinese nation. The task for detailed research is to understand the process through which each group of ethnic people developed.

Indeed, for the reasons outlined above, Professor Fei Xiaotong has continued to carry out his projects to "study the development of border areas and ethnic group areas" and "the rise and development of the ability to come together as the Chinese nation" and other such large-scale projects. When he took us with him to the Wuling area in 1991, it was to carry out this project. In 1994 Qiu Zeqi and I returned to Wuling and selected Laifeng in Hubei and Longshan in Hunan as two village societies to study. The photographer Zhang Zudao joined us on that trip. He was already seventy-two years old but still very active, and he carried his photography equipment and accompanied us. This was his third visit to these areas on Tujia projects, making his way again through difficult terrain that he had covered before. All the photos used in this article are his work.

Three generations of Chinese sociologists, anthropologists, and ethnic researchers have worked for forty years doing on-the-ground research in the homelands of a single ethnic group to understand the overall development of their ethnic heritage and their social and economic life. While not all their methods or the main issues they have studied are exactly the same, such pursuit of truth by intellectuals in service to the nation and the people are indeed part of the same great effort.

This article is by Pan Naigu. It was published in Old Photos *no. 25, pp. 121–129.*

My Father: A Civil Engineer's Life Before and After Liberation

Introduction

This story of a highly educated man, Tan Bingxun, who was devoted to his country, is typical of many others of similar backgrounds. This memoir was prepared by his oldest child, his daughter Tan Tianjian, who was clearly devoted to her father. She notes that her father hoped to "save the nation through science." This was a common outlook among many professionals in technical fields, who after 1949 continued to believe that their competence was in itself sufficient to strengthen China and were less attuned to the political campaigns carried on as early as 1951–52. Tan Bingxun saw enough of the flaws in the Guomindang leadership that his own views began to change.

Ms. Tan's simple analysis of her father's character discloses why he was vulnerable to those political campaigns. Introverted but completely devoted to his country, Tan Bingxun was not well prepared to deal with the harassments of repeated "anti-rightist" campaigns. Besides his family background, Tan had many connections with the Guomindang elite including Jiang Jingguo, Jiang Kaishek's son. Devotion to the nation was by no means enough, and Tan was a ready target for institutional leaders at the various universities where he was assigned. Among the details of their family life, Ms. Tan notes the many times their family moved, sometimes as promotions and sometimes as punishment, even before 1949. The family's last move, to Xi'an, came after his entire "unit" was relocated, as was sometimes done out of national security concerns. So this privileged family experienced instability and sometimes peril, even before the broad-scale campaigns of the 1950s began.

The Text

Our father left us about fifty years ago, but my feelings about him have not weakened. They have actually have grown stronger. These memories and my regrets are often bound up with each other. They bring tears and cause my heart to break.

Our father, Tan Bingxun, was born into an important family in Jinan, Shandong, but by our grandfather's time the family had begun to decline. Our grandfather was a lawyer, yet he didn't have great hopes for his children. After my father graduated from high school, he left home and set off on his own for Tianjin, and he examined into China's first institution of higher education, Beiyang University.[4]

While he was still a student, Father displayed his talent and independent nature. He believed in devoted effort in self-development and in using science to save the country. Not only did he concentrate on his major in architecture and engineering, he also showed concern for the welfare of the nation and building a just society. His thinking was very deep and his personality introverted. He wasn't good at socializing with other people and often looked as if he were somewhat indifferent, but he had a deep love of our country and

4. This is today's Tianjin University.

was enthusiastic in pursuing its salvation.

Father graduated from university in 1931. He then was appointed to an important position in the Public Works Office in Qingdao. During this same year an article he wrote, "Some Thoughts on Building the Nation's Defenses," was published in *Da Gongbao*.[5] This article was well-received by people in all circles, and later, Qingdao Middle School put it in their teaching materials. One of the Guomindang's founding fathers, Huang Fu, was also impressed with the article, and he called Father to Beiping. He regarded Father as a capable, talented younger person and presented him to a number of upper-level members of the Guomindang. So after this, Father was given important work and was promoted.

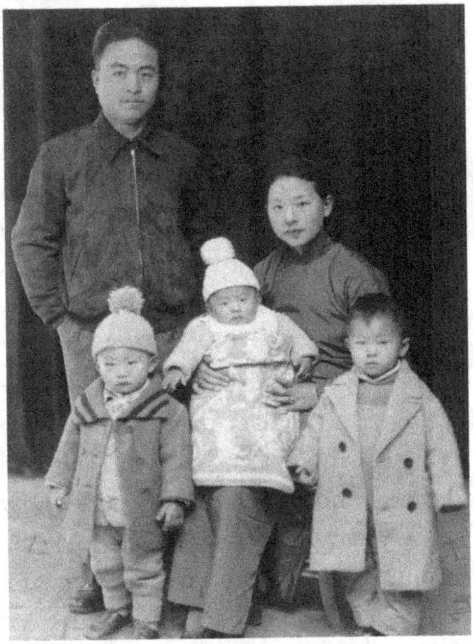

Photo 1: Tan Bingxun's family in Jiangxi, 1938.

In 1932 Father was made supervisor of the Qingdao waterworks. During this time, he became acquainted with the younger sister of a university classmate, very young and beautiful, who then was studying fine arts and was also a member of the Qingdao women's basketball team. This was our mother, Wang Shuqin. They fell in love and were quickly married.

In 1933 Father was transferred to Beiping and shortly was made head of the Beiping Public Works Office. At that time, he was just twenty-five years old and was the youngest department head.[6] By

5. *Da Gongbao* was a major newspaper during the 1930s and '40s, perhaps the closest thing to a truly independent journal in this period in China. To get something published in this paper was quite an achievement.

6. "Beiping" is the same city as Beijing. This new name, meaning "Northern Peace," was adopted after Jiang Kaishek's government established Nanjing as its capital in 1927–28.

this time the Northeast was already occupied by the Japanese, and Beiping's situation was growing tenser day by day. The Japanese puppet government's activities were getting more extreme every day, but there was little that patriots could do about these conditions.

By 1935, because of construction needs at Lushan, Father was sent there as head of the city's Management Office, a position he held until 1938. During these years Father renewed the famous sights, planted many trees, went forward with a number of new construction projects, and overall, made an important contribution to the development of Lushan.[7] Always following national affairs closely, Father became ever more indignant over Japan's aggression. In order to express his feelings about China's humiliation by the Japanese, he planned and built a new bridge called "The Bridge of Humiliation." These characters were carved into the main arch of the bridge. This bridge quickly became recognized as a landmark in Lushan.

At this time Father was also in contact with several high-level Guomindang people. He had a very good relationship with Song Meiling and regularly participated with her in meetings of the advisory committee of the Lushan Management Office. Once when we met Song Meiling along the road, she said I was lovably fat, and we got a picture with her holding me, and maybe saying something like "here's a solid girl!" During the Cultural Revolution thirty years later, this photograph and those few words became the "proof of my guilt."[8]

In 1938 Father was sent to Jiangxi as head of the Highway Department; this position was at Ganzhou. My own first memory of Father was during this period at Ganzhou. One day our family bought two children's bicycles. My little brother and I were very happy, and our father taught us how to ride. I learned how to ride very quickly, but I couldn't get on or off the bike. Father would always lift me off. From that time on, I had a special feeling toward bicycles. I

7. Lushan became a resort for well-to-do Chinese and foreigners during the late nineteenth and early twentieth centuries. Tan's assignment in Lushan might have been partly a reward to this capable young engineer, but he also had the skills to refine the infrastructure of a town that had been built piecemeal.

8. Song Meiling was Jiang Kaishek's wife. For Tan Tianjian, author of this account of her father's life, the problem was that her father was influential enough to have interacted personally with Madame Jiang.

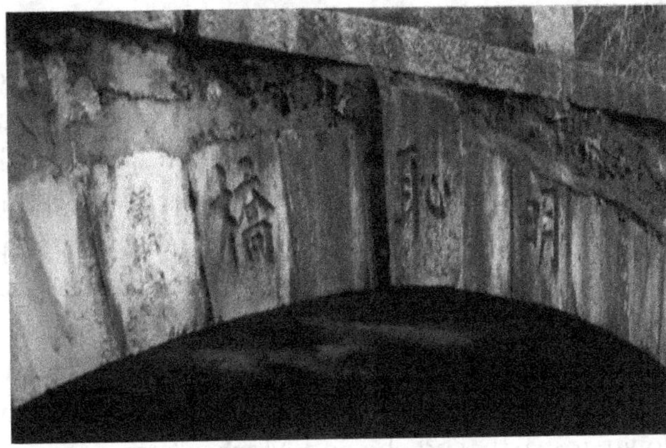

Photo 2: The "Bridge of Humiliation" in Lushan.

thought riding a bike was like flying, it was just so beautiful. When I was in middle school in Qingdao, I rode my bike to school every day. Every Sunday, Father would teach me how to wash and take care of the bike. Even now, in my retirement, I still like to ride my bicycle to go shopping, and it brings back those unforgettable memories of my father.

At this time Jiang Jingguo was the administrator in Ganzhou.[9] He had just returned from the Soviet Union and was very determined to get things done. Father was often in contact with him through his work. They were about the same age and had the same interests, and they enjoyed talking with each other. They both advocated resolute resistance against Japan and wanted to rely on science to save the country, so in time they became good friends.

In 1942 we went with Father to Chongqing. In order to avoid the Japanese air-raids, we soon moved to an area outside the city called "Mountain Cave." It was beautiful there. Except for a few houses, everything was open fields and mountain forests. Near the

9. Jiang Jingguo was the son of Jiang Kaishek, from the father's original marriage. Jiang Jingguo was one of a number of young Chinese who went to the Soviet Union for his education. His father's political orientation makes Jingguo's long sojourn in Russia seem unlikely, but so it was! The younger Jiang of course went to Taiwan, where he eventually became president (1975). In this last role he distinguished himself as the first Chinese leader to assent to a genuinely representative democracy when he ended the "period of tutelage," the stage of development laid out by Sun Yatsen to prepare the nation for political participation. For all the years that Jiang Kaishek ruled mainland China and Taiwan, he never terminated this as a façade for authoritarian rule.

top of the mountain was a church school called Holy Light School, where my brother Tianlu and I studied. After school my classmates and I liked to play in the fields, and we collected specimens of all kinds of plants. We especially enjoyed climbing the mountain, and we liked to slide down at a certain spot. We did this so often that we created our own groove! This became our favorite place.

During this period Jiang Jingguo's home also was in a suburb of Chongqing. On some Sundays Father took us to their house to play. We would play for the whole day, and I remember that we spent the night there once. Jiang Jingguo's wife, Jiang Fangliang, was very down-to-earth, and she and Mother became very close. They talked about almost everything. Jiang Fangliang spoke the Ningbo dialect, and she was interested in anything that involved China.[10]

For me, those days in Chongqing were probably the happiest in my young life. But for Father this period was quite different. At that time, he was head of the postal transport section in the Communications Bureau and was also a member of the Central Economic Construction Committee. He was responsible for transporting materiel, and his work kept him very busy. Usually he could only come home once a week, and sometimes he couldn't return home at all. That was the most difficult period in the Anti-Japanese War and also was one of the worst times in Father's work life. Besides these concerns, he was dissatisfied with corruption in the Guomindang at this difficult period of the war. He even wrote an article criticizing people for these problems, including the head of the Communications Bureau, who was so corrupt he couldn't do his job.

At that same time an underground agent of the Communists managed to buy off fuel from the postal transport section and get it sent to the liberated areas. The Bureau head used this situation to accuse Father and called on him to pay for the lost fuel. He also curtailed Father's efforts to have the British and the Americans investigate the matter. This episode had a serious effect on Father and took him to the brink of a mental breakdown. One day when Father

10. Jiang Jingguo married a Russian woman during his student years in the Soviet Union. Theirs was a happy union with several children. Jingguo's experience in the Soviet Union and his marriage were key features in his relative openness, especially as compared to his father.

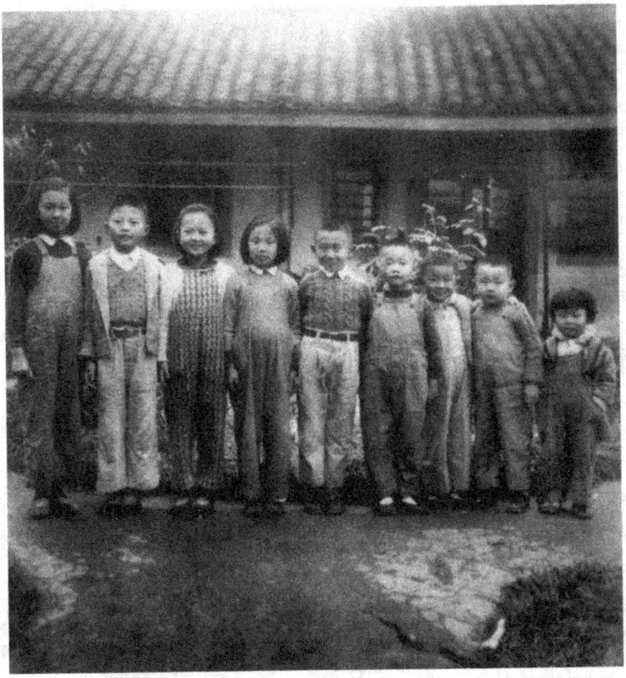

Photo 3: The four children of the Tan family at their home in Chongqing, with their playmates, five children in the family of Chen Cheng, a prominent Guomindang leader. 1944.

came home he felt very depressed, and after dinner he took me on his knee and said, "Jianjian, you must avenge me." When he said this to me repeatedly, I could see that he felt so helpless and dejected. This was hard for me to accept because I had always thought of him as strong and self-confident, but suddenly I realized he had a weak side. There were times when he couldn't deal with things that came up.

After victory in the War of Resistance, we went once again to Beiping, with Father serving as chief of the Beiping Public Works Bureau. During this time his spirit became much better and he recovered his old vitality. He had time again to take us on Sunday outings to Beiping's historic places or out in the country. I remember one time when he was especially happy. He told us about the project he had been working on to redesign the entrance and gate to the North Lake Park, and he wanted us to go see it together. After seeing the new gate, we walked along the lake shore lake and then enjoyed a cup of tea near the White Pagoda—a beautiful sight and a lovely memory!

Not long after we returned to Beiping, we moved into a house that Father designed himself. It was a Chinese-style courtyard house

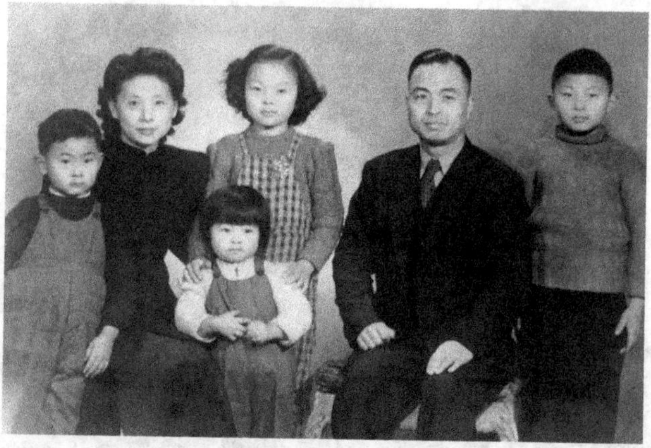

Photo 4: The Tan family in Beiping in 1946.

that included elements of Western design. We children lived in several rooms down the right side of this courtyard. All along the left side of the house were rooms where books were stored. Throughout his life, Father loved to buy and read books. He labeled them all and arranged them carefully on their shelves. This was his own library.

During the latter part of our time in Beiping, Father became deeply involved with a group of intellectuals headed by Hu Shi, who started a study association.[11] They discussed ways to achieve democratic government in China. Their goal was to build democratic consciousness among the people and to propagate the idea that China must follow the path of democracy. The group met at our house sometimes as well as at other homes. Father had realized it was not enough to hope to save the country through science. We needed to seek a way to build a democratic government. So he was really enthusiastic about this project and regarded it as a key to the country's future.

At the beginning of 1948 Beiping organized an election of

11. Hu Shi was a major figure in China's twentieth-century history, the country's most important liberal thinker. Hu was educated in the United States at Cornell and Columbia universities, and he earned a PhD in philosophy at Columbia. Returning to China in 1917, he joined the faculty of Beijing University and began to write for progressive journals such as *New Youth*. He remained a supporter of the Republican government and served as that government's last ambassador to the United States. He moved to Taiwan in 1949 and remained there for the rest of his life. This study association was representative of Hu's ongoing efforts to foster democracy in China.

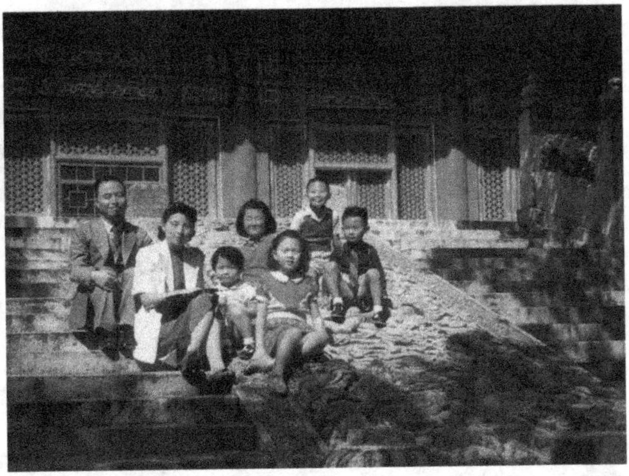

Photo 5: The family on an outing in Beiping, 1947.

representatives to the National Assembly. Father was extremely interested in the election and believed it was the dawn of democratic government. He took Tianlu with him to vote, but then he realized that the Guomindang people were buying votes and trying to steal the election. He felt the situation was hopeless.

In 1948 as the civil war grew more intense, Fu Zuoyi's army withdrew to protect Beiping and set up a temporary airfield at the Temple of Heaven. They wanted to cut down the trees, but Father opposed this, arguing that the trees were the result of long years of growth and should not be cut. He went there himself to intervene and firmly resisted these plans. Not long afterward, Father was transferred to Shanghai.[12]

In Shanghai we lived in an ordinary apartment building. Father was a consultant to a construction company, and my siblings and I went to school. It was 1949 and Father faced a difficult choice. There were three options: the first was to go to Taiwan, the second to go to the United States, and the third to stay in China. He seriously

12. This episode came when the Republican government was struggling to save itself in the Civil War against the Communists. To this day, visitors to the Temple of Heaven should thank Tan Bingxun for saving the "ancient" trees in that great complex. The trees are magnificent, and the Temple of Heaven along with many other temples of all kinds are the best places to see China's oldest and most beautiful trees. Tan's reassignment in Shanghai almost certainly was punishment for his stand regarding the trees.

considered going to Taiwan, and Jiang Jingguo urged him several times to do that. But Father thought there might not be much of a future there. Taiwan also might be liberated before long, and he was very familiar with corruption in the Guomindang. So he decided that Taiwan was not the best choice for us. He thought the best choice would be the United States, where some of his classmates and friends had relocated. Going to America would provide the most flexibility, but relocating there would take a lot of money. We were already a large family, with five children. Father said he didn't have enough money for that move, and if we got to the US and didn't have any money, what could we do then?[13]

Our house in Shanghai was very near Ba Jin's home. One day Father said that workmen had shut off the water to Ba Jin's house because he hadn't paid the water bill. Father said it was tragic that such an important writer was not able to pay his water bill. A bit later, Father suggested I write a letter to Ba Jin to let him know of my great respect for him, so I did that. What I had not imagined was that, not long afterward, I would get a note back from Ba Jin. It was short and simple but very cordial, and I was greatly moved.[14]

In 1950 father was invited to become a professor in the Civil Engineering Department at Shandong University, and we moved to Qingdao.[15] Both Father and Mother were very happy to move to Qingdao because it was Mother's hometown. Father was relatively happy with his work there. Besides his classes he began to do some research. He was extremely hard-working. Sometimes when I got up in the middle of the night I saw him still working on an article. Father's relationships with the students were also very good. One

---

13. Countless others weighed the same factors as did Tan Bingxun at this time. Had Tan imagined what lay ahead for him, he probably would have accepted the riskier possibility of leaving with his family for the United States.

14. The writer Ba Jin has been greatly admired by the Chinese. Born in Chengdu to an upper-class family, he left home during his student years and went to Shanghai, where he spent several periods of his lifetime. This particular period did not provide much income even for a well-known writer. His response to young Ms. Tan was typical, and one of many reasons why people loved and respected him.

15. The main campus of Shandong University is at Jinan. Apparently this department in Qingdao was a branch of the university, or perhaps the campus has been moved over the years since 1949.

New Year's Eve, he invited all the students who didn't go home for the holiday to come to our house to see in the New Year, so seven or eight of them helped make *jiaozi*.[16] We all remembered this happy occasion.

In 1951 the Party Center promulgated a "Directive on Launching Reformed Thinking and Organizational Rectification in Schools," calling for teachers and workers at schools and all university students to hold study sessions in order to root out the small numbers of anti-revolutionary people among older intellectuals. In 1952 Shandong University commenced this reform movement. Father naturally became a target. At first he didn't realize how serious the matter was, but as time went on things became increasingly tense, and all he could do was write one statement after another. He became downhearted, and his colleagues no longer dared to associate with him. This thought reform movement was later linked to the Three-Anti and Five-Anti Movement, and Father was asked about financial matters during the time he was working in Beiping.[17] Father saw no way to recall the details of accounts from memory alone. There was no way to present it all clearly. When he said this, the work group said he was being dishonest and had a bad attitude.

It was probably in the spring of 1953 that something happened we could never have imagined. Very early one morning, at the height of this political movement, Mother realized that Father was missing, and we discovered that he had tried to hang himself in the garage. But we were able to rescue him!

Then I happened on the note that Father had written, a small piece of paper he had placed under a stone. He had only written a

16. *Jiaozi* are a simple snack food, extremely popular in China and served in thousands of Chinese restaurants outside China, where they are normally called "dumplings" on the English menu. There is a range of fillings, including meat only (almost always ground pork), meat and vegetable greens, or greens only. *Jiaozi* are usually boiled, actually boiled repeatedly to produce "wet *jiaozi*" (*shuijiao*). "Pot stickers" (*guotie*) are another form of these dumplings, dry-fried on a griddle or skillet.

17. The Three-Anti and Five-Anti Movements were directed against intellectuals and former business people whose pre-1949 interests were opposed to the revolution. It was assumed that many such individuals were corrupt and might pursue their self-interest in their new positions. Tan's case seems typical, except that there was nothing in his background to suggest corruption.

few lines. Mother burned the note when she discovered it and told us that under no circumstances should we tell anyone that Father had tried to commit suicide. We should just say that he was sick, because at that time suicide would only compound the crime. It defied the goals of reform! I noticed once when Father was asleep that there was a deep red ring around his neck.

Even though Father's note was burned, I had already seen it. The note said he didn't want to reject what he believed or do anything that would contradict his own heart. I have never forgotten these words of my father. I have never told anyone about the contents of his note, but now I want to write the truth about Father. I think I finally understand him. He was unwilling to go against his conscience and would rather die than bow his head. Now I realize how difficult and painful it was for him to make that choice. We five children were still quite young. I was the oldest, just starting high school, and my youngest brother was twelve years younger. Mother was simply a housewife without the means to earn a living for our family.

In 1954 I passed the entrance examination for Fudan University's Foreign Languages Department. Father was very happy, and he saw me off at the railroad station when I left for Shanghai. During the years I was studying in Shanghai, he looked after me in many ways, and he wrote me a letter almost every week. He encouraged me to let him know what I was thinking about. When I told him I enjoyed reading books outside of class, he told me:

> Your enjoyment of books is really a very good habit. There are two ways to approach reading: one is to read in depth, and the other is to skim through a book. The classics and literary works are rich in content and should be read thoroughly, and most books by contemporary writers can simply be read quickly. . . . Shakespeare's works belong to the first kind, and popular propagandistic plays and novels belong to the second kind.

On May 13, 1955, People's Daily published "The First Set of Materials Regarding the Hu Feng Anti-Revolutionary Group," and then added a second and third set.[18] Thus began a nationwide struggle to criticize and investigate the Hu Feng anti-revolutionary group, which supposedly had expanded into a serious opposition movement. Once again Father became an important target of this campaign. One day a few school leaders came to search the house. They started in the evening and searched until the next morning. Finally it seemed they didn't find anything, and they took a few books and articles, and also took quite a few photos from our home.

In the autumn of 1955 Father was taken into custody. He was finally released a year later, with the conclusion that "we cannot conclude that he is anti-revolutionary, so he need not go to court." He had become very thin, his spirit was not good, and he had developed high blood pressure. He told Mother that life in jail had been

18. Hu Feng (1902–1985) was a long-time dissident in Communist China. This is a pen name; his original name was Zhang Guangren. By the 1950s Hu Feng was perceived by the authorities as the head of a literary and cultural "clique" advocating that one's personal viewpoint does not need to be sacrificed to an orthodox view of the arts in order to be valid literature for the socialist cause. Like many others among the two thousand or so individuals at first implicated in this movement, Tan Bingxun probably had never met Hu Feng.

From Hubei Province in central China, Hu Feng studied literature at Beijing University and Qinghua University, and then he studied in Japan during the late 1920s. There he became a Marxist, and he was expelled from Japan for radical activities. After returning to China he became closely associated with Lu Xun, the paragon of modern Chinese literature and culture. After Lu Xun's death in 1936, Hu Feng consciously sought to be seen as Lu Xun's successor, and he succeeded in this effort. Hu Feng edited journals and series of publications through which he tried to foster younger writers who agreed with his point of view. He nurtured younger writers carefully and gave much of his time to critiquing their work both before and after it was published.

Hu Feng made one major enemy during his earlier years, Zhou Yang, who became an influential literary theorist after 1949. Zhou supported Mao Zedong's views on literature and other forms of cultural expression, the latter expressed most forcefully at the Yan'an Forum on Literature and Art in May 1942. Mao argued that all literature is a form of political action and must display the proletarian point of view. Hu Feng countered this view by saying a writer retains his own personality and should express it in his or her work. To Mao and his followers, especially Zhou Yang, this was a form of "subjectivism," opposition to which was the motivation for their actions against Hu Feng. Hu Feng was imprisoned from 1959 to 1980, and finally was exonerated of the charges made against him. For fuller understanding of "the Hu Feng clique" in China's recent cultural history, see the excellent study by Kirk A. Denton, *The Problematic of Self in Modern Chinese Literature: Hu Feng and Lu Ling* (Stanford, CA: Stanford University Press, 1998).

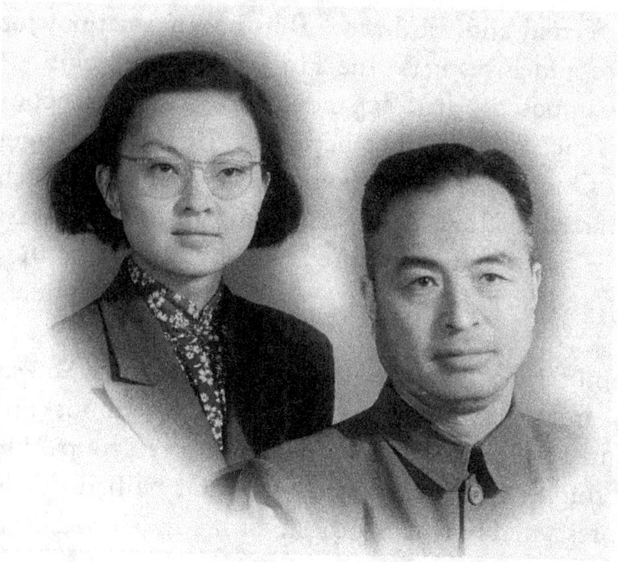

Photo 6: Tan Tianjian, author of this article, and her father Tan Bingxun, taken in Xi'an, 1957.

very difficult. They often ate mushy vegetable leaves and were terribly hungry, and roundworms even came out of their mouths because even they were hungry!

Father returned to his original school, Qingdao Technical College. Then the entire school was transferred, and Qingdao Tech's Environmental Department was sent to the Metallurgical Institute in Xi'an, so we moved to Xi'an. After this move to Xi'an, Father's health improved greatly and his morale also gradually improved. We were assigned a good apartment in first-class faculty housing. From our window we could see the Qinling Mountains to the south.[19]

In April 1957 the central government issued "Instructions on a Rectification Movement," laying out ways to implement an open rectification. They sought the opinions of people within the Party, and they also wanted to hear from those outside the Party and the masses of people. If there was knowledge about how to get things

19. The Qinling Mountains are an important feature of Chinese geography, dividing north from south and affecting rainfall north of this mountain range. The name for this group of mountains translates as "Qin territory," a reference to the original extent of the early state of Qin, whose ruler first conquered and organized the Chinese empire as *Qin Shi Huangdi*, "the Founding Qin Emperor." The modern province of Shaanxi includes territory south of these mountains that is rather different from that to the north.

done, people should talk about it fully and openly. There would be no crime for speaking out, nor for hearing people's opinions. At the beginning of May this movement began on a nation-wide scale, and the Metallurgical Institute joined in too. Big-character posters expressing opinions openly were put up, and the entire institute was excited by it all.[20]

The leaders mobilized everyone to speak out in meetings large and small. They invited Father to speak at the "democracy lectern," but at first Father wouldn't agree because he was worried. Then they asked him repeatedly, and after reviewing Chairman Mao's own speech about this movement, Father agreed to speak. He talked about questions involving intellectuals. He also spoke on dividing the three powers of government, and said that the legal system should be independent. He talked about how the life of peasants in villages was very difficult. He spoke about his visit to Caitai Village to investigate conditions there, and he took his data with him to support his views.

This movement began to change before the end of May. On May 15 Chairman Mao wrote "The Situation Is Beginning to Change," and pointed out that "the bourgeois right wing is now attacking." On June 8 the Party Center issued "Instructions on Organizing Our Strength to Strike Back against the Right Wing's Outrageous Attacks," and then a nation-wide anti-rightist struggle began. Later, Father told me that some formulations in Chairman Mao's original talk and his formal documents did not agree. He gave me two documents to look at, and he said, "I presented my views out of good intent and hoped that the Communist Party would govern well." On this I agree with Father because what he said was true, and I also thought, how did "speaking out is not a crime" become "speaking out can be a crime"? Before long, Father was designated a member of

20. The campaign introduced here is generally referred to as the "Hundred Flowers Movement," a term inspired by the intellectual life of the late Zhou period, fifth to third centuries BCE, when the variety of movements in thought were called the "Hundred Schools of Thought." The reference to these schools as "flowers" also harks back to that time. Tan Bingxun was right in being very hesitant about expressing his opinions, and his fate was worse than many, who even during that relatively early time were "sent down for re-education." "The "Great Leap Forward" also began in 1958, its goals economic rather than political.

the right wing, and then a bit later on it was decided he was of the extreme right wing.

That autumn, school began again and I went back to Shanghai. Later I learned that Father had attempted suicide again and was rescued again. Finally, in June of 1958 he was arrested. For the first few months he was in a jail in Xi'an, along with political prisoners. Mother went three times to see how Father was getting along, each time taking one of the children. Each time he looked worse, each time thinner than before. He wouldn't talk about conditions in jail, but only asked how the children were doing. Then as he was going to be moved from Xi'an to Tongchuan, Mother took our second younger brother, Tianju, to see him the last time. Father only asked about Tianju's schooling, and he said it might not be possible for him to go to university, but that he should try to obtain higher education if later there was an opportunity for him.

Father was then transferred to the Cuijiagou Coal Mine for labor and "reform" at Tongchuan.[21] During this period he was allowed to write letters home. He asked that the family send him newspapers to read. After getting permission, Mother sent newspapers every week. In spring 1959 a strange thing happened: Mother received a packet of newspapers back from the coal mine with a note saying, "This person is not found; return to sender." Mother thought there had been a mistake, and she mailed the papers again—but with the same result. So Mother wrote a letter of inquiry, and the reply came, "Tan Bingxun has committed suicide."

Upon learning of Father's death, Mother and Tianju immediately went to the Tongchuan Mine to find out how it had happened. They saw the place where Father was buried—just a mound by the side of a road. In summer of that year when Tianlu and I got back to Xi'an, Mother said she wanted to go to Tongchuan again to see the grave and to meet some of the leaders there. They were polite, and they took us to see the grave. Mother said the location was not at all suitable and asked that the grave be moved to a more peaceful and secure place. She also asked for a stone marker with the name engraved, so the grave could be located at a later time. The leaders

21. Tongchuan is a county about fifty miles northwest of the city of Xi'an.

agreed but decided to move the grave the next night. They didn't want anyone to know about it. They only allowed Mother alone to be involved in this entire process. Later when they saw that Mother would have difficulty climbing the hill, they agreed that our brothers could go along to assist her. So I waited alone at the place where we were staying—a night of foreboding.

We wanted to leave the next day, and the leaders agreed to put up a stone marker as soon as possible. Just as we were about to leave, Mother said she would like to visit the grave again to say good-bye. So this time the three of us went together; it was a very peaceful spot up on the hill, with lots of small trees all around. Mother picked some wild flowers and placed them on the grave, and she began to cry furtively. Mother always had a strong disposition. She had never cried openly in front of us, but this time she couldn't stop the flood of her tears.

In July 1979 the Shaanxi Provincial Public Security Bureau issued a "Revised Statement on the Arrest and Education through Labor of Tan Bingxun." This redress of Father's case made me reflect deeply and feel no end of sadness. In Qingdao early in 1990, Tianlu began gathering up all the traces that remained of Father's life, and I started helping him put them in order. I began to think of ways to prepare reminiscences of him and decided simply to recall Father as I remembered him, to write about Father as I knew him, and to put down my intimate feelings about him.

When I finished this article for publication, I decided to use what I wrote in 1990 in the first draft of my reminiscences of Father's life as the conclusion here:

> Father's life was one of struggle, one of bright accomplishment, and also a life of tragedy. Throughout his life he pursued truth, and showed concern for the country and for the people. He worked hard exploring ways to help the country through science and through democracy. He worked to engineer China's construction during the War of Resistance and to nurture human talent. He left a legacy that cannot be forgotten.
>
> Although Father left us early, his love for our country and

our people, his pursuit of truth, his ceaseless struggle, and his unbending spirit will always live in our hearts and admonish us to go forward. He lives deep in our memory.

This is by Tan Tianjian. It appeared in Old Photos *no. 52, pp. 17–1.*

From Soldier to Handyman

Introduction

Gao Malong, the subject of this affectionate memoir by his daughter, triumphed throughout a life of intermittent hardships. Orphaned during his early teens in Yunnan province, he was allowed to join the local army with the help of a sympathetic relative. His hard work and developing abilities through his military career brought Gao the stability that was absent in his early years. After 1949 he endured a number of frustrating and unfair episodes that would have discouraged most people. But Gao Malong always had an answer and maintained a positive attitude, the qualities his daughter most admired. This man's life provides a thought-provoking contrast to that of Tan Bingxun.

The Text

It's early in the morning, and again I dream of my father. As I wake up, I hear his deep country accent: ". . . the best season of the year is spring; the best time of day is morning. . . ." When I was small, I often woke up hearing my father recite poems, and now, as if it's another spring when flowers are blooming, I ask, "Father, where are you?"

Father was kind and optimistic, full of vim and vigor, and people found it hard to believe he was an old man past seventy. But after a medical crisis, unfortunately he left us forever on March 9, 1998. That day, I was a thousand miles away in Beijing. He had a great love of life and left his family so much love, but sadly, he had to close those kind eyes.

On January 6, 1998, as I hurried back to Beijing after a few days with him, when I was quite a distance from the door of his house, as I suddenly turned my head, my eyes met his and I heard him murmur, "Ah, no need to see me off!" This was the way we always said goodbye. Who could have known that would be the last time I would hear his voice!

Father's name was Gao Guochen. He joined the army when he was thirteen, and when he was sixteen he went off to fight the Japanese with the Yunnan Sixtieth Army. His service took him to more than ten provinces, and he fought in the famous Battle of Tai'erzhuang. After the victory over Japan, he stayed in Jiangxi. Following Liberation in 1949, he changed his name to Gao Malong to commemorate his old home, Malong in Yunnan.

Father was born in October 1921 in Tongquan Town, Malong County, Yunnan. When he was seven, because his father's pig strayed into the local tyrant's yard, his father was beaten to death. Father went to live with an uncle and had to drop out of school and learn how to tend cows at home. When he was thirteen, his mother died of poisoning from eating some wild greens. Father was left all alone and even had to wander the streets.

In October 1934, a local Yunnan general came to Malong to recruit troops. A relative in the area, trying to get Father into a good situation, asked a member of the military committee to take pity on this young orphan, and they accepted Father.

From this, Father began his life of hardship as a soldier. Right at the beginning, there was the problem of finding a uniform that would fit such a young boy. Father adapted a uniform, taking tucks in some places so he could wear it. An older soldier taught Father how to sweep the floor, how to prepare water for the officers to wash their faces in the morning and their feet in the evening, how to give them a little more to eat at mealtime, and how to wash the dishes and wipe the tables. After a bit, Father was sent to the bugle corps to learn how to play the instrument, but he didn't have the strength to blow hard enough to play well, so he was returned to regular duty.

After the Marco Polo Bridge Incident of July 7, 1937, the leader of the Long-Yun Provincial Army of Yunnan ordered the transfer of all the patriotic forces of the province to form a regular army and join the fight against Japan. Father's unit was affected, so they departed

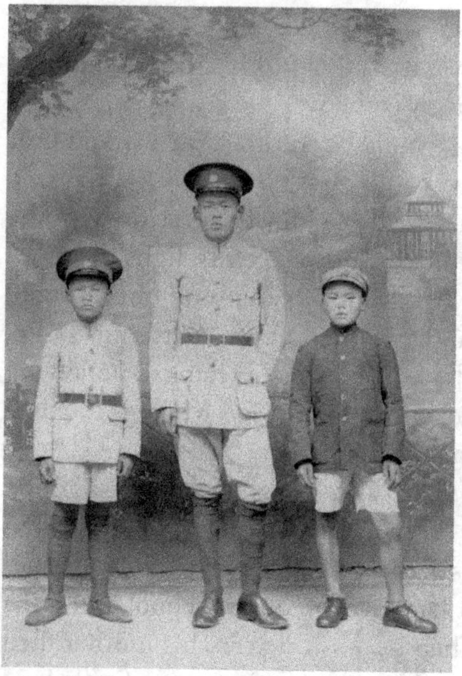

Photo 1: Father (on right) as a fifteen-year-old member of the Sixtieth Army. He was the youngest and the shortest soldier in the unit. Photo from 1936 at Jianshui, Yunnan.

Jianshui County in Yunnan as part of this larger army.

By the winter of 1937 this Sixtieth Army had finished its preparations. Before leaving, the division commander issued ten orders: (1) Do business in a fair way. (2) Compensate for any goods you damage or destroy. (3) Respect crops that the masses are producing. (4) It is not permitted to beat anyone. (5) Return all borrowed items. (6) It is not permitted to misbehave with women or to abuse private individuals. (7) The camp must be properly cleaned before deployment. (8) Follow orders in all situations. (9) It is not allowed to block or misuse vehicles. (10) You may not drop out of ranks. Violations will be dealt with by military regulations.

Early one morning before the army left on its new assignment, all the troops gathered for morning instructions. A soldier of the 1083rd Regiment was executed by firing squad. He had taken something from a civilian and had been reported by a witness. This execution took place in the public square of the county seat in Pu'an County, Guizhou.

In the spring of 1938 the Sixtieth Army was ordered to reinforce Tai'erzhuang.[22] When they arrived at Xuzhou, the bridge over the

22. Tai'erzhuang was notable in that it was a successful engagement for China's armies. As suggested in the text, Tai'erzhuang is near Xuzhou, in northern Jiangsu. Xuzhou was and still is a major railroad junction for both north-south and east-west routes. Up to this time the Japanese had advanced as much as they pleased, and they anticipated little

Grand Canal had been damaged by enemy bombing so that trains couldn't cross. The only thing to do was to have the troops, carrying lighter packs and equipment, cross the bridge on foot. At six o'clock that morning, as the troops advanced quickly, they began to exchange fire with the enemy. Japanese aircraft, armored vehicles, and tanks all joined in. The Yunnan Army could only rely on the courage of the troops and the decision not to carry much gear as they fought hand-to-hand with the enemy, and they handled the Japanese easily this time, showing them the strength of the Yunnan troops.

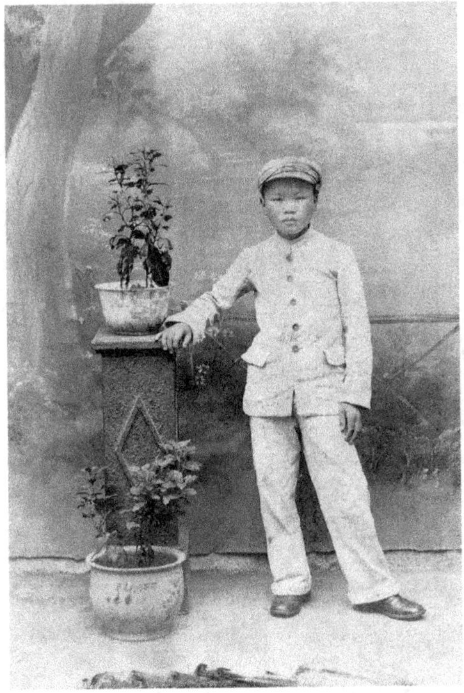

Photo 2: Father in 1937, stationed at Jianshui, Yunnan.

By this time Father was serving on the staff of the 541st Brigade of the 183rd Division of the Sixtieth Army, and because the telephone lines had been destroyed by enemy gunfire, he had to carry the orders of the staff leader to each of the regiments. Father told us that during these days at Tai'erzhuang, he had to climb over piles of dead bodies. In cutting across the Japanese blockade line, under his feet were decomposing bodies, and the stench and countless flies filled

---

difficulty. Chinese military leaders chose this situation to try to stem the tide, and they were successful. This major battle became a great victory for the Republic of China and its leadership, including Jiang Kaishek. The battle offers important evidence that the Nationalist troops did not lack imagination and spirit. Tai'erzhuang involved hundreds of thousands of less sophisticated Chinese troops and tens of thousands of better-equipped Japanese forces. Success for the Chinese at this point in a long war limited Japan's future capabilities and must be regarded as an important factor in the ultimate defeat of Japan.

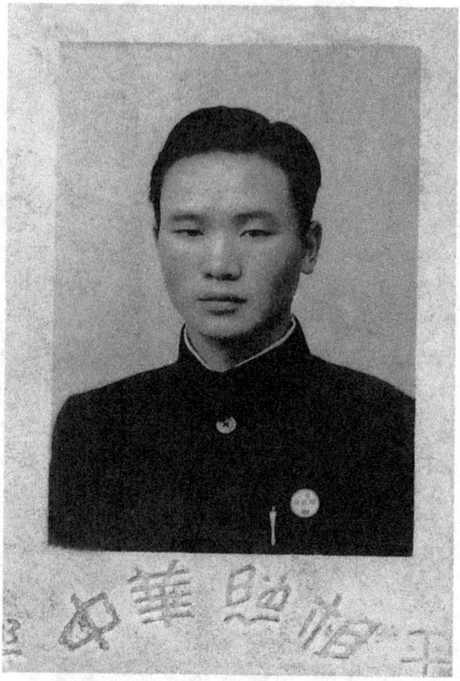

Photo 3: This photo is from 1945 after victory over Japan. At that time Father was located at Jiujiang, where this photo was taken.

the air—a tragic scene! But to carry out his duty, he had to take this in stride. The 1082nd Regimental commander Yan Jiaxun was killed in battle at the front in this struggle against the Japanese. Father had served with him throughout this time and held fond memories of him. Father felt fortunate to have survived eight years of the War of Resistance.

Father would talk of his experiences of that period from time to time, but we knew almost nothing of this forgotten history, so we couldn't respond appropriately. All Father could do was keep the sadness stored deep in his heart.

In 1957 when Father was working in Shangrao (Jiangxi), he answered the call for cadre to go to the countryside. He went with 160 other leaders to Ma'anshan Village in the Tianzhu Mountain and Forest Area. The village only had eight households, so not much farming was done, and their life was very difficult. About 60 percent of the families had no housewife, as most had died giving birth or from other gynecological disorders.

Father looked positively on and wrote of his experiences in the mountains and how he had further matured through that discipline. He said, "In this 'green ocean' I have had deep feelings, have gained much knowledge, have strengthened myself physically, and I have learned how to distinguish poisonous snakes and treat physical problems with herbal medicines."

During the Great Leap Forward it seemed best for Father to

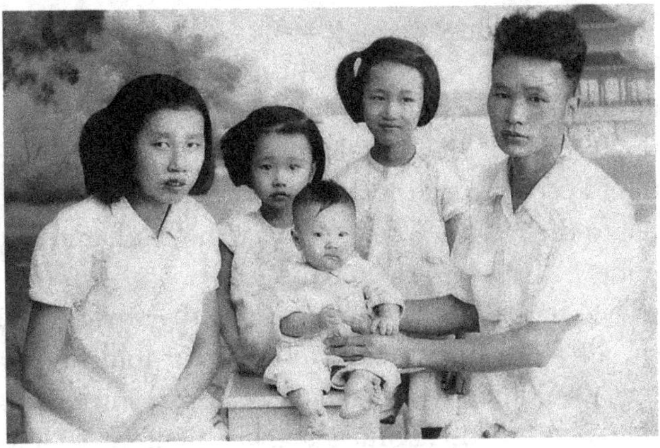

Photo 4: Our family in 1957 just before Father went to the countryside. Photo at Qianshan, Jiangxi.

continue to work in the mountain area; he wasn't able to come home very often. He said, "We work day and night, nearly driving ourselves to death. If you followed me as I move bags of grain, you'd see us making the 'output per *mou*' of three thousand or five thousand *jin*—or some newspapers say people are getting ten thousand *jin* or even twenty times ten thousand *jin* per *mou* (1 *jin* = 1 pound). We have a ten-*mou* plot (1 *mou* = 1/3 acre) as an experimental field and have asked people to come and take a look, to build public support. Twenty *dan* (50 kilograms or about 110 pounds) of millet, with sixty or seventy people taking turns picking—from south to north and then again from north to south. They change the pickers but don't change what's been picked, and then call it twenty or thirty thousand *jin*. Actually it's still the same few baskets of millet! This is just the way it is in this Great Leap Forward.[23] The farmers sweat year after year, and day after day they might get three to five *fen* (cents)." Father said that the Great Leap Forward wasted the wealth of workers. During the Mid-Autumn Festival in 1959, each person got half a *jin* of sweet-cakes. This was what the cadre working in the mountain area finally got for their work that year.

Father told a story about how during the great drive to make

23. Gao Malong's experience was only one such example during the Great Leap Forward, which while claiming to achieve great gains in production was actually failing to increase production in most parts of the country. The result was disaster and widespread famine, with the political result that Peng Dehuai and others challenged the leadership of Mao Zedong, who had promulgated the Great Leap policies.

steel, a decision was made to build an earthen forge in the valley, and in order to provide the fuel this furnace would need, they cut down all the trees for miles around. They even cut more than twenty trees that were at least a hundred years old, from the Yu Family Village in the forest district. These camphor trees and scholar trees were essential to the *fengshui* of the village. This Yu Family Village was also the home of Father's leader, Mr. Yu.

When Brigade Leader Yu told Father to take sixty men and cut all these trees, Father once again told him what the effects of that would be, and all the older people in the village, including the eighty-year-old Qi Shuashua, knelt before Yu and asked him to reconsider this decision. Brigade Leader Yu simply said, "I am a member of the Communist Party, and this is what the Party has decided. I can't change it." Everyone in the village, the elders and the ordinary people, embraced the trees and shared in this anguish over their loss. Brigade Leader Yu himself joined in cutting the trees, and these hundred-year-old trees were brought down in a single morning.

In March 1963 Brigade Leader Yu himself was sent down and returned home. Everyone refused to let him come into the village. They tore down his old house and refused to have anything to do with anyone in his family. Finally, Brigade Leader Yu had to leave his position. He ran out of money and starved to death at his mother-in-law's home.

During the Cultural Revolution, Father suffered persecution and pain, and he also faced the heavy burden of looking after us. He took the lead in starting an electrical machine repair shop, but when it became well-established, it was taken over by the neighborhood association. Father then started a bicycle repair business, braving the hot sun, wind, or rain day after day. He went to the far suburbs of the city and put up his booth, and then after a while he built a shop, thus taking care of us while keeping himself out of the weather.

From notes and money order receipts of his that we saw a few years before his death, we learned that over several decades, Father had sent contributions to the "Hope Project" to support education in poor rural areas. This reminded me of one of the lessons Father taught us: "Money is no more than dirt; virtue is worth a thousand." Father was staunchly upright, yet he was also a person rich in feelings.

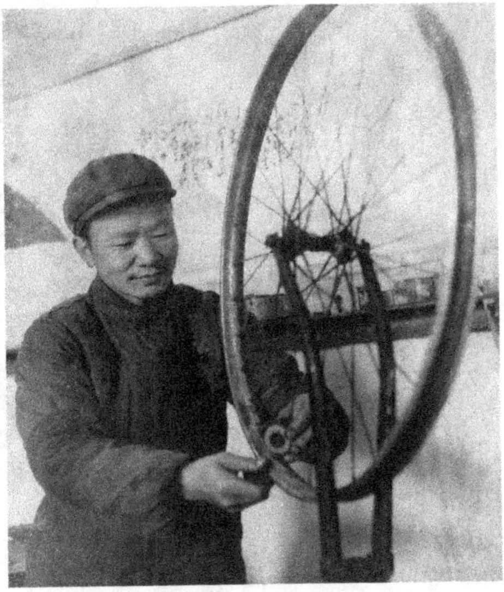

Photo 5: During the Cultural Revolution, Father opened a bicycle repair shop to take care of the family.

In the winter of 1980, at the age of sixty, Father had several bundles of porcelain bowls inscribed with the words, "Having left my old home forty years ago, I spare no effort to come back, even traveling a thousand *li*," and went back to see his friends and family and learn firsthand of what had happened in their lives.

Again in November 1996, when he was past seventy, Father returned alone to his old home in Yunnan and again made a contribution to the "Hope Project." For this, he received the highest commendation from local government: "He has come a thousand *li* to help. We of his old home place are deeply moved." A number of television stations and newspapers also reported these efforts in support of education. After returning to Nanchang (Jiangxi), he recounted a number of times the changes in his old home place over the decades, and repeated that old line, "I left home when I was small, and came back as an old man." He also thought things over and came up with a plan that he wrote out. He enthusiastically suggested that we take advantage of the natural resources of his old home area to create a woodland park. We could give some money to support it, and our family could also help to develop the park with our own labor.

Father always watched over us very earnestly and would recall hardships in his effort to teach us. If he saw that one of us had left a bit of food, he would observe, "When you see the plate filled with food, you know that every bite comes through hard work."

Father was a hard-working person. One summer, Mother gave him fifty *yuan* to buy a nice pair of leather sandals. But Father came home with a pair of plastic sandals he had bought out on the street

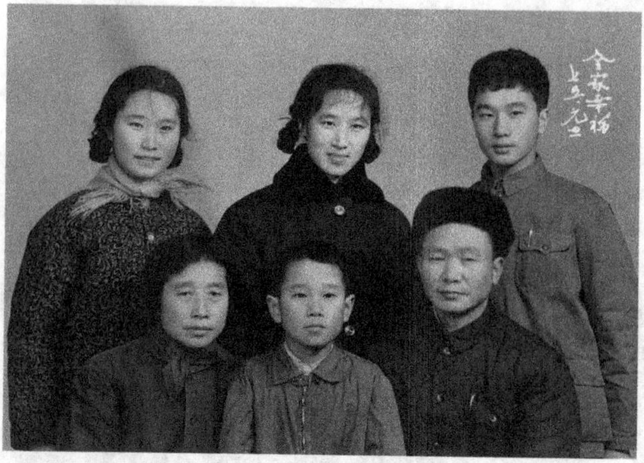

Photo 6: Our family on New Year's Day, 1975.

somewhere. Mother got awfully angry, but Father, like an old urchin, was secretly pleased with himself.

Once Father recited for us the "draft of a will" that he thought of on the "Six-Six" holiday of 1986, which said:

> I have worked hard throughout my life, have seldom been sick and have had no unusual diseases. Upon my death I wish to donate my body to a hospital for research purposes. It may be cut apart or used in any beneficial way. I ask no fee for this.

He then wrote this up into an article that he sent to a newspaper and asked them to publish it.

Father was seventy-five before he quit work. He often said, "The saddest thing is to be idle." During his senior years he wrote a long memoir, which he divided into a number of sections. He sent some parts of it to newspapers, offering them for publication. He regarded Yunnan as his "first old home" and Jiangxi as his "second old home." When I pick up these thick sheaves of his writings, my tears flow and my heart aches as I savor the details of his life again, and I am again filled with great respect for him.

"If you don't sleep through dawn, you'll hear the crows cawing all around." This is a poem that Father often recited. We grew up amidst the sound of his accented voice. Father loved springtime, and to this day I see his smiling face and hear that thick country accent in my dreams. Father was a book, full of memories and not easily understood!

This memoir is by Gao Yulin. It appeared in Old Photos *no. 32, pp. 113–120.*

FIVE

Campaigns

A Struggle to "Seize Power" in a Remote Mountain Village

Introduction

The "Four Cleanups" Movement (or "Four Cleans"), more formally known as the Socialist Education Movement, was the last nation-wide campaign before the Cultural Revolution. It began in 1962 and ran until 1965 and was largely directed toward rural areas. After the failures of the Great Leap Forward, Mao Zedong had been rather quiescent. This campaign marked his return to activity. Mao hoped to restore high ideals to rural life. The four areas to be "cleaned up" were work scores, accounts, warehouses, and property. (Throughout the rural economy during earlier years, points were assigned for the number of hours a person worked. The "work score" was the total of points accumulated, which would be converted into a monetary value to arrive at a person's wage. In this article the amount of pay for a work point is given as .012 *yuan*—no more than a few pennies at most.)

Another aspect of the "Four Cleanups" movement was to send students to rural areas to learn through practical experience. In the late stages an effort was made to extend the Socialist Education Movement to cities. By that time the momentum of inter-Party and leadership struggles was leading toward the Cultural Revolution, and the Socialist Education Movement came to an end. Objectively, and as this writer suggests, the movement was not very successful.

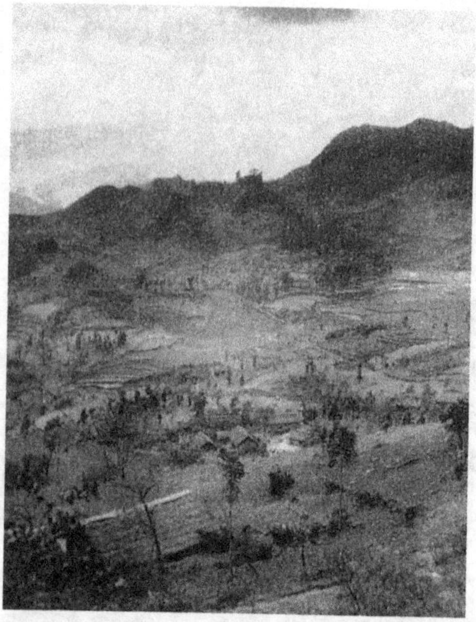

Photo 1: A bird's-eye view of Yuji Village in its mountain valley (photo by Ma Dazheng, June 1965).

This article also describes efforts made to address concerns about the "Four Elements," a second set of categories concerned with socioeconomic classes. These were landlords, rich peasants, counterrevolutionaries, and "bad elements." In some parts of China differences between landlords or rich peasants and ordinary peasants were not great. But such campaigns need individuals to target. It is noteworthy that the one specific situation presented here concerns a form of maternity leave—an issue that still is of concern in far more developed societies, including our own. The writer, then a young researcher, reflects on the implications of the report that was made and the photos he took during his team's involvement with these villagers.

The Text

Everyone who lived through the Cultural Revolution is unable to forget that powerful slogan, "seize power." This so-called seizure of power meant to "seize the power from those in power who are taking the capitalist road." But this brief narrative is about something that happened two years before the "Cultural Revolution" began, from my own experience of a struggle to "seize power" in a small, out-of-the-way village in Guizhou.

In 1964 I had just finished eight years of university studies and finally left my alma mater, Shandong University. I moved on to the Institute of Nationalities Research in the Philosophy and Social Sciences Division of the Chinese Academy of Sciences. The Philosophy

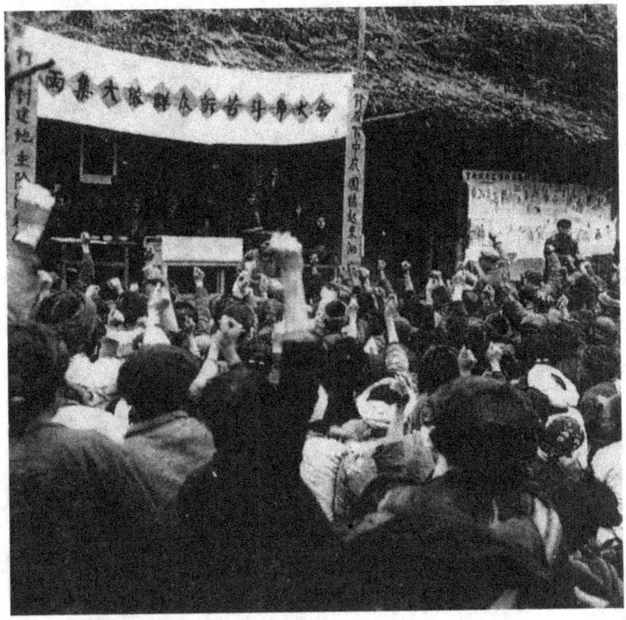

Photo 2: The chair's platform at the Yuji brigade mass meeting to "speak bitterness," i.e., to decry injury at the hands of those in authority (January 3, 1965, photo by Ma Dazheng).

and Social Sciences Division, located on Jianguomen Street in Beijing, was the "highest echelon" of studies in the social sciences. This was what society believed, and this was certainly the feeling of those of us who entered the division. From the autumn of 1964 until summer of the next year, under a unified leadership, the Institute of Nationalities Research, from the head on down to ordinary researchers—a few dozen people—set ourselves up in Guizhou Province, in Qinglong County, Jichang District, Yuji Commune, to participate in the Four Cleanups Movement there. I was assigned to this commune, the Yuji Production Team of the Yuji Production Brigade, as a member of the Four Cleanups Work Team.

From the time we entered the village, we sought to advance the processes of "taking root," developing social assistance groups for the poor, seizing power, and making careful analysis of the Four Elements. We tried to help the villagers develop production, re-establish Party branch organizations, and consolidate socialist positions. In the end, both the "denouncers" (this was called mass power) and "those being denounced" (this was called the object of revolution) all lined up in tears to see the Four Cleanups Work Team that organized to "struggle" against each other in the village. I participated in

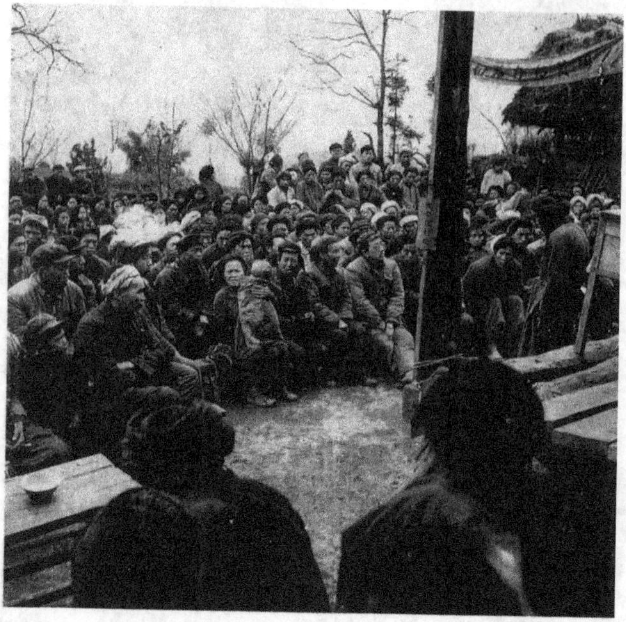

Photo 3: Luo Deming being "struggled against" (denounced) at a mass meeting. (Luo stands in front of the platform with his head bowed.) Sitting in the front row, wearing glasses and old army clothes, is the writer of this article, Ma Dazheng.

the entire process of the Four Cleanups in Yuji Village. I entered the village as a member of the work team, but by the time we left I had been promoted to a group leader, the lowest-ranking official. By the standard of that time, I performed well. I picked out a few old photographs from my own albums that present some actual parts of this historic scene. It is good that I still have them on hand. I also have a document entitled "The First Battle in the Yuji Seizure of Power (A Brief Report on Seizure of Power Meetings)" which I helped to draft. Looking at the photos and the briefing makes us remember and reflect deeply.

Yuji Village is deep in the mountains of southwestern Qinglong County. Here there are rows of mountains, bubbling springs, green trees, verdant bamboo, and lush growth all around. There are no real roads, only goat paths and mountain tracks for transportation. From the village to Jichang District you have to cross three mountain ranges—a four-hour round trip to walk. Yuji is a village of several ethnic groups, with Buyi, Yi, and Han all living in the village, a total of 221 people. At that time, Yuji was called a fortified mountain village. Yuji Village was managed by a commune committee, and the home of the secretary of the Yuji Brigade Party branch

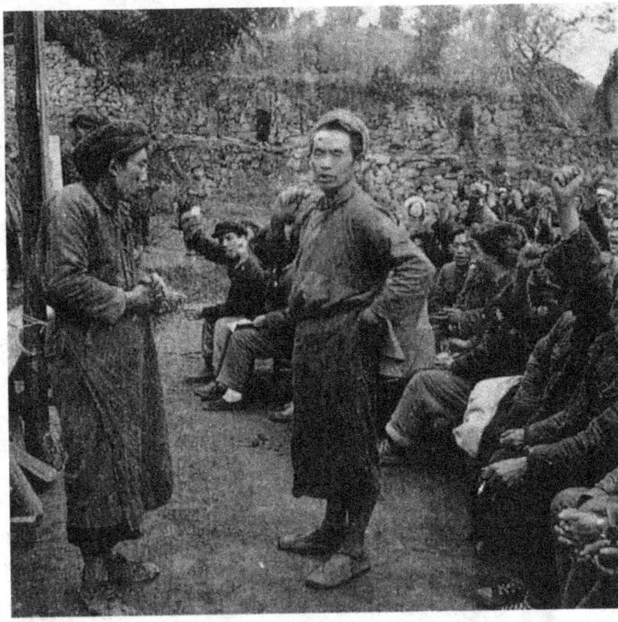

Photo 4: Fan Zhongchun, chair of the Poor and Lower-Middle Peasant Association, struggles against Luo Deming (photo by Ma Dazheng, January 3, 1965).

organization was also in the village. It was the political center of a commune and a brigade.

According to the brief report from that time, "The meeting place was set up in the middle of Yuji Village," and "at about nine o'clock the mass of people entered the meeting place. At this time a detachment of the militia organized a patrol group to maintain order in the meeting place and assure the safety of the meeting."

The targets of struggle at this meeting were Party Secretary Luo Deming (Buyi) of the Yuji Brigade, and Yuji Production Brigade Assistant Brigade Leader Ma Changlun (Han). The leader of the meeting was the preparatory committee of the Poor and Lower-Middle Peasant Association, established a few days earlier under the supervision of the Four Cleanups Work Team. The object was for the Association to "seize the power" of the Party Branch Secretary and the head of the Production Team.

From the gesture of the peasant in the middle of the participants, things were indeed the same as described in the "Brief Report": "A poor lower middle-class peasant and other peasants, using their earth-shaking strength, struggle against the degenerate element's representative of the landlords and rich peasants, Luo Deming and

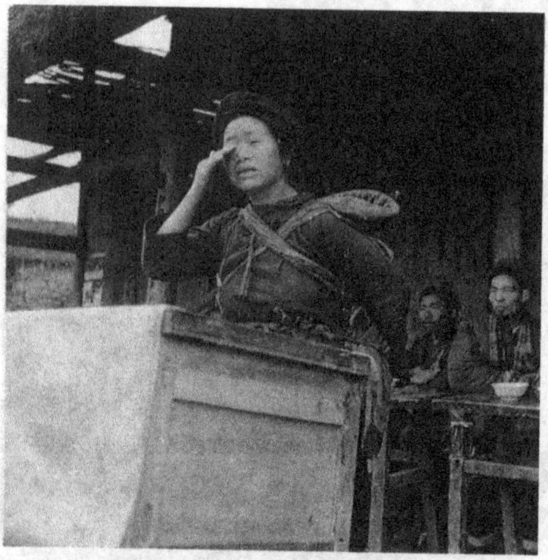

Photo 5: Huang Yongzhen, a poor peasant, weeps as she makes her complaint (photo by Ma Dazheng, January 3, 1965).

his accomplice, Ma Changlun. The people are shouting over their great victory."

From the materials brought to light before this struggle meeting, Luo Deming's major "crimes" fell into two categories. One was that he didn't distinguish clearly among different subgroups of peasants as to their land holdings, and that he led those who were better off in building new houses for their families. Another problem concerned his work style, which was sometimes gruff and rigid, especially during the time of the Great Leap Forward, and in economic matters he sometimes did not distinguish between what belonged to all and what was private, or used public things for his private benefit.

What was Luo Deming thinking as he stood there in this photo being criticized? It is hard to guess, but it seems likely he was so shocked that he couldn't think or speak. After criticism, he was recognized as one who had a good attitude, "hanging his head and acknowledging his crimes." Hence after this struggle session, he was removed from his official position but was allowed to keep his party membership. I recall that on the eve of our work team's departure from the village, we went to his house to say good-bye, and with tears in his eyes, Luo Deming said, "Chairman Mao has saved me," and "Your work team has given me a new life." I believed he spoke sincerely and I was touched.

Fan Zhongchun and Huang Yongzhen were two activists after the work team came to the village. Fan Zhongchun, who became chair of the peasant association, was criticizing Luo Deming with all his righteous indignation. Huang Yongzhen shed painful tears as she recalled her own grievance during the Great Leap Forward.

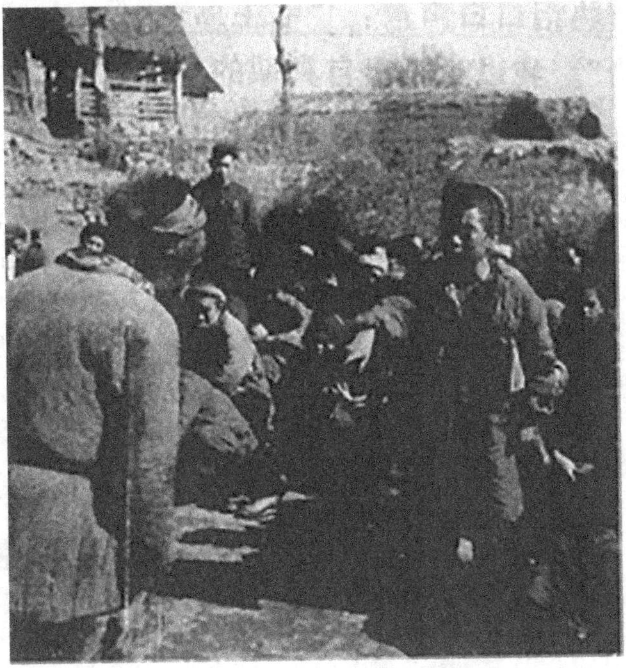

Photo 6: Che Mingyi, a poor peasant, challenging a rich peasant (photo by Ma Dazheng, January 21, 1965).

She had just given birth, and before a month had passed, the brigade called upon her to go back to work in the fields. Naturally, the person who gave this order was Luo Deming. During the later period of the movement, Huang Yongzhen joined the Party and also became assistant secretary of the Branch Party organization.

Analyzing the Four Elements was an important part of struggling against the enemy. In the Yuji Brigade, during the time when class background was being determined, there were three landlord families, one bankrupt landlord family, and one rich peasant family. When the Four Cleanups Movement began, these five households had a total of thirty-one people, male and female, old and young. Except for the rich peasant family of Li Zhaoguang who had more able workers and so were called "prosperous," the standard of living of these landlord households was really not distinguishable from the majority of families in the brigade. At that time in the Yuji Brigade, the daily work point was evaluated at .012 *yuan*. Even though this was the 1960s, there is nothing else to say but that this was a poor mountain district. Che Mingyi had originally been a hired farm worker, and the work team brought him along as an activist so that by this time he was associate chair of the Poor and Lower-Middle Peasant Association and a militia company commander, and had developed tremendous authority in the village.

Such struggles for power occurred countless times during the Four Cleanups Movement. No matter how sincere were those who organized these struggles for power, the triumphs or defeats on both sides of these struggles were inevitably shaped by the larger historical background. Although the pages of history have already been turned, a few old photographs can stimulate people's memories and their reflections on history.

This is by Ma Dazheng. It appeared in Old Photos *no. 4, pp. 1–6.*

A Shaanxi Peasant's Critique of the Economy During Mao's Great Famine

Introduction

The first thing one might think after reading this critique is that peasants are not just farm workers doing the best they can. Hu County, Shaanxi, is about fifty miles west of Xi'an, a major city, although at that time this county seat did not interact so closely with the city as in more recent times. The peasant Yang Weiming and his colleagues attracted attention at the time because of their astute observations and wise suggestions.

The concerns that Yang Weiming and his colleagues addressed were the policies that had produced the disaster of the "Great Leap Forward" and the "Three Bad Years" of 1959 to 1962. This most terrible famine resulted from natural disaster, bad farming methods, falsified crop reports, and bureaucratic cover-ups. Estimates on the number of people who died range from twenty-five million to as high as forty-three million. A great majority of those who died, ironic as it might seem, were the peasants who produced the food. They were tied to grain quotas, and whatever they produced had to be shipped away to feed city people, who were guaranteed a ration allowance.

By 1962 when Yang and his associates produced their critique, there was perhaps some improvement. But these peasants in Hu County still rightly had concerns. The initial enthusiasm for their ideas reflected many people's readiness to hear new ideas.

When Yang's critique reached supreme authority, however, it was enough that Mao criticized it, so Yang and his friends were scorned and disgraced. Many others among our subjects were scorned and disgraced. Note that their fate differed from most others whose stories are translated here. Their case "went to the top," where support ended and disgrace began. Many others among our protagonists ran afoul of local authorities who in some cases had for years been seeking an opportunity to bring them down.

The Text

This is the fortieth anniversary of the article written by the peasant Yang Weiming, of Hu County, Shaanxi, entitled "Some Thoughts and Feelings on the Present Situation." Yang and two associates who were also Party members, Brigade Secretary Jia Shengcai and Brigade Chief Zhao Zhengao, wrote this article, which was also known as "A Single Leaf Reveals Autumn," that stirred the entire Party and enraged Mao Zedong! In 1962, at the Expanded Political Bureau Session at Beidaihe,[1] Mao said that this article has one thing to say:

> From a single leaf we know autumn is coming, and this applies everywhere. A single leaf tells us about autumn, and it could also tell us about winter and even more importantly it tells us about spring, and about summer too . . . every class can talk about having hope. These comrades from the commune near the Hu County seat also can talk about hope and can talk about the hope of their units and their leaders. . . . Have we replied to the letter from these three comrades from Hu County? Communist Party members cannot remain indifferent to such questions.

1. Beidaihe is a seaside resort in Hebei province. Its beach on the Gulf of Bohai is regarded as one of the best on that stretch of seacoast. Hence it became the site of an annual summer retreat for the Communist Party's central leadership.

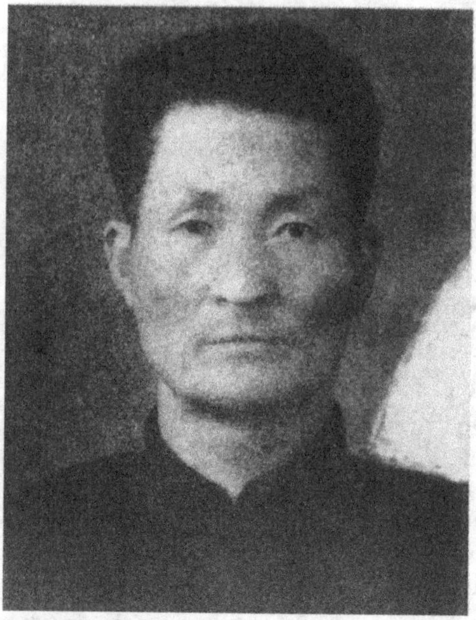

Photo 1: Yang Weiming.

Yang Weiming was born in 1922 in Hu County, Shaanxi. As a boy he attended a private school for three years, but he received no further education because his family was poor. He participated in underground work for the Communist Party and joined the Party in February 1949. In May 1949 he became an assistant leader of his village but in winter of that same year he decided on his own to leave the Party, and returned home. In 1957 he rejoined the Party and became the accountant for his production brigade. In 1962 his "A Single Leaf Reveals Autumn" drew severe criticism from Mao Zedong, and the criticism and persecution continued until 1968, when he could no longer stand the disgrace and took poison to commit suicide at age forty-six. In 1979 he was rehabilitated and exonerated.

In early 1997, I saw a reprint of Yang's article appended to *A Brief History of Agricultural Cooperatives in Shaanxi*. I was amazed by every aspect of this lengthy article. It took our country's affairs as a personal responsibility and showed deep feeling for people's suffering. Using plain and candid language, the article addressed issues forthrightly and maintained a levelheaded outlook. It showed close investigation of the issues, exactness in its judgments, and depth of thought. Among such writings by Party members and people outside the Party during the period from the 1950s through the 1970s, for the level of its thought and judgment, "Some Thoughts and Feelings on the Present Situation" is an exemplary work. The article is divided into thirteen parts, some with such witty titles as "The Belt," "Going Too Far" and "Not Going Far Enough," "Using the Back Door," and "Two Sets of Tracks."

In his foreword the writer makes his purpose clear:

> Having worked at the village level for more than eight years, so many feelings come together as I look back on this experience, especially in the present difficult times. . . . This piece about my thoughts and feelings is not written to "report good things" to my superiors, but rather to "report worries." This is more important now than just reporting good news. As these thoughts and feelings have come to me, it is like "good medicine making one's mouth bitter" or "words that offend the ears."

The author also states candidly,

> The thoughts and feelings in this article are a matter of personal opinion, or may be seen as "a single leaf revealing autumn, and it being the same in other places." This might also be regarded as "looking at the world from the bottom of a well" or the error of "seeing everything through a bamboo tube."[2] But I can say with confidence that what is presented here is actual fact based on my personal experience.

The article says,

> We have already acknowledged that "the difficulties are extremely serious." Just how serious are they? As to the villages, if we compare conditions before collectivization and the present, people's seething resentment has replaced widespread praise, withering production has replaced fertile production of grain, and hunger has replaced prosperous peasants and adequate food. A village economy about to lose its productivity has replaced the flourishing scene of yesterday. How can this darkness be returned to light?[3]

2 Both of these are common Chinese expressions to say that someone has a very limited view of the world. Here they are used to show humility in dealing with major issues.

3 The development of agricultural cooperatives began in the early 1950s, with a second stage to create communes launched during the "Great Leap Forward" in 1957–58. Yang

What should be done? Yang said,

> The situation seems extreme, but it should be easy to overcome the difficulties. The key is whether we can break away from the "resolute Yan'an spirit" and respond to actual conditions quickly.[4] We can dare to consider such policies as "making all materials the same" and "free markets," and replacing "reform" with "regulation" in small- and middle-scale industry and commerce. In agriculture we can choose either a collective or independent approach according to the resources of the people!
>
> In recent years we have been going in the direction of "retreat," and while we have seen some good effects, we still have a long way to go. We should "adjust" policy and planning of the national economy thoroughly! And we should stop only when we have conquered the difficulties.

This discussion was directed at the modest eight-character slogan on the national economy: "Adjust, rectify, consolidate, and improve."[5]

He used a belt as a metaphor for the condition of the Chinese economy. If the belt was too loose, things wouldn't work, but things also wouldn't work if the belt were too tight: "If it's too tight, people will be gasping for breath! If we consider our national economy at present, is this gasping a symbol of the situation? Do we have a tight belt constricting our whole body so that it seems we can't

---

Weiming here refers to this later stage. His comments are addressed to the disastrous results that occurred when neither the level of development nor—far more importantly—the productivity they were supposed to achieve actually came about.

4 The "Yan'an spirit" refers to attitudes that developed after the Chinese Communists arrived at their new base in northern Shaanxi in 1935 following the Long March. Their outlook was no longer affected by Soviet principles but came to be based on practical experiences. This new spirit was refreshing, vital, and largely successful at first, but by the time Yang Weiming's article was written, it had been subsumed in an increasingly centralized system and lost its earlier effectiveness and appeal.

5 The policies expressed in this eight-character slogan were used, and themselves adjusted, during the "Great Leap Forward" from 1959 to 1962. They are broad in meaning and thus have been used with some variation in other campaigns in more recent years. In the conditions that Yang Weiming and his colleagues were addressing, problems were already far out of control. Yang Weiming thus called for bolder approaches.

move? Indeed, we do, and only if we loosen the belt a bit can we breathe and move comfortably." Thus, concern about the state economy brings questions:

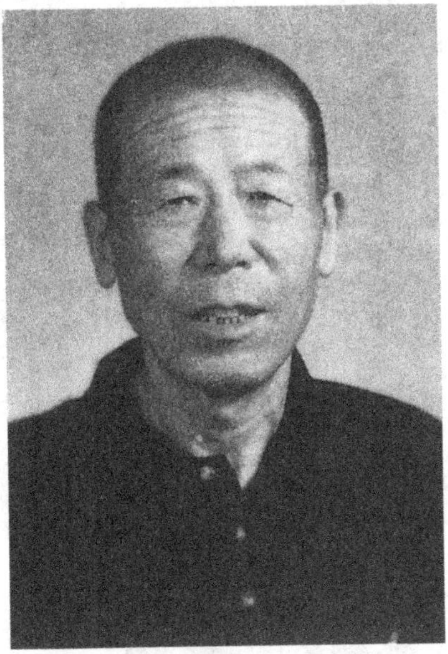

Photo 2: Zhao Zhengao.

> The state economy is like a human body. The freedom for small and middle industry and business to produce and the freedom for villagers to trade outside their assigned farm work are like arms and legs without a body, the arms and legs have nothing to depend on. Without arms and legs, the body cannot do anything either. The two are interdependent and must work together.

This idea was very similar to the idea of "opening up and invigorating the economy" of twenty years later.

Yang Weiming made an unusual and very clear judgment of China's overall situation:

> I think our country is a "poor and blank" country, and in trying to build on such a weak foundation during the period from 1949 to 1955, did we really succeed in the task of building under New Democracy?[6] The answer has to be

6 Mao Zedong began in the early 1950s to use the expression "poor and blank" to describe conditions in China and saw this characteristic as positive rather than negative. He used this expression in a talk in 1958 made upon launching a new phase in cooperative agricultural organization, lauding an early commune in Henan Province—where the "poor and blank" character of the people had presumably contributed to their success. The concept of a "New Democracy" was first presented in an article by Mao in 1940. The idea was that all classes in the society would cooperate to build a modern society under the leadership of the Communist Party. While the idea of New Democracy remained and continued in use as a goal, Mao himself had lost his earlier enthusiasm. Instead, his

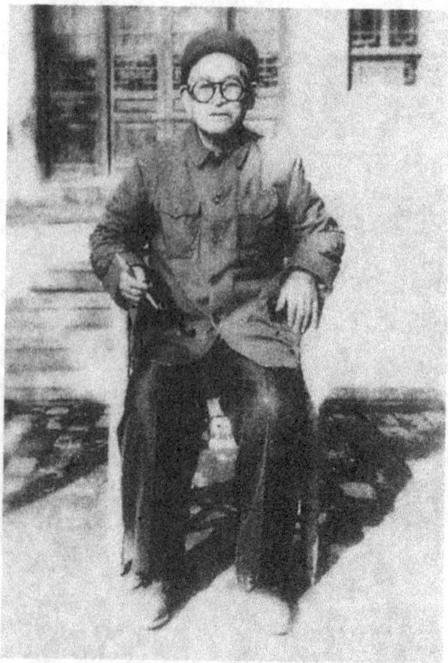

Photo 3: Jia Shengcai.

"No." Furthermore, no matter how hard you try, it is simply impossible to build a great industrial nation in a backward agricultural country of six hundred million people.

Some people said our socialist construction needed to take two steps (from New Democracy to Socialism). So if we didn't make the first step very well, could we then take the second step well? Normally it would take twenty or thirty years to build New Democracy. If we look at it from this angle, what we tried to do was clearly over-enthusiastic and violated objective rules.[7]

The theory he put forward on "the initial stage of socialism" is exactly the analysis of more recent years. All of the observations, judgments, and policies of "the theory for the initial stage of socialism" of the 1980s could have come from his ideas.[8]

---

original intended successor Liu Shaoqi (1898–1969) emphasized New Democracy, especially the role of the existing bourgeois business and industrial groups who had already begun to contribute to China's economic development. This disagreement became a wedge between Mao and Liu by the early 1960s and was a basic cause of the Cultural Revolution by the mid-1960s.

7 Many people in China came to believe that major policies for development had been overly ambitious and in particular that it would be impossible to build a strong industrial economy on an inadequately developed agricultural base. Mao Zedong was blamed for excessive enthusiasm as early as 1959, when the Great Leap Forward policies led to great difficulties, most importantly to the combination of reduced grain production and exaggerated reports of the amounts of grain available.

8 Soon after consolidating his position as China's new leader in 1978–79, Deng Xiaoping expressed the view that the policies of the 1950s had been incorrect in their theoretical analysis and had led to the disasters of the "Three Bad Years" of 1959–62. In contrast to the Maoist approach followed in China's first socialist decade, Deng emphasized that correct Marxist theory provided that some members of society must become wealthier than most, and that the larger society's wealth would develop from this source.

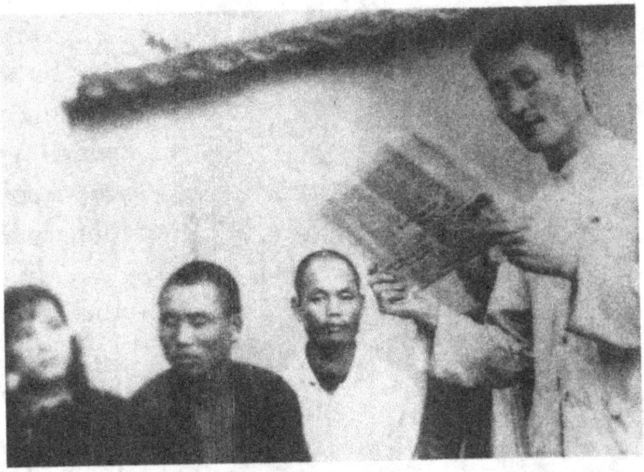

Photo 4: Yang reading a newspaper to his fellow villagers.

Clearly, Yang was passionate in thinking about the nation; the policies were in error and had brought disastrous results. He thought the pattern of social organization had gone too far:

> If a man wants to go to a certain village that is ten *li* away, if he walks fifteen *li* he has gone five *li* too far. Then if he goes back three *li*, he is still two *li* from that village. If he then goes back ten *li* he's still a few *li* short of where he wants to be. Ultimately, what he needs to do is to go back five *li* from that first trip when he overshot, and he'll be at the village, right where he wants to be. It seems that the serious difficulties we are now experiencing are a question of going five *li* too far, so the way to overcome the problems is a matter of just going back five *li*.

So to what degree did we need to move in order to "go back five *li*"?

> In agriculture we could use the principles of collective methods, individual methods, and the masses' own resources. This is like raising chickens to get eggs, which is better than killing chickens to get eggs; or taking drastic measures to deal with the situation, which is different from ineffective remedies. We need a basic approach, not just stopgap measures. This is the point to which we need to retreat.

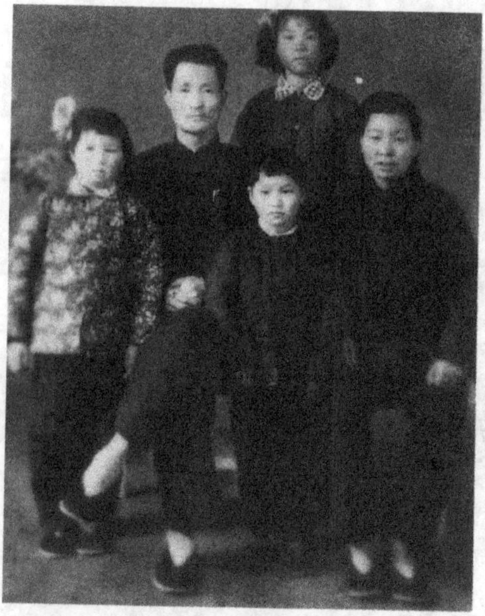

Photo 5: Yang and his family. His upright bearing reflects confidence.

What we needed to do was to "step back from the resolute Yan'an spirit," by which Yang meant to put good policy before revolutionary fervor.

He thought that once the economy had fallen into "complications," it had many "symptoms of the sickness," among them "food shortages" and "tensions in supplying the market." These symptoms had caused problems such as "going through the back door," "speculation and profiteering," "incidents of theft," and "relocation because of famine." Other effects included "a dispirited outlook," "resentment," "low morale on the job," and even "a deterioration in friendships" and other such spiritual phenomena associated with the symptoms.

The writer also laid out the idea of an open market for material items (grain, cotton, oil) and using "the principle of value" to adjust problems in supplying markets and to stimulate farmers' initiative in the production of sideline products. Small and middle-scale industrial and commercial enterprises "should use moderate 'regulations' rather than 'reform' as a method." A system of democratic amassing of goods was his approach and what he recommended for China's peasants.

Among his ideas, Yang Weiming used the image of "parallel tracks," a colorful way to represent the relationship between above and below in a process of "democratic accumulation":

> With parallel tracks, trains can go in both directions—one east and one west, without obstructions. But if there is only a single track, efforts to have things move in both directions

> will result in conflict and mutual obstruction. If the trains can move in turn, as they change directions and start back, the process doesn't need to stop. . . . To expand the will of the masses of people is to "fill up the track" and "accumulate up to the maximum." Then a process can begin to put things in order. People can make their own decisions, leadership is linked to the people, and policy gets through to the lower levels.
>
> When the system is working well, the two trains are each moving in their own direction without any obstructions. When "the will of the masses" and "current policy" develop contradictions, we must make sure that "the will of the masses" is accumulated up the line as quickly as possible. Thus can we ensure that errors in current policy are corrected in a timely way. We need to have open movement in both directions so that these two can interact well. Otherwise, we'll have "two trains going toward each other, unable to avoid each other, and getting damaged in a collision."
>
> With this process working, the trains from opposite directions can turn around and change directions, and start again without a break. . . . So we may say: "If the west-bound train doesn't come in, the east-bound train will be empty. And if there is no east-bound train, the west-bound train would be useless." This is the interaction of cause and effect to replenish our resources.

The writer continued:

> We are a people's democratic country, and if we call ourselves this, our democracy should be a 100 percent democracy. "Democracy" means a high level of "concentration of the people's will." Democracy and "concentrating the people's will" should interact and not be seen as two opposing things!

If ours was not "a hundred percent, honest-to-goodness democracy" that "concentrates the people's will" to a "high degree," how could we speak of "a system that coordinates the people's will"? Yang understood that this was actually a question of the political system. Yang presented an objective critique throughout his article, from practice through theory, on how the feelings from below were not reaching the leaders above. Those above paid no attention to the will of the peasants, nor to the feelings of the nation as a whole, and formulated policy that departed from practical concerns and brought the nation and the people into a disastrous situation. Yang's discussion presented a systematic way to resolve the problems.

"Some Thoughts and Feelings on the Present Situation" was mailed on May 10, 1962, to the commune, county, district (*diqu*), provincial, Northwest region, and Party Central levels, and at the same time was sent to the *Shaanxi Daily*, the propaganda department of the County Committee, the Xi'an City Committee of the Party, and to responsible people in the counties of Weinan, Hua, and Tong'guan.[9]

The initial response to the article was positive. In June, Yang Weiming received a letter from Managerial Commissioner Wang Shijun, which said as follows:

> Your discussion shows a good understanding of our situation . . . and it is very beneficial for guiding our tasks. I thank you for your concern about the great affairs of our nation. I have put your article with other letters containing suggestions and distributed it to all the relevant activities and comrades for their reference in studying issues of concern, and I hope you will correspond often and maintain this contact, so that we may often hear responses on the practical situation of peasants.

Also during June, Xi'an Mayor Liu Geng gave the article to the Provincial Committee's office and told them to "deliver this to the Provincial Committee for study and execution." At the beginning of

9. Hu County is about fifty miles west of Xi'an, the capital of Shaanxi province. The other three counties to which Yang sent his article lie to the east of Xi'an—so he clearly intended to get the attention of local leaders in an extensive area of Shaanxi.

July an assistant chief in the Shaanxi Propaganda Bureau also wrote to Yang Weiming and said, "The opinions and responses to the current situation presented in your letter all have been sent on to the Provincial Committee and to Party Central, and I have asked the Party to consider and deal with them." He encouraged them and the opinions they put forward: "Whether they are correct or not, we should respond to them promptly and send them on to the Party for study. This is the responsibility of Party members and also is their right." Yang Weiming's article was reprinted in several provincial- and central-level publications and was clearly regarded as stimulating and worthy of discussion.

During this period Yang Weiming was discovered by higher levels within the Party and was employed by the Xianyang[10] Region Committee as a research consultant and a reporter for the Northwest Region publication *Northwest Construction*.

These early responses show that Yang's analysis seemed accurate and that it represented the feelings of a great majority of the Party and the people.

But after August 6, when Mao Zedong at the Beidaihe meeting sternly criticized Yang's article as expressing class determinism, the situation was quickly reversed.

By the middle of August, the Shaanxi Provincial Committee began to review "Some Thoughts and Feelings on the Present Situation." Soon after that, officials from provincial down to county level called in Yang Weiming and his associates four different times for talks. Each time they were told that their article had "erroneous viewpoints."

On October 11 and 19 the Shaanxi Provincial Committee twice sent reports to the Northwest Region headquarters and to Party Central that "Some Thoughts and Feelings on the Present Situation" "is a flagrant and relatively systematic reactionary outline calling for the restoration of capitalism" and that the three Party members' proposals "in substance would be the revival of the capitalist road. . . ."

---

10 Xianyang is a smaller city west of Xi'an. Today the Xi'an airport is nearer to Xianyang than to Xi'an itself. The Founding Qin Emperor (*Qin Shihuangdi*) built his capital in this area.

On October 18 the Provincial Propaganda Section made a report to the Provincial Committee giving the opinion that the three Party members' suggestions and proposals "are a spontaneous force of capitalism calling for a relatively thorough, systematic reaction within the Party," and they "bring together opportunistic positions and viewpoints within the Party since the beginning of the socialist revolution and socialist construction, and present a thoroughly bourgeois political outline for restoring capitalism."

Then on October 29, Shaanxi Provincial Committee First Secretary Zhang Desheng, speaking to a major meeting of the provincial committee, said "the three Party members from Hu County have asked us to 'initiate a retreat from the resolute Yan'an spirit' and revert to going it alone. This is absurd and completely reactionary."

Yang Weiming and his associates were compelled to undergo self-criticism.

From then on, the words of one person [Chairman Mao] closed the hearts of all and caused this tragedy.

In 2001 at the time of Qingming Festival for paying respect to the dead, I went to the bank of the Lao River in Hu County to pay my respects to Yang Weiming's spirit in heaven, but his grave had been covered over by a chemical plant. There was no place for him in the world of humans when he was alive, and after his death he couldn't hold a spot of yellow earth. This causes us deep regret!

I once wrote as follows:

> The spirit of a people is not simply the accumulation of the mass of the people but is the cup of wine brewed from a bucket of water and grain, or a piece of raw steel forged from a variety of ores. Gu Zhun was like this, and so was Yang Weiming.[11] Fortunately, people today understand the value of Yang Weiming and remember him with respect forty years later, for the insight shown in "Some Thoughts and Feelings on the Present Situation."

11. Gu Zhun (1915–1974) was an economist whose views conflicted with the Maoist leadership and who continued to advocate policies opposed to those of the leadership. He was jailed during the Cultural Revolution and died while incarcerated.

On July 31, 2002, I received a telephone call from Liu Gaoming, director of the Hu County Library. He called to tell me that important county officials had agreed to name the library for Yang Weiming. In the past there have been cases of a library's name being decided by money, power, or an influential person. But Yang Weiming is the first ordinary person to have a library named for him.

This article is by Lu Yuegang. It appeared in Old Photos *no. 27, pp. 32–42.*

The Geology Bureau during the Cultural Revolution

Introduction

Many of the stories included among our selections describe what happened to individuals during the Cultural Revolution. As its writer says, this article offers insight into the processes through which the Cultural Revolution worked its way into organizations, in this case a government bureau in Beijing, the national capital. As noted, the source for this set of photos was Panjiayuan, the name of an area in the southeastern part of Beijing that is also used to identify the "dirt market" there. This market provided sources for a number of similar articles in various publications over decades after the Cultural Revolution (perhaps these may be identified as down-to-earth sources).

As things moved along during the Cultural Revolution, the groups that arose tended to become splintered over significant or imagined differences. That tendency showed itself in developments in the Geology Bureau within the first year. Compare this article with the armed struggles described in the piece about the Cultural Revolution cemetery in Chongqing.

The Text

At the market for old goods in Panjiayuan, in Beijing, I discovered an old album that contained about a hundred photographs of different

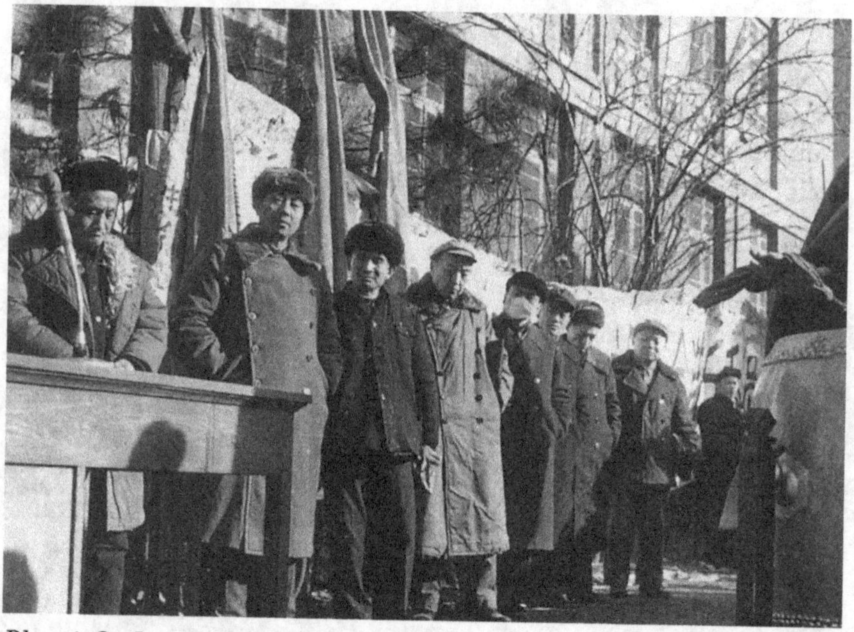

Photo 1: On January 1, 1967, the rebel group in the Geology Bureau organization called a meeting to announce the establishment of "the Geology Bureau Revolution and Rebellion Committee," and to tell the crowd that He Chang'gong and the other leaders had been removed from power.

sizes, the great majority of them taken in the courtyard of the Geology Bureau organization. This provides a detailed record of the process of the Cultural Revolution in the Geology Bureau. The content touches on the struggle for power to establish the Revolution and Rebellion Committee. They celebrated a first-month anniversary and a great union of all factions after half a year—from January to September 1967. The events that occurred in the courtyard of the Geology Bureau organization during this period epitomized what happened throughout Chinese society during this same period.

Photo 1 was taken in the courtyard on January 16, 1967, at the general meeting to establish the Revolution and Rebellion Committee. Besides this photo there is another one with a different perspective that shows the larger scene at the meeting. Written across the top of the photo is "the general meeting to establish the Revolution and Rebellion Committee in the Geology Bureau." At the time, the earliest reorganization through a struggle for power occurred in Shanghai (January 8, 1967), and the next place was in Shanxi (January 12, 1967). The new political power established by the rebel group in Shanghai was called "Shanghai People's Commune," and in Shanxi Province they simply used the name "Shanxi Province Revolutionary Rebel General Command Section" to seize power. The earliest to call themselves a "Revolutionary Committee" were those in Heilongjiang Province (January 31, 1967). During the last part of February, under the direction of Mao Zedong and the Party Center, "Shanghai People's Commune" changed its name to "Shanghai City Revolutionary Committee." After that, throughout the nation, in every region and every unit, the new general name was changed to "Revolutionary Committee."

This explains why the new political authority in the Geology Bureau was first called "Revolution and Rebellion Committee." This also indicates that the struggle within the Geology Bureau was rather early. Actually, on January 16 the *Red Flag* editorial writer promulgated Mao Zedong's most recent instruction. "To seize power from a small group of capitalist-roaders within the Party is a revolution in which one class overthrows another class under the dictatorship of the proletariat." The struggle for power in the Geology Bureau occurred on the same day. On January 22 the *People's Daily* published its editorial calling on people to "open a broad struggle for power throughout the nation to achieve full victory in the great historical duty that Chairman Mao has given us." This was exactly one week after the struggle for power within the Geology Bureau, so you can see that the rebel faction within the Geology Bureau was somewhat ahead of the new wind that would blow.

In this photo album there is a brief text to explain each photo, and from this we learn that the eight people standing in this photo are the original leaders of the Geology Bureau. They are, from left: He Chang'gong, Xu Guangyi, Liu Hansheng, Cui Zhendong, Song Ying, Zhang Tongyu, Li Xuan, and Zhao Chengfeng. At that time, He Chang'gong was vice minister of the Geology Bureau and Party Secretary, and Xu, Liu, Cui, Song, Zhang, and Li all were vice ministers; Zhao Chengfeng was Associate Director of the Bureau's Political Section. Actually, all of these individuals had rendered outstanding service to the People's Republic and had been high-ranking

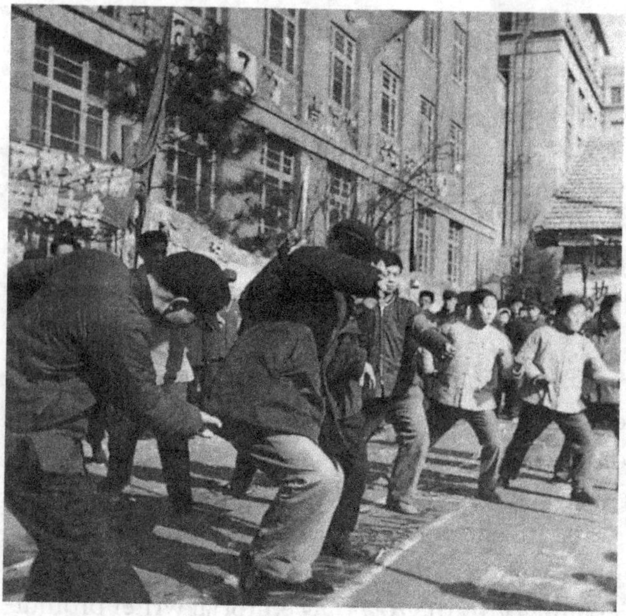

Photo 2: On February 16, 1967, the Revolution and Rebellion Committee of the Geology Bureau organization met to celebrate the "one-month anniversary" of the seizure of power, and people at the meeting performed a "Loyalty Dance."

officials for a long time. For them now to be brought down in the space of a single day was truly not what anyone would have imagined! You can see this from their dejected, perplexed expressions in the photo. Liu Shaoqi said that at the beginning of the Cultural Revolution, the awkward plight of leaders was "old revolutionaries meeting up with new questions," and although this was a somewhat self-mocking comment, it was also very much to the point.

As can be seen from the photos, these men accused as "those in authority taking the capitalist road" were not made to wear tall caps, did not have to hold up signs or to kneel, nor did they have to be "jet-planed."[12] They were only made to stand there. At that time this was already a kind of benign "special treatment." Xu Guangyi in particular (second from left) shows this lenient treatment. As before, he was wearing his otter-fur winter military cap and his wool military overcoat, with his head held high. He did not lose his military dignity. Xu was a career army officer and had served as a logistics

12. "Jet-planed" refers to the common practice in struggle meetings during the Cultural Revolution, in which a target person was forced to bow deeply with two arms twisted from behind on each side by two revolutionary rebels. The person's body was said to look like a "jet plane," hence the name.

chief of staff, and in 1955 he had been awarded the rank of brigadier general. For "capitalist roaders" to get this kind of "benign treatment" was very likely related to the nature of this meeting. If this had been a special struggle session, the situation would have been very different. With the passage of these thirty years, the youngest of these men would now be an elderly person more than eighty years old. Who knows how many of them have survived to this day?

Photo 2 was taken one month after Photo 1, specifically on February 16, 1967. On that day the rebel group held a meeting in the Geology Bureau courtyard to celebrate the "one-month anniversary" of their struggle for power. It seems a bit strange that there would be a celebration of a one-month anniversary, but this is what the caption of the photo states. There would seem to be two possible explanations as to why they didn't want to wait to celebrate. One is that they couldn't contain their happiness over their seizure of power. The other is that celebrating this great change would help to establish the position of the new authority in the bureau.

As before, this photo was taken in front of the main building of the Geology Bureau, but what are the people in this photo doing? At first glance it looks like they are performing *taijiquan*, but on such a glorious day, who would be in the mood for that kind of activity—and even less, in this kind of gathering? Is this some kind of exercise during a break in work? Not only do these motions not suggest this, but again it makes no sense that there would be a break in work at such a meeting. After repeatedly considering what this could be, it is most likely that they were performing a "Loyalty Dance," dancing to show their "loyalty" to Chairman Mao. During the early phase of the Cultural Revolution, this dance was a popular group activity. It could be performed by a few people, by hundreds, or even by thousands dancing together. Accompanied by the music to "Always Loyal to Chairman Mao," they danced gracefully together in set movements. Like "Request for Instructions at Morning" or "Evening Report," the "Loyalty Dance" was a form of "class" that people would do at that time. Once I saw a photo of a Red Guard Propaganda Brigade performing a "Loyalty Dance" at Tian'anmen Square. The movement and the style suggested a performance and was not so straightforward and natural as this group at the Geology Bureau.

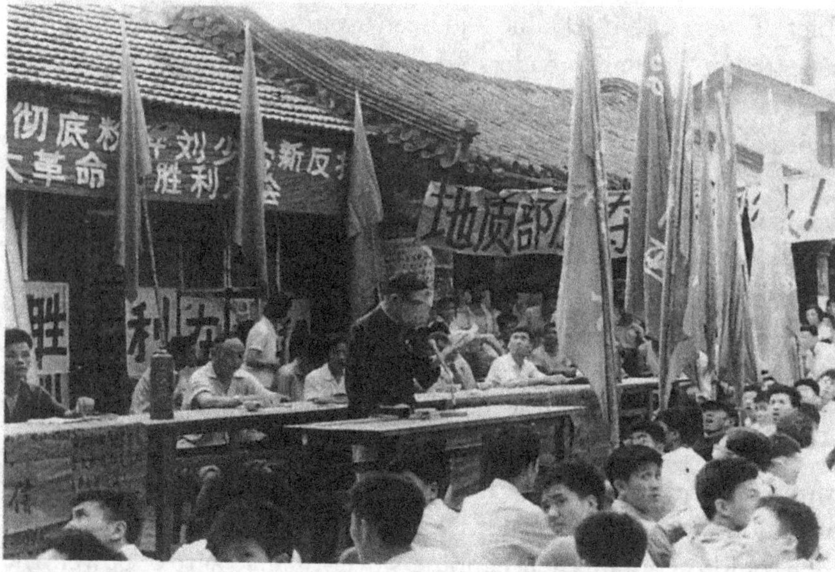

Photo 3: On July 16, 1967, the Revolution and Rebellion Committee of the Geology Bureau Organization met to celebrate six months of their seizure of power. Here a representative of the revolutionary organization talks at the meeting.

Photo 3 shows the meeting that the rebel group convened for the six-month celebration of their seizure of power. The major theme of this meeting is shown in one of the signs, which says "The Seizure of Power in the Geology Bureau on January 16 is Awesomely Good." The names of those seated along the main table and the person who is speaking in this photo are written in the original photo collection, and we may assume they were the leaders of the rebel group. Although citizens who perform public activities that occur in public places are not covered by principles of confidentiality that apply to people in private, based on principles that we all understand I have omitted their names from this article. As activists during the Cultural Revolution, they echoed the calls of Chairman Mao and the Party Center and reflected the changes of a particular time. But still later after all the cases were overturned, I suspect the behavior of these people should be investigated thoroughly, but I still couldn't say whether they would be judged as "the three kinds of people."[13] This

13. In July 1982, the CCP designated "three kinds of people" as politically untrustworthy based on their behavior during the Cultural Revolution: persons who followed

reflects the words of Luo Jinbao in the movie *The Little Soldier Zhang Ga*: "Don't look at the thrills of the present hubbub, but fear it later when they bring out the list of culprits." Just as He Chang'gong and his associates could not have imagined they would be brought down in a single evening, the rebel leaders, after their success during this period of revolutionary fervor, probably couldn't imagine what their lives would hold for them in the end.

When Photos 4 and 5 were placed in the album, only a brief line of explanation was included: "Expressing support for destroying the capitalist roaders." Who was it that wanted to "destroy the capitalist roaders"? This is not explained in the album. Both these photos were taken in the auditorium. In Photo 4 it appears that as they debated on the stage, the two sides broke out in dispute, and that another person off the stage is waving his arms and yelling out slogans—the atmosphere is a little tense. Photo 5 comes right after Photo 4, and in it, a man from one side has just wrestled someone from the other side down off the stage. You have to think the photographer quickly pressed the shutter button, and when you figure in the weakness of light in the room, this influenced the exposure level of the photo. We still must thank the photographer's quickness because only such quickness could provide people of later times this truer reflection of the Cultural Revolution.

Photos 6 and 7 also should be taken in sequence. In Photo 6 a number of people are busy decorating the meeting-place, with everybody dashing about, smiles on their faces. The scene has a festive air. Then after a time (it might have been the next day), all of the rebel groups in the Geology Bureau gathered in front of the building for an "assembly to lower our flags so as to follow closely Chairman Mao's great strategic plan in its glory" (see Photo 7). This "lowering of the flags" was a kind of symbolic language for the different factions in this larger unity. In the process of struggling for power, the various rebel groups had become more violent the longer their factional struggles continued. In some places this even turned into large-scale military struggle. From July to September

---

the Cultural Revolution radicals to gain power, persons who had serious factional thoughts, and persons who had records of being violent.

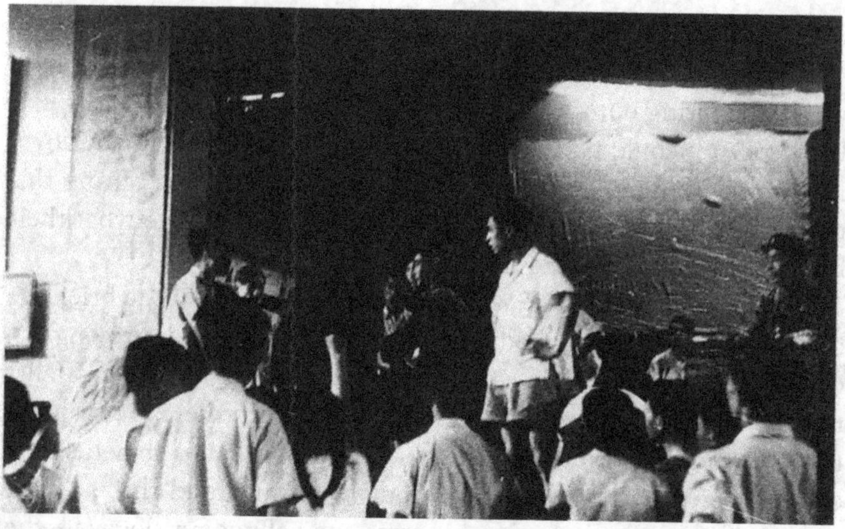

Photo 4: In July or August 1967, at a meeting of "The Army to Destroy Capitalism," two of the mass groups got into a dispute.

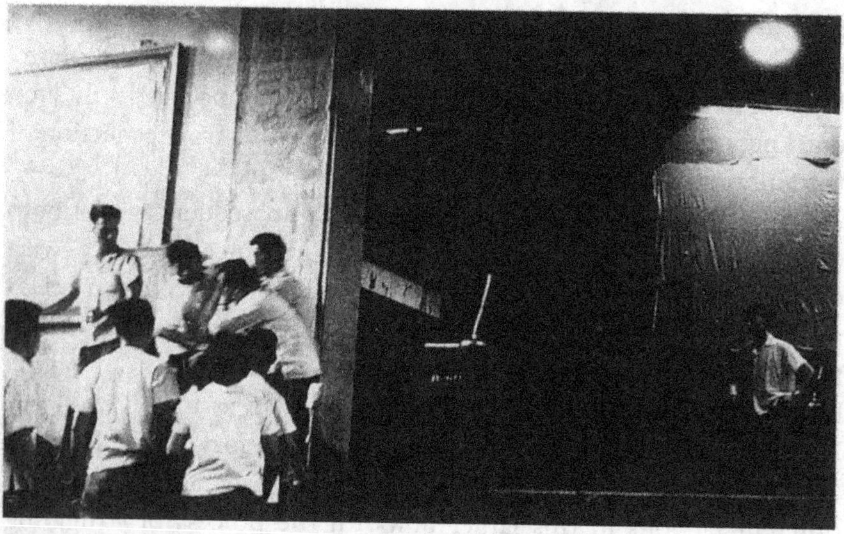

Photo 5: At the same meeting as in Photo 4, a member of one of the struggling groups tries to push a member of the other group off of the stage.

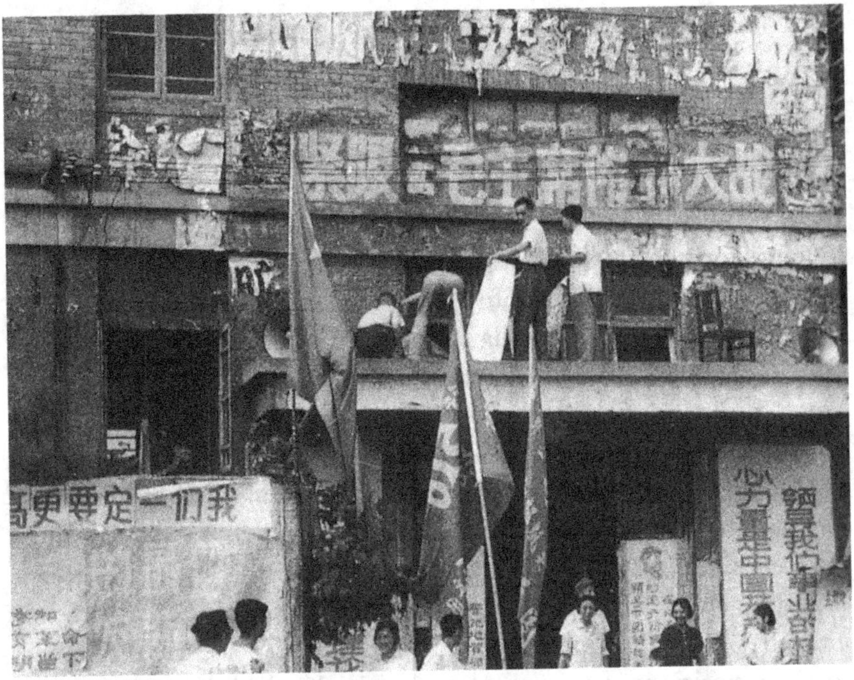

Photo 6: In August or September 1967, the revolutionary organizations of the Geology Bureau prepare for a "Gathering for the Glorious Lowering of Flags."

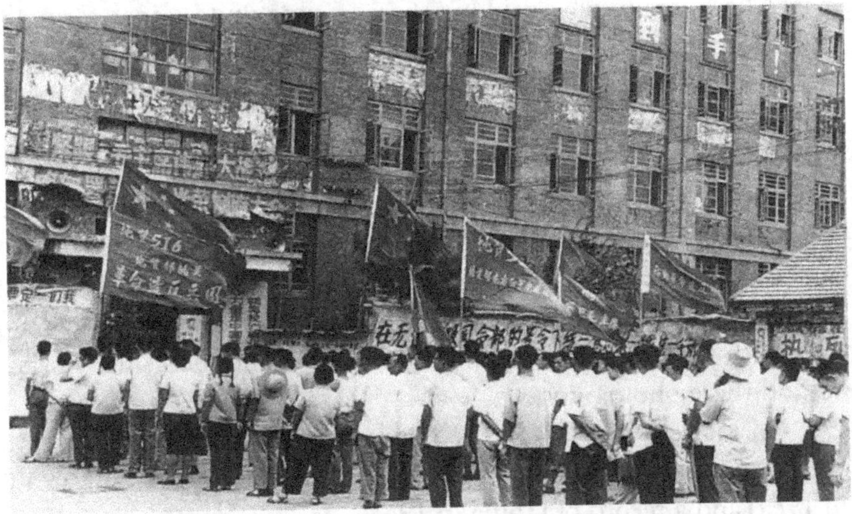

Photo 7: In the same setting as in Photo 6, the different revolutionary groups in the Geology Bureau Organization assemble for the "Gathering for the Glorious Lowering of Flags."

1967 Mao Zedong several times said, "We only want the two factions to be mass revolutionary organizations and to work out a great unity under the principles of revolution." On October 17 the Party Center issued a "Notice on Systematic Implementation of a Great Revolutionary Alliance."

The great alliance of the groups in the Geology Bureau was following along in this background. It is worth noticing that, judging from what those in this photo are wearing, it appears to be still summertime and is probably August or September. This again indicates that in pursuing the themes of the Cultural Revolution, those in the Geology Bureau wanted to be in the lead and not just follow along with what was happening elsewhere. In Photo 7 you can see four flags of different rebel groups: the "Revolutionary Rebel Military Corps," the "East Is Red Revolutionary Rebel Military Corps," the "Protect Chairman Mao Military Corps," and the "Logistics Revolutionary Rebel Military Corps." But from these names of the different rebel groups, people today find it difficult to see the differences in their positions. For example, what is the distinction between a "Revolutionary Rebel" and an "East Is Red Revolutionary Rebel"? "Protect Chairman Mao" and rebels responding to the call of Chairman Mao—again, why couldn't these groups unite under a single flag? Only the name "Logistics Revolutionary Rebel Military Corps" provides meaningful information to let people know that this rebel organization was formed to do logistical service.

This group of photographs made in the courtyard of the Geology Bureau in 1967 clearly reflects the social and human environment of that time. The kind of information these photos provide is much more than one small article such as this can include. I hope the people shown in these photographs, and especially the photographer with the perceptive eye, might come out and tell us the stories behind these photographs.

No matter what, all people living at a later time can see from these photographs a record of the courtyard of the Geology Bureau in 1967, and a record of China in 1967.

This is by "Mo Ming" (a pseudonym). It appeared in Old Photos *no. 14, pp. 91–99.*

Cultural Revolution Cemeteries in Chongqing

Introduction

Strife among competing Red Guard factions occurred throughout China during the Cultural Revolution. Sichuan and its largest city, Chongqing, saw some of the most intense struggles. Competing groups took up arms as the intensity of conflict increased, and struggles became armed battles. This article describes the reopening of a cemetery for Red Guard victims in these struggles.

This article was preceded by a note written by the *Old Photos* editorial staff, as follows:

> At the beginning of 2010, the media publicly opened what the city government had approved in late 2009—the cemetery for victims of military struggles during the Cultural Revolution located in Shapingba City Park (in suburban Chongqing). Under the name "Red Guard Burial Park," this was made a city-level cultural protection site. He Shu's article introduces the current circumstances of the cemetery and the relevant background.

The Text

In January 1969, in an area within the Shapingba District's public park, a cemetery built by the people themselves was completed. According to the calculations of Chen Xiaowen, who studied the progress of the cemetery for a long time, it contained a total of 113 graves. (But according to another researcher who carried on the investigations after middle school teacher Chen Xiaowen retired, there were more graves than he knew about, and the total is more than 1,300.) Among these, besides some 92 graves that had no markers, or on which the inscriptions could no longer be read as of the early 1990s, 345 people were buried there. But then if you add those in graves too obscure to study, an estimate of about 400 would not be too high.

These victims appear to be more than actually participated in the major military skirmishes in Chongqing. A small number were simply unfortunate bystanders. Among the 196 victims for whom there is reliable information about their ages, the largest number were less than 20 years old. There were 69 in this group. The next largest group was 21 to 30 years old; there were 66 of these. The youngest of the entire number was 14, and the oldest was 60.

Of the 299 for whom there is reliable information about their occupations, the largest number were workers—176 of them. Next were students, with 104, and among these the largest number were ordinary middle school students—82 of them.

The largest number of those buried in this cemetery worked in factories. For example, in the section for the "Eight-One Military Group" that manufactured guns in their machine tool factory, 31 people are buried there (and among these are middle and elementary school students from schools associated with this factory). Of those from the section for the "Eight-One Military Group" that produced vacuum equipment for tanks and armored vehicles, 37 people are buried there. From the Number 102 Copper Factory that made copper products for the defense industry, their "2307 Revolutionary Worker Rebel Group" has 15 buried there. There are 10 in another section of the cemetery for the Jiangling Machinery Factory, also of the "Eight-One Group" that produces ammunition. The "8/15 Battle Group" of commerce workers has 16 people buried in their section. The section for the general command group called the "Gold Category Group" has 8 dead. With all of these victims representing the various worker and trade groups, it is clearly not accurate to call this a "Red Guard Cemetery."

"Red Guards" specifically refers to mass organizations among students, and really is not a very clear concept. During the early period of the Cultural Revolution, the name was applied mainly to the sons and daughters of high cadres, this younger generation known as "Old Red Guards" (this was mainly in Beijing), but it also came into use for Party and Government groups elsewhere, organized by officials, following the style of Beijing. Then it came into use by others in many different places—groups rebelling against Party and Government, seizing power or trying to. The name became fashionable and largely lost its original meaning. So this writer has always advocated that the cemetery discussed here should be called "Cemetery for Victims of Military Struggle during the Cultural Revolution" or more simply, "Cultural Revolution Cemetery."

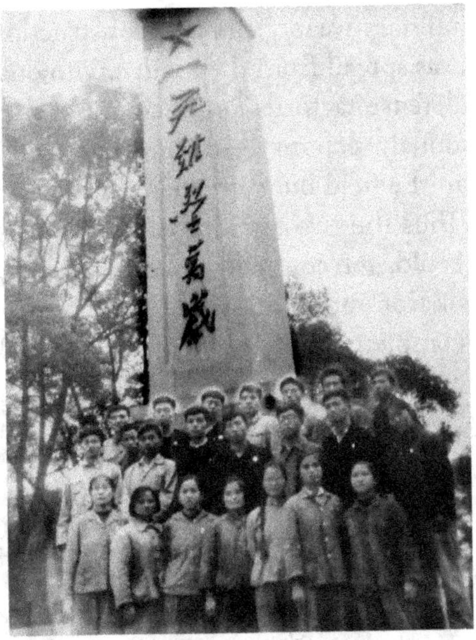

Photo 1: Students of the Machinery Department, Chongqing University, at the "Martyrs' Monument" to their classmates who fell in armed struggle. November 1967 (Photo contributed by Huang Zhaoyan).

The mass-organized Cultural Revolution opposition groups in Chongqing as of early 1967 quickly divided into two large factions, the "Eight-One Faction" and the "Fight to the End Faction." Both sides claimed they were defending "Chairman Mao's Revolutionary Line" and both were convinced their opponents were being "controlled by manipulators walking the capitalist road" and had sold out to the "Liu (Shaoqi)–Deng (Xiaoping) Line." Each said the other was a tool for "restoring capitalism" and was of the "Guomindang Reactionary Clique." Because of the atmosphere in society and habits of thought growing out of education in class struggle and the revolutionary tradition, the two sides would say of each other "they have no room to make adjustments on issues of the Party line." "Everything they do is something reactionary; if you don't resist it you won't be able to defeat it." "Democracy does not yield to the reactionary faction, not one bit or even half a bit." "Mercy toward the enemy is brutality to the people." Thus, treatment of the other side must include suppression, not allowing them to give an opinion, tearing up their slogans and big-character posters, smashing their propaganda vehicles and broadcasting stations. All this would arouse conflict and create struggle. From fighting with fists and throwing rocks, they said, "We can build up to knives, cudgels, steel rods, and beat them to death. Finally we can use modern military weapons."

Chongqing served as a production center for conventional

military weapons in the "construction along three lines" policy.[14] This was spread from the main city to outlying areas in several large-scale defense factories producing and storing large quantities of conventional weapons—some of these still had not been supplied to troops in the field but were being readied to send into the war in Vietnam. Thus the two factions each had their strengths and ultimately both could get completed weapons either from the factory storerooms or from army arsenals. Those that weren't quite complete could be completed with just a bit of work. Besides these sources there were even homemade weapons (like the Chongqing University group's "Eight-Fifteen style" machine gun and hand grenade). With so many weapons floating around among the mass organizations, large-scale fighting could progress very quickly, and it became a scene of out-and-out "civil war." But in this "civil war" both sides were fighting for the same leader—Mao Zedong—and for the same "revolutionary goals: fighting to the finish for the 'Great Cultural Revolution'!" At the moment of death, they all uttered the same quotation: "Remain resolute to the end, do not fear sacrifice, overcome all difficulties, and fight for victory!"

During the period of large-scale military struggles in Chongqing, the Shapingba District was controlled by the "Eight-Fifteen Faction." Thus, in the part of the cemetery that is in Shapingba Park,

---

14. "Construction Along Three Lines" was a broad development policy launched in 1964, a response to defense concerns against both the Soviet Union and the United States. Following the "Sino-Soviet split" of the early 1960s, the American presence in Taiwan, and the growing involvement in Vietnam, the Chinese leadership sought to decentralize military-oriented industry and develop the areas of China that lagged behind the earlier industrialized parts of the country. This was an intelligent policy under the circumstances, one that probably grew out of the leadership's experiences on the Long March, which passed through some of the areas where development was sought.

Chongqing became a major focus of this program, and the presence of the defense industry described here was a direct effect of "Construction Along Three Lines." As noted here, these new local industries produced weapons that were not yet available at the front in Vietnam, where Chinese forces were again resisting the Americans as they went into action in a neighboring country, as in Korea barely a decade earlier.

"Construction Along Three Lines" now seems a precedent for the "Development of the West" policy carried out in recent years under President Hu Jintao. While there are other aspects of the goals of China's leadership since the turn of the century, this ongoing attention to less developed areas shows due concern for these areas and how their development can affect residents there and larger economic achievements.

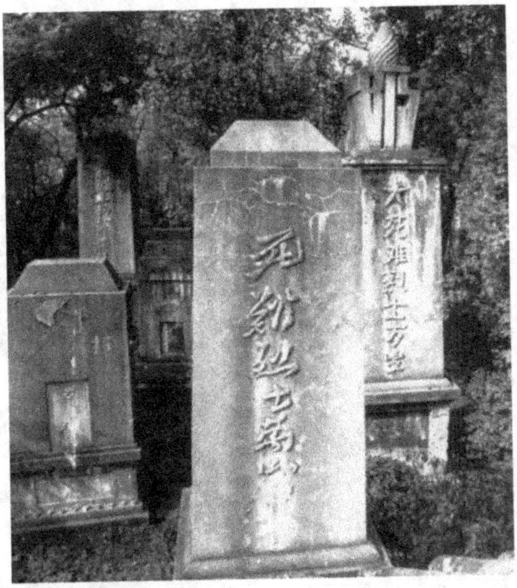

Photo 2: A Cultural Revolution slogan refurbished for visual clarity during the early 1990s, in Shapingba Park (photo by Liu Qingfeng).

the victims buried there are mostly of the "Eight-Fifteen Faction" or related groups. Many of their stones are inscribed with a torch symbolizing "the Eight-Fifteen rebel revolutionary spirit." The handle of the torch is made up of the three numbers "815." This design was worked out by Zhou Ziren, editor in chief of *Eight-Fifteen Struggle News*—the organ of Chongqing University's "Eight-Fifteen Struggle Group." This design was chosen for its simplicity, clarity, and ease of drawing, so it became the representative design of the "Eight-Fifteen Faction."

There are two major styles among the gravestones at the cemetery. One is old-style gravestones and the other is a style modeled on the Monument to the People's Heroes in Beijing. From the words on the tombstones, you can see clearly the sources from that time—the "rebel group" and education in the revolutionary tradition. Besides the stones that have Cultural Revolution slogans (such as "revolution is not a crime, rebellion makes sense"; "one's head can be taken off, blood can flow, but Mao Zedong Thought cannot be overcome"; "we can endure beatings and struggle; we pledge our lives not to lose sight of revolution"), generally they are inscriptions like Mao Zedong's to heroes who died in the Anti-Japanese War or the Civil War between the Communists and Nationalists, or are well-known lines from Mao's poems such as "long live suffering martyrs," "eternal glory to suffering martyrs," or "when we sing the stirring 'Internationale,' Heaven sends a hurricane to help us." And there were other similar lines.

Among the graves, the inscription on Chongqing's Number 29 Middle School stands out. . . . Here is the complete text:

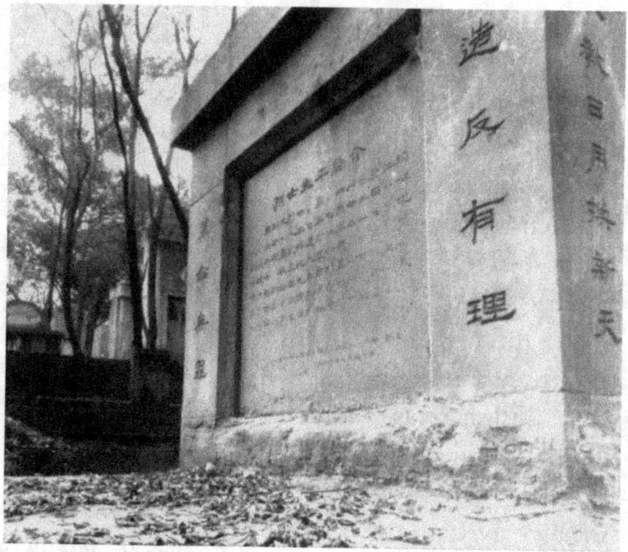

Photo 3: Cultural Revolution slogans on a monument in Shapingba Park (photo by Liu Qingfeng).

Blood enriches our country to produce strong grass, and winter sets the great earth to bring forth spring flowers.

Chairman Mao's most loyal Red Guard soldiers, the very best soldiers in our Maoist struggle group Zhang Guangyao, Sun Yulou, Ou Jiarong, Yu Zhiqiang, Tang Mingyu, Li Yuanxiu, Cui Peifen, and Yang Wuhui—these eight martyrs who gave up their last drop of blood to defend Chairman Mao's revolutionary line in those August days when blood and fire were exchanged used up the glory of life to light the way on the road of life for those who will come after.

You, our comrades who died in distress, when we think of you we are filled with courage, our strength is unending, we remain determined, we can break through all difficulties and struggle on to victory.

Wherever disorder challenges the Red Flag, righteous blood hastens the blossoming of heroes.

Beloved comrades, today we have used military struggle to welcome the cheerful red cloud.

Speak out boldly for all that you strive for, thrill to the sight of a world gone red.

Your hot red blood soaks through our Eight-Fifteen's reddest red flag of rebellion. Ah, this great flag, we hold it high. Your ardent blood has permeated the flaming torch of our 815—ah, this flaming torch, we hold it high!

We might lose our heads, and blood might flow, but Mao Zedong Thought will never be lost. Your urgent words of alarm have brought together thousands of volunteers as an earth-shaking roar for justice. Your heroic stature is like rich green pines and poplars, standing upright like Red Crag Ridge and the peak of Gele Mountain.[15]

Eternal glory to you, martyrs of our Mao Zedongist fighting group! Eternal glory to you, martyrs of our 815 Revolutionary Group!

Chongqing Revolutionary Rebel Fighting Academy
(originally Number 29 Middle School)
Mao Zedongist Fighting Corps — October 1967

The group of eight students of Number 29 Middle School in Chongqing, shown below, all died in the intense fighting of August 1967. Brief information on each follows:

Zhang Guangyao, third-year student at Number 29 Middle School and a "Mao Zedongist Military Group member" was one of these victims, falling on August 3, 1967. Because of an internal military struggle at the Number 29 school involving the Mao Zedongists and the "815 Group" from the Petroleum School, a member of the "815 Group" was hit by a bullet and lay on the athletic field, and no one went to help. Several girls from the Petroleum School ran to the headquarters of the Maoist Struggle Group looking for someone to assist the wounded youth. Zhang Guangyao volunteered and went out to help. In the words of his classmates, "it looked like he was turning to catch a basketball." He was shot dead by the opposing group.

Ou Jiarong, Yu Zhiqiang, Tang Mingyu, Li Yuanxiu, Cui Peifen,

15. These place-names are commonly used as references in Chongqing. Red Crag was headquarters for Zhou Enlai and other Communist Party leaders for activities in Chongqing. There is now a museum at this site. Gele Mountain is a major ridge in the area. See also "House of Democracy" in this collection.

and Yang Wuhui were helping the Financial Section of the "815 Group" move their office, and at the entrance to the Finance Club they met up with a struggle group from "Rebel to the End," who abruptly shot them dead. They did this as some of these girls were sucking on popsicles. Yang Wuhui was one of the two youngest students in this entire cemetery. She was just fourteen years old.

Sun Yulou, another student from Number 29 Middle School, at this time was serving as a representative of the Chongqing Revolutionary Rebel Command Post stationed on the south bank of the Yangzi River. On August 22, 1967, "Rebel to the End" attacked the "815 Group" position on New Street south of the river, and while protecting others as they retreated, Sun Yulou was hit by a bullet and killed.[16]

This cemetery also has the grave of Ai Shuquan, a higher middle school student at Number 3 Middle School. At that time Ai Shuquan was a member of the "Service Group to Resist Large Forces" and a leader of the Military Struggle Team. The inscription on his grave says the following:

> The martyr Ai Shuquan, male, from Tianjin, belonged to a worker's family. He was an excellent member of the Communist Youth League and a fine soldier in the "Service Group to Resist Large Forces." On August 3, 1967, he made a heroic sacrifice to defend Chairman Mao's Revolutionary line. He was twenty-one years old.

Xu Ruodan was a female student, and a friend of Ai Shuquan. Unfortunately, she died on July 16, 1968, when she decided to commemorate Mao Zedong's swim in the Yangzi River by swimming in the Jialing River, and she drowned.[17] In her diary she wrote her

16 At this point the original tells the story of another martyr, Jiang Pijia, who was assigned to provide security for a central government investigation team. Jiang was eighteen years old and the son of a professor at Chongqing University. Although he was at the assigned place, when he was killed he was merely leaning against a tree to rest and was hit by a stray bullet. The writer notes that the investigation team had not yet arrived in Chongqing and were to stay at another place. Jiang's story again shows the randomness of death during those days and the irony in treating him as a glorious martyr.

17. Mao Zedong's swim in the Yangzi River at Wuhan on July 16, 1966, displayed his

innermost thoughts after Ai's death, and also the scene at the cemetery in Shapingba Park:

> August 3, 1967:
> I woke up after napping and went to the entrance of the Red Expert Building, where I heard that Ai Shuquan had died. I thought someone was lying to me. Then a bit later, our classmate "Fishing Boat" came in, carrying Ai Shuquan's body, and all of us were so saddened that our cries of grief reached to heaven. . . .
>
> Later Ai Shuquan's body was laid out on a table in the meeting room. He was dressed in a military uniform, with a copy of "Quotations from Mao Zedong" clasped in his hand and with badges of the Chairman on his breast and as an armband. His eyes were set and his lips open as if he would say, "I couldn't fight to the end in the Cultural Revolution, and couldn't fight all of our enemies wherever needed. I died but I have not closed my eyes."

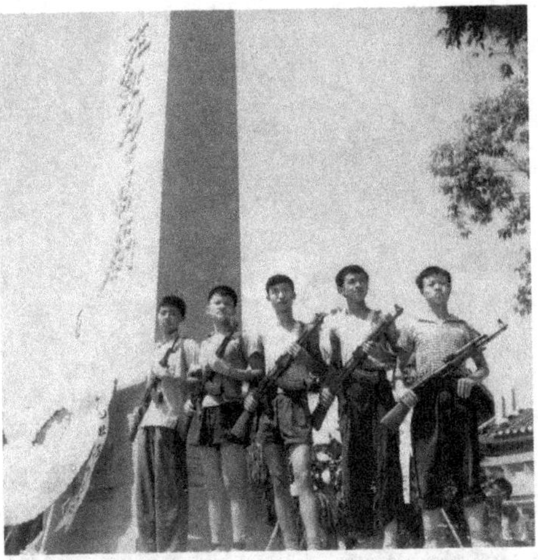

Photo 4: After the cessation of military struggle in 1967, at the just-completed "Martyrs' Monument" to fallen students of No. 29 Middle School Maoist Military Struggle Corps, these classmates hold submachine guns. They wanted to express their determination to "wipe our eyes, carry on the martyrs' will, and pledge to carry on with their great plan" (photo contributed by Guang Li).

---

vitality at age seventy-two. It was intended as a symbolic action to show his fitness to continue leading China's revolution—this after the setbacks and opposition Mao faced at the beginning of that decade in the failure of the Great Leap Forward and the "Three Bad Years." The Cultural Revolution began very soon after Mao returned to Beijing following this feat. From his youth, Mao like many modernist activists had emphasized the importance of physical fitness, a great contrast from traditional intellectuals. Mao's swim is commemorated with an annual re-staging on this date at Wuhan.

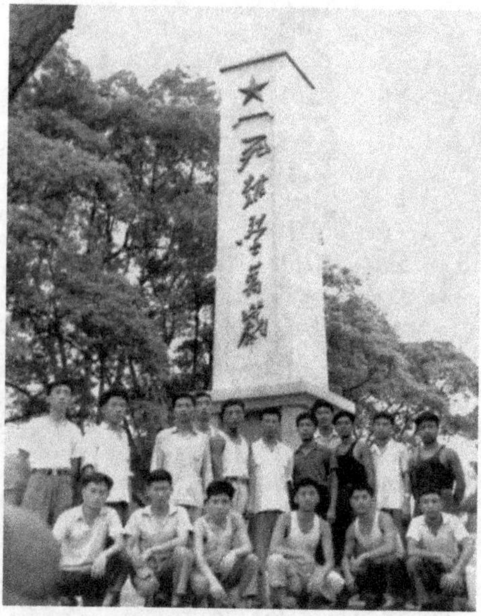

Photo 5: After the armed struggles concluded, classmates from the Metallurgy Department of Chongqing University gathered at the "Martyrs' Monument" of their classmates who had died (photo contributed by Huang Shunyi)

August 4, 1967:
Got up early this morning and went to our headquarters, and as I approached I heard people crying. They were Ai Shuquan's mother, older sister, father, and younger brother grieving. On hearing such utter grief, who wouldn't also be moved to tears? Ai Shuquan was one of our group's leaders and also a leader of the Revolutionary Alliance and a member of the Alliance's Standing Committee. He often came to spend time with our classmates, always wearing a smile. . . . In his work he was always on top of things, doing everything conscientiously and positively. . . . His death was a terrible shame. I heard that his older brother is still coming, so this will be the last thing before he is buried.

August 6, 1967:
Ai Shuquan's older brother has arrived; he is a student at Tianjin University. When he saw his younger brother's body, he grieved deeply for a long time. He stroked his brother's body and placed his commemorative badge and his armband on the martyr's body. . . . Seeing this scene, who could keep from tears! The strength of this grief and pain is more than enough to cause my pen to seek a blood debt from our enemies who would destroy the Party!

> August 8, 1967:
> Yesterday the memorial gathering was held, and everyone in the martyr's family spoke. Friends of the family who knew him throughout his life talked about how he lived. He joined in rebellion very early in order to defend Chairman Mao's revolutionary line. He was beaten as an anti-revolutionary, called a thug and condemned as a cow's ghost and snake spirit. Through all these difficulties he never bowed, and on behalf of the revolution he went everywhere and rode every wave. He said, "I want to give everything to the Party." Now, he has fulfilled this promise. . . .
>
> Today all of us had deep pain in our hearts as we buried him in the martyrs' cemetery at Shapingba Park. We pledged to him that we would carry out his mission and see the Great Proletarian Cultural Revolution through to the end. To seal our pledge, each of us picked up a handful of dirt and tossed it on his grave.
>
> When one martyr goes down, ten thousand Red Flags will fly high. The death of a hero even more calls forth a courageous mass, and the martyr's blood was not shed in vain. We want to give the springtime of our lives to protect Chairman Mao's revolutionary line.

(These excerpts are from Xu Ruodan's diary, which was provided by her older sister, the writer Xu Jun.)

In those times people could sing and cry, and now we sigh and feel the tragedy of it all. In every grave in this martyrs' cemetery there are many stories like these.

During those years in Chongqing, at every school and every factory, in important places on streets or in gardens (for example, the City Gymnasium, Chaotianmen Landing, the head of the Jialing River bridge, major traffic circles, and other places)—"martyrs' cemeteries" were built in all of these places. But later all of them were leveled.

During those years of large-scale fighting, the "Eight-Fifteen Military Struggle Group" of Chongqing University lost a total of twenty-four people. After the fighting ended in 1967, the university started a "martyrs' cemetery" in a grove of pine trees on the campus.

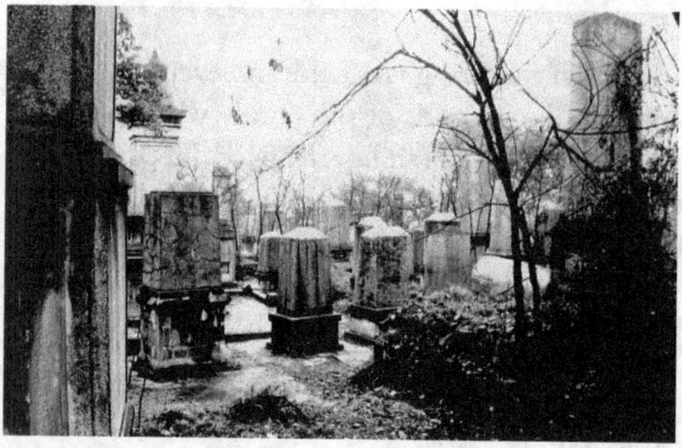

Photo 6: A corner of the mass grave of those who died in the armed struggles of the Cultural Revolution. Shapingba Park in Chongqing (photo by Liu Qingfeng).

A commemorative monument was placed at the front of the cemetery, on which the names of these twenty-four martyrs were inscribed, and these words from Mao Zedong's writings were inscribed in the same style: "Long live these martyrs who died in difficulties." A vertical panel for poetry was also placed at the entrance to this burial park, with a line from Mao's poem in memory of Yang Kaihui and Liu Zhixun: "I lost my beloved proud poplar and you your willow; they flew on light wings to the great beyond."[18] In those years, many of the "Eight-Fifteen Group" would stop by this memorial garden when they returned to the school to reflect on the past, pay their respects to their classmates, and show that they cherished the memory of these fallen classmates.

Not long after the Cultural Revolution ended, these cemeteries were leveled, and these "martyrs" or "wronged youth" fell into a situation in which their bodies could not be recovered and permanently laid to rest.

It was fortunate that the "martyrs' cemetery" in Shapingba Park was not destroyed. For many years, from time to time there would be

18. This line is from "Reply to Li Shuyi," written in 1957. Li Shuyi was a woman friend who, like Mao, had lost her spouse, Liu Zhixun, in early revolutionary activity. Mao's first wife was Yang Kaihui, daughter of a professor for whom Mao had great respect. Yang Kaihui was killed by the Nationalists in 1930, and Liu Zhixun in 1932. The reference to the two kinds of trees comes from the family names: Yang (杨) means poplar, and Liu (柳) means willow. This is not the most common character for families named Liu (刘).

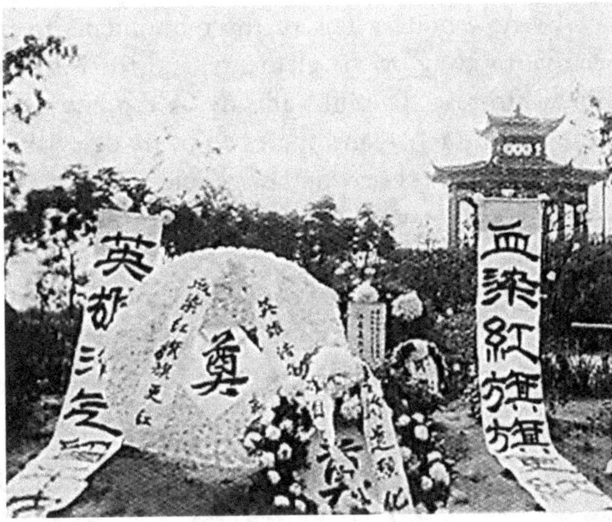

Photo 7: The first grave for Jiang Pijia, at Mt. Pipa. See note 3. (Photo contributed by Zeng Zhong).

news that it too was to be leveled, but this always aroused public opposition. Because it was in a suburban area and not close to the city of Chongqing itself, the number of people interred there was large (more than one hundred), and several hundred others were buried there too, so family and friends also spoke up. Talk of development never resulted in action, and those seeking development avoided unwanted troubles.

The last version of this cemetery was created forty years after these events, at the end of 2009. The Chongqing city government included this place as a City-Level Unit for Cultural Preservation, even though its name was the not quite accurate "Red Guard Cemetery," and when all was said and done it was formally designated a historical and cultural landmark. For many years Mr. Ba Jin called for a "Museum of the Cultural Revolution" to be established. Here at least we have an actual "Branch Museum."[19]

19. Ba Jin wrote essays on many subjects during the years after the Cultural Revolution. These were collected in Random Thoughts (*Suixiang lu*) (*Beijing, 1987*). The essay "A 'Cultural Revolution' Museum" (*"Wenge" bowuguan*), pp. 819–23, recalls Ba Jin's fear that China might again experience such mass hysteria with a repetition of social trauma and personal tragedy. He urged that museums be established so that everyone would remember that period and do everything possible to avoid a repetition. As the writer of this article says, the Chongqing cemeteries meet Ba Jin's hopes for these memorials. Other museums or memorials have been launched elsewhere in China.

By now quite a few of the monuments in this cemetery have been worn away by the elements, and the inscriptions on them are difficult to read. It might already be too late to preserve or recover these cultural relics, and there can be no delay about redoing the engravings. It isn't clear whether people will give attention to the need to preserve the graves, but their concern should include reflection on that time in history. . . .

This is by He Shu. The article was published in Old Photos *no. 70, pp. 83–97.*

A Diary of the 1976 Tian'anmen Incident

Introduction

This essay provides details on the people's response to the death of Premier Zhou Enlai on January 8, 1976. By early April of that year—around the Qingming Festival, the traditional Chinese holiday to honor the dead—a crowd that spontaneously gathered in Tian'anmen Square in central Beijing in memory of Zhou turned into a rally against the Gang of Four, the radical faction headed by Mao's wife, Jiang Qing, and indirectly against Mao's Cultural Revolution. In response, the ailing Mao ordered a crackdown on the movement. The number of victims of this crackdown is unknown. In comparison with the Tian'anmen Incident of 1989 caught by CNN and other media in the West, few in the West know about this Tian'anmen incident.

This was the first of a number of major events for China in 1976, the last of which was the death of Chairman Mao Zedong. Zhou Enlai had been a moderating force during the excesses of the Cultural Revolution, and it was widely believed that he had often acted to reduce if not eliminate destruction, hence the groundswell of sadness and sympathy at his death. This initial response to Zhou's death enters into most textbook discussions of the fateful year of 1976, but the repeated memorials in 1977 described here with personal diaries and other documents are not often noted.

The Text

Not long ago I was rummaging through an old cardboard box and discovered a diary of mine from thirty-three years ago, with quite a number of photo negatives still tucked inside. The photos were taken during the Qingming Festival in 1976.[20]

For the people of China, 1976 was an eventful year. During that year, three of China's leaders—Zhou Enlai, Zhu De, and Mao Zedong—died one after another, the "April Fifth Movement" broke out in Tian'anmen Square, the Tangshan earthquake hit, the "Gang of Four" was arrested, and other momentous events happened.[21]

Of these important events, the most memorable for me is the "April Fifth Movement," the outburst that had an enormous influence on China's political situation.

After Premier Zhou's death on January 8, 1976, everyone in the country was immersed in cherishing his memory. At the Qingming Festival, in order to express their grief over Premier Zhou and also to express their dissatisfaction with the "Gang of Four," masses of people went to Tian'anmen Square to lay wreaths of flowers and put up poems and songs opposing the "Gang of Four." So in this way a movement got started to mourn Premier Zhou and oppose the "Gang of Four." At the time I was an ordinary worker who enthusiastically joined this movement and took many precious photographs.

Although the "Gang of Four" always found ways to attack Premier Zhou by innuendo and continuously sought ways to limit people's activities to mourn him, masses of people found ways to go ahead with their mourning. The No. 1 Shoe Factory where I worked set up a memorial hall for the Premier in our main dining hall and held a memorial gathering there. At the same time people mourned Premier Zhou, their dissatisfaction with the "Gang of Four" intensified.

---

20. The term *Qingming* actually means "pure brightness." This is the day in early April when people pay their respects to those who have died, from earlier generations or, in this case, quite recently.

21. Zhou Enlai (1898–1976), China's premier, died on January 8; Zhu De (1899–1976), the Standing Committee Chairman of the People's Congress of China, died on July 7; Mao Zedong (1893–1976) died on September 9.

Today, thirty-three years later, publication of these photographs and parts of my diary will add evidence for that episode in history.

January 15, 1976—a clear day
This evening, four of us—Zhao Kexin, Lu Long, Dong Changwei, and I—went together to Tian'anmen Square to pay our respects to Premier Zhou, and we also hoped to see the hearse that carried his ashes. First, we left our bicycles at Dong Changwei's home near Xidan and then walked to Tian'anmen Square. All along the street we saw lots of people. When we got to the Square, there was already a sea of people. Masses of people had set out wreaths for Premier Zhou, laid layer upon layer around the Monument to the People's Heroes. And the small pine trees beside the monument had small white flowers placed in them. In the vast crowd of people in the Square, there were so many groups of two or three people who, like us, had decided to come. Then there were also large groups from the largest units, organized by the units themselves. Everyone formed a great long line in order to reach the area beside the Memorial to the Heroes and pay respects. All around the Memorial you could see people in tears, expressing their sorrow and their loyalty to Premier Zhou.

The four of us chose a spot on the north side of the monument and first stood there in reverence for a moment. Then we moved closer to express our loyalty to Premier Zhou. My testimony was this: "Beloved Premier Zhou, your spirit will be passed on for countless generations and your achievements will remain eternally. We resolve to follow the road you walked until we reach the world of communism." As the four of us swore our faithfulness, all of us shed tears. Finally, although we were reluctant to go, we left Tian'anmen Square.

After this and all the time until the Qingming Festival on April 4, people went to Tian'anmen Square in droves. Although there were so many people, everything was very orderly. With so many people walking and others riding bicycles, there could have been accidents, but there was almost no bickering about such inconveniences. As the expressions of grief for Premier Zhou continued, people's feelings changed from grief to righteous indignation. Poems and writings mocking and criticizing the "Gang of Four" were copied and passed on among the masses of people. The "Internationale" became the most popular song in the Square.

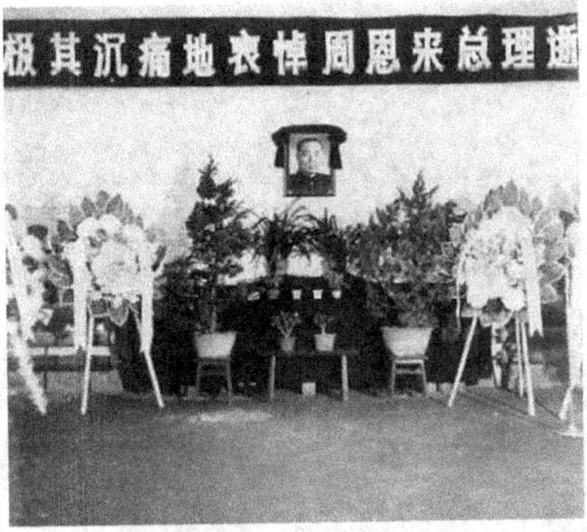

Photo 1: In the factory where the author worked, workers turned their dining hall into a memorial hall to mourn Premier Zhou.

Toward evening on April 4, I went to the Square again with several friends from our factory. This time a vast crowd of about two million gathered there. I heard someone recite a poem, "Reporting to the Premier for Instructions," in a loud voice:

> There's a bridge across the Huangpu River, which is rotting and shaking. The bridge looks like it's toppling. Please tell us, is it breaking down or is it burning?[22]

In perfect unison, the crowd erupted together: "It's burning! It's burning! It's burning!"

At 7 o'clock that evening the loudspeaker in Tian'anmen Square broadcast the voice of Wu De, First Secretary of the Beijing City Committee, who said, "Today at Tian'anmen Square there have been bad people who have carried out destructive anti-revolutionary

22. In this poem or ditty the sounds *jiang*, *qiao*, and *yao* are repeated several times: *jiang* represents Jiang Qing, Mao Zedong's wife; *qiao* is the last syllable in the name Zhang Chunqiao; and *yao* is for Yao Wenyuan. These were three of the four in the "Gang of Four." The Huangpu River is in Shanghai. It is used in this ditty because all the other members besides Jiang Qing were from Shanghai.

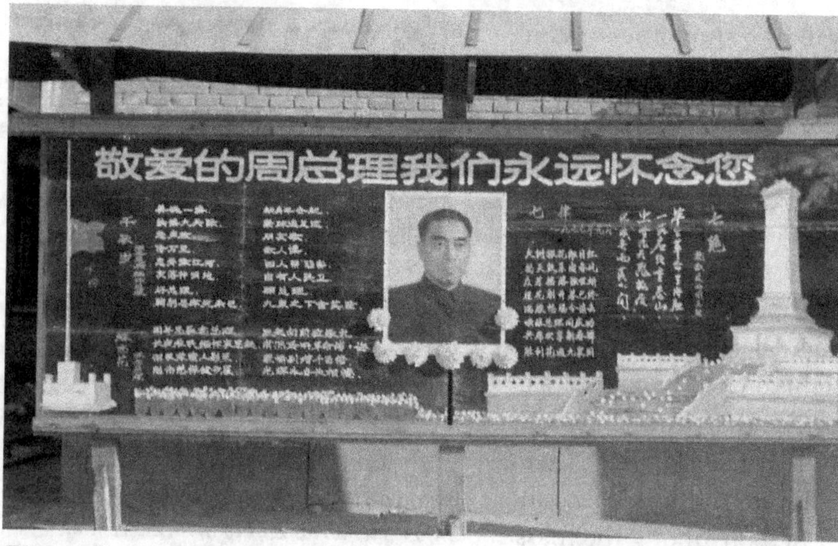

Photo 2: Poems dedicated to Zhou Enlai are written in chalk on the factory blackboard. The line on the top reads "Esteemed and Beloved Premier Zhou, We Forever Miss You"

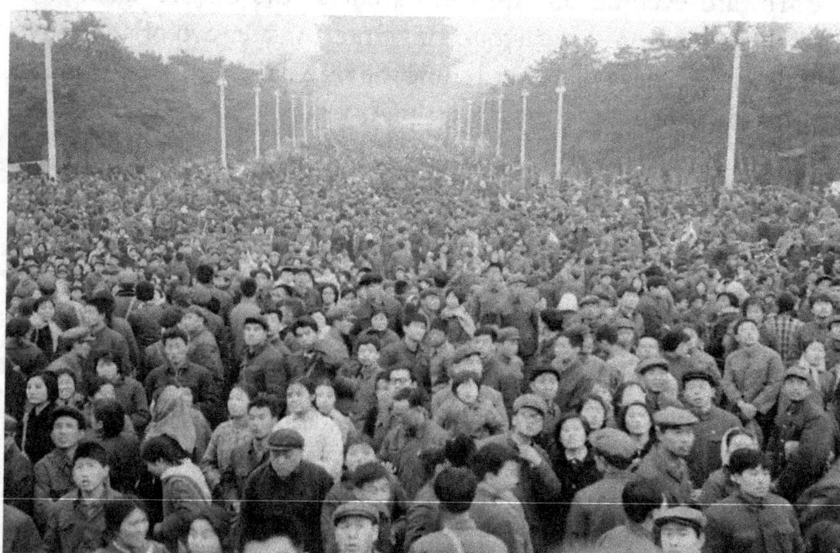

Photo 3: The mourning crowd in Tian'anmen Square with Zhengyang Gate in the background—an early fifteenth-century structure that, with the approval of Zhou in 1965, survived the demolition of Beijing's old city gates

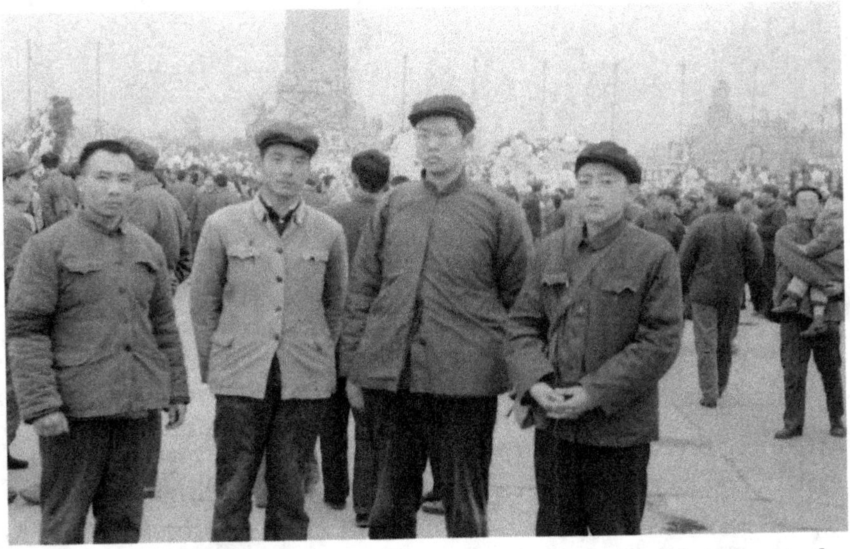

Photo 4: The author, first on the right, and friends in front of the Monument to the People's Heroes.

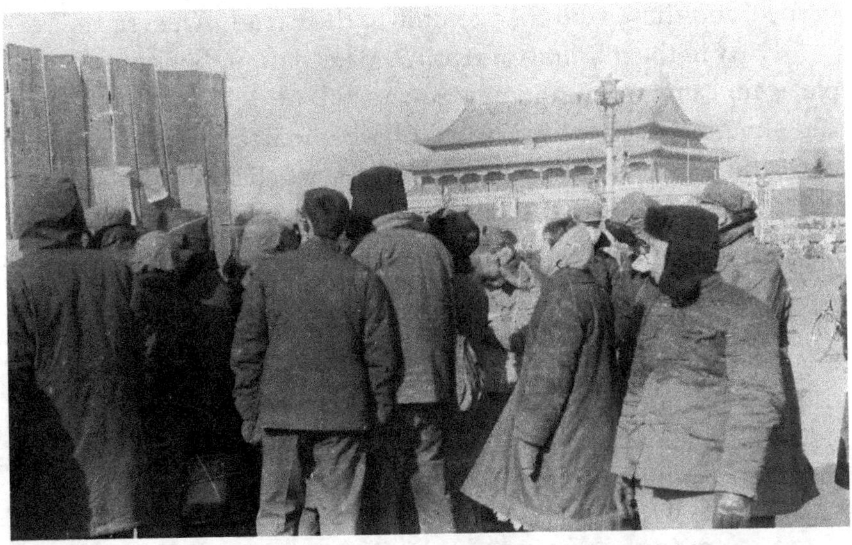

Photo 5: Viewing commemorative poems and notes in Tian'anmen Square.

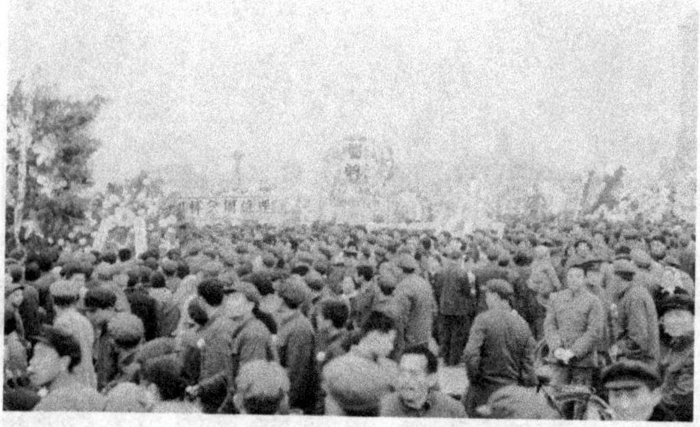

Photo 6: The mass of people who came to pay their respects to Premier Zhou, most of whom had come previously.

activities. The revolutionary masses should leave the Square immediately and not accept their deception." Wu De's brief speech was repeated continuously. We assumed that the authorities wanted to break off these activities. Not long after Wu De's pronouncement, we left Tian'anmen Square.

That same evening, the wreaths were forcibly destroyed or burned, and those who tried to protect the wreaths were seized.

Early in the morning of April 5, many tens of thousands of people again came to Tian'anmen Square, full of righteous indignation and shouting insistently, "We still have our wreaths, and still have our comrades!" Then those same militia people, police, and troops began a serious clash. In the confusion, the vehicle with the loudspeaker was destroyed, and the small building of the workers' militia headquarters was set on fire.

At 9:30 a brutal crackdown began; the authorities sent out ten thousand militia, three thousand police, and five battalions of garrison troops, armed with cudgels and surrounding the square on all sides, beating and dispersing the masses of people who remained in the Square and arresting some.

April 7, 1976—a clear day

Today I tore up and burned my diaries of the past few months. They were not of great importance. Although great national events are not recorded on paper, I trust myself that I will never forget them. From now on I will simply record everyday matters in the diary. Today's broadcast reported

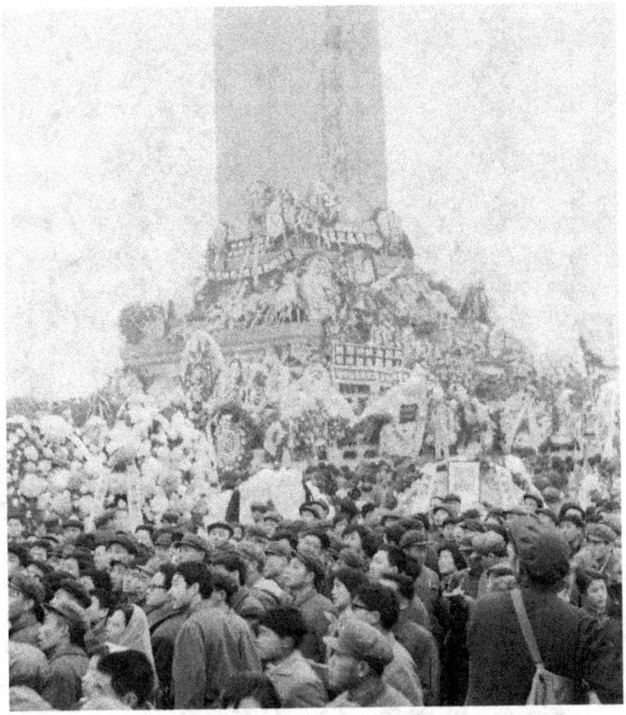

Photo 7: In front of the Monument to the People's Heroes.

two major decisions by the Party Central. It reported on the experiences of anti-revolutionary events in recent months at Tian'anmen Square. At the factory there was a major meeting in which everyone supported these decisions.

After the popular activities to commemorate Premier Zhou were declared to be anti-revolutionary, Beijing was enveloped in fear. Every work unit began investigations to find out who had gone to Tian'anmen Square during the period of grieving Premier Zhou, whether or not they had copied poems, or taken photos, or done other such things. I had gone to the Square with my friends several times and not only had copied poems but also had taken photos—a good many people knew this. In order to get through this crack-down, I had to turn over several hundred of the poems and songs that I had copied in the Square as well as some negatives of photos that were blurred and unclear. I completely destroyed everything in the diary that might invite trouble. The negatives that were rather

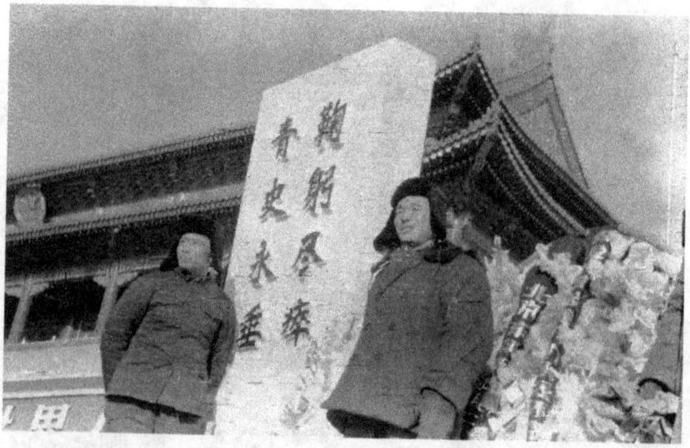

Photo 8: In front of Tian'anmen, the Gate of Heavenly Peace itself.

clear, I tucked away in a good hiding place. The photos published here are some of those.

January 10, 1977—cloudy

During the past few days, huge crowds of people have been going to Tian'anmen Square to commemorate the first anniversary of Premier Zhou's death. They take wreaths or baskets of flowers, write poems, pass out leaflets, paste up slogans—all sorts of means to express their feelings. There is lots of activity there. Especially yesterday, as I walked along Chang'an Road, I saw a very brave group of people carrying wreaths to the gate at Zhongnanhai and presenting their petitions.[3] There were six items in these petitions:

1. Support Hua Guofeng as chairman
2. Call for the appointment of Deng Xiaoping
3. Reverse the verdict on events at Tian'anmen
4. Hold a public trial of the Gang of Four
5. Dismiss Wu De (First Secretary of the Beijing City Committee)
6. Establish a memorial to Premier Zhou

The crowd of petitioners tried to enter Zhongnanhai, but then the Liberation Army guard closed the gate. (It is said that this was the first time the gate to Zhongnanhai had been closed.) The spontaneous action of this crowd could not easily be controlled. No matter what the leaders did to deal with their demands, the crowd still meant to see their demands met. If Party Central does not respond to the petitions of the masses of people before the Qingming Festival, there could be a crisis at the time of the festival.

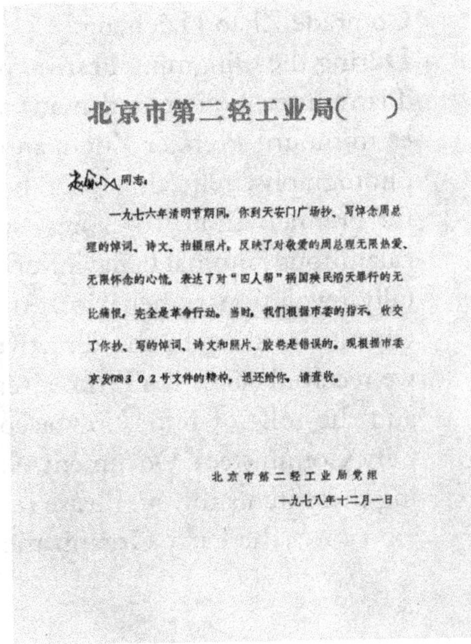

北京市第二轻工业局(　　)

[illegible]同志：

一九七六年清明节期间，你到天安门广场抄、写悼念周总理的悼词、诗文、拍摄照片，反映了对敬爱的周总理无限热爱、无限怀念的心情，表达了对"四人帮"祸国殃民滔天罪行的无比痛恨，完全是革命行动。当时，我们根据市委的指示，收交了你抄、写的悼词、诗文和照片、胶卷是错误的。现根据市委京发(78)3 0 2号文件的精神，退还给你，请查收。

北京市第二轻工业局党组
一九七八年十二月一日

Photo 9: The letter sent to the author by the Party Group of the Number Two Light Industry Office of the City of Beijing, dated December 1, 1978

In October 1976 the "Gang of Four" had been broken up, and millions of people took to the streets in Beijing, applauding this "Second Liberation." During the period of the first anniversary of the Premier's death, the masses again took to the streets and presented their demands, and over the next few years almost all their demands were realized. In July 1977, Deng Xiaoping was formally reinstated; in December 1978, the Third Plenum Session of the Eleventh Central Committee [of the CCP] announced that the activities of April 5, 1976, in the Tian'anmen incident were indeed revolutionary behavior, and that incident indicated mass support for the Party to break up the "Gang of Four"; at the end of 1978 Wu De was removed as First Secretary of the Beijing City Committee, and as chief of the City Revolutionary Committee; and on November 21, 1980, the Supreme Court of the Ministry of Law tried the "Gang of Four."

In December 1978, I received this letter from the Party Organization of the Beijing Number Two Light Industry Bureau:

Comrade Zhao Haicheng:
During the Qingming Festival period of 1976, you went to Tian'anmen Square and made written records of speeches to mourn Premier Zhou, and poetic writings. You took photographs, reflecting your boundless love and feelings for Premier Zhou. To express such deep anger toward the calamitous criminal behavior of the "Gang of Four" was totally revolutionary behavior. At that time, it was an error when, according to the directive of the City Committee, we received (seized) all those items, the poems, the photos, and the rolls of film. Now according to the spirit of the City Committee's Document No. (78) 302, we are returning these items to you. Please receive and inspect them.

The Party Organization of Beijing Number Two
Light Industry Bureau
December 1, 1978

This is by Zhao Haicheng. It was published in Old Photos, *no. 65, pp. 24–32.*

On Taking the Photo *A Petitioner*

Introduction

The idea that an ordinary person could appeal to authority, even to the emperor, was established in Chinese tradition. At the entrance to the Imperial Palace stands a pair of *huabiao*, ornamental stone columns with "wings" near their tops. Legend says a commoner was allowed to write a petition stating a grievance or offering an idea concerning statecraft and place it at the base of one of these columns. Occasionally an emperor would indeed respond to such a petition. The contemporary practice grew out of this tradition although the petitioners shown here used a different form to present their grievances.

Here photographer Li Xiaobin tells not only how he took a photograph that eventually had great influence, but also describes the concerns a photographer faced if his work were to be published and have some influence in the larger society. Both Chinese and foreigners were for many years ordered not to take pictures of bridges, railroad stations, or other subjects that might compromise national security. (Chinese regulations closely followed those applied in the Soviet Union, one of the many everyday Soviet influences in China.) To authorities in China, political embarrassment was as important as security concerns. Hence Li's concern about even getting his film developed. Security concerns became somewhat less important by the early 1990s, but as Li observes, it is still rare to see these kinds of photographs. While materials from the "bad old days" of the Cultural Revolution may be acceptable inasmuch as they are supposed to show that those conditions existed only in the past, any expression of dissent is still unwelcome. On the other hand, local demonstrations staged for many reasons are now almost rampant in China. Thousands of such popular actions are staged every year.

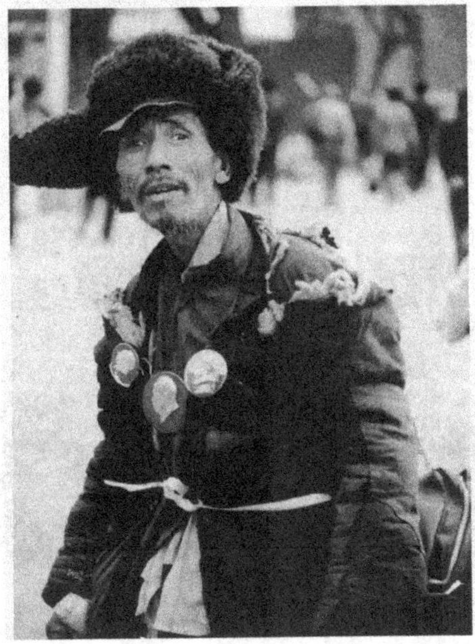

Photo 1: A Petitioner, *photo taken in late November 1977, on the streets between Tian'anmen and Duanmen, Beijing.*

This piece is clearly related closely to the preceding one, in which a worker during these same years and several colleagues joined many thousands in paying their respects to Zhou Enlai soon after Zhou's death in January 1976.

The Text

This photo, *A Petitioner*, was taken in November 1977. At that time the "ten years of turmoil" had come to an end, but the pain they had

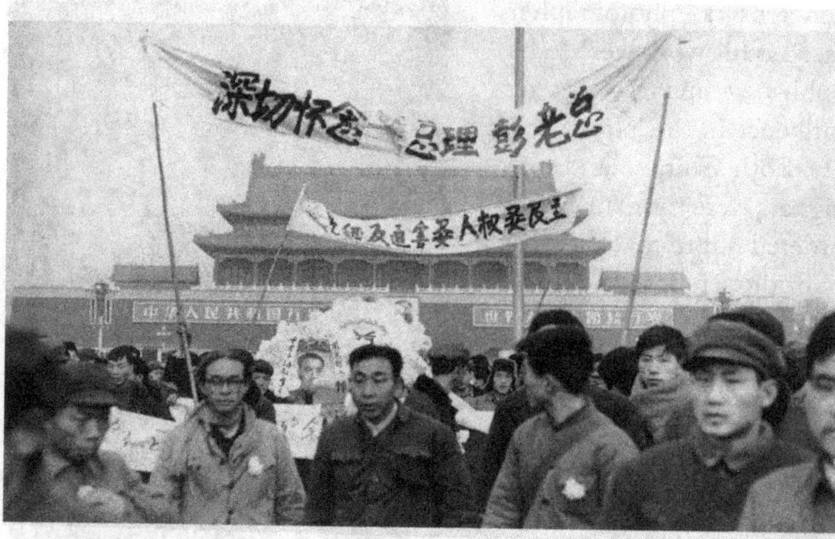

Photo 2: Crowds are gathering in Beijing's Tian'anmen Square in commemoration of the second anniversary of the death of Premier Zhou Enlai, January 8, 1978. The banner in the background reads in part: "Anti-persecution; [we] want human rights and democracy!"

brought to society and to the people had not yet ended.[23] Because of the influence of "the two whatevers," most of the trumped-up cases against "rightists" were still unresolved, and new cases were still being brought forward.[24] Those who had been hurt found themselves in a world of difficulty, and they could only try to present petitions at some level of authority. Beijing is the capital, so a vast number of petitioners streamed into the city, searching for one last ray of hope. Petitioners surged all around Tian'anmen Square, Zhongnanhai, and at the entrances to all the organizations concerned with security and investigation. I lived on a street right beside Tian'anmen Square where many of these offices were concentrated, so I was well acquainted with what the petitioners were going through.

23. The phrase "ten years of turmoil" is a common expression in China referring to the Cultural Revolution decade, 1966–1976

24. The slogan "the two whatevers" appeared in editorials in *People's Daily* and *Red Flag* in early February 1977, as themes in the leadership of Mao Zedong's successor Hua Guofeng. Hua sought to be faithful to Mao and to continue his ideas and policies.

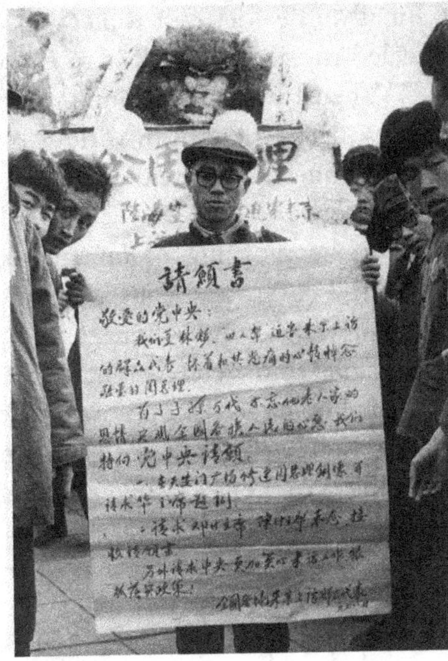

Photo 3: A man holds a petition asking Deng Xiaoping and other government leaders to support the petitioners.

After *A Petitioner* was published, analysts with various journals commented that the photographer had shown "political consciousness," "good observation," or "concern for our people's fate," but I wouldn't say that. What moved me to take the photo at the time was a professional photographer's reaction to this petitioner, a kind of spontaneous reaction. I was moving toward him on my bicycle, and I spotted him just as he emerged from the surging mass of people. I felt a strange urgency and nervousness. Actually, my hands were shaking just a bit. I realized that the lighting conditions were just right. I quickly set my bike down about fifty meters from him, got the camera set up and waited behind a large tree. When he came into the frame, I took three quick exposures.

I thought the three shots had captured his expression and that surely I had a good photo, so I jumped on my bike and took off. I sensed that some people out on the street thought I shouldn't take a photo like this, that they resented me for it. In those days you might have your camera taken from you, or your film, for such an action. That wouldn't have been at all unusual! So as I rode along I was also looking to see if anyone was following me. To avoid a problem, I took a roundabout route and after making sure that nothing would happen to me, I relaxed and went back to my home, which actually was not far from where I took the photo.

I also thought a long time about getting the film developed. A private photo service might mess it up. A state outfit might confiscate it. Finally, I decided just to keep it for a time. By February of 1978, I steeled myself and took it to the China Photo Agency to develop.[25] When I went to pick up the photos, it was hard to conceal my feelings. I had thought of various possibilities and how I might respond to questions. I rejoiced when I got the photos back without any problems.

On the morning of January 8, 1978, thousands of petitioners filled Tian'anmen Square carrying white flowers and signs denouncing the "Gang of Four" and honoring the memory of Premier Zhou Enlai and Peng Dehuai.[26] They made a huge demonstration to "fight hunger, resist oppression, advance democracy, and advance human rights." Several well-known photographers were present, including Wu Peng, Wang Zhiping, and Ren Shimin, who were serving as editors for a large-scale special publication called *The People's Grief.*

During the years 1977 to 1980 the masses of petitioners were at their peak. I probably took thirty rolls of film during that period. Besides the subjects I've already described, I also took photos of Democracy Wall at Xidan in 1979, the activities of Renmin University students to return to school and to their places in university dormitories, the demonstration in Tian'anmen in January 1978 of students from Yunnan demanding that they be allowed to return to

---

25. Many amateur photographers during those years believed that a roll of film should be exposed all at once and developed immediately—otherwise the film would go bad. Li's decision suggests his experience as a photographer.

26. Peng Dehuai (1898–1974) was a prominent Chinese Communist military leader. His long military career began during the 1920s, when at first he was associated with the left-wing Guomindang. By the late 1920s he had joined the Communists in their Jiangxi base in the south. He became Minister of Defense after the founding of the People's Republic and was the commander of the Chinese People's Volunteer Army during the Korean War. He first showed differences with Mao Zedong in his intention to follow the Soviet example and make the army more professional—rather than primarily a politically oriented body. Peng's biggest disagreement with Mao came in the summer of 1959 at the widely known "Lushan plenum" where Mao's Great Leap Forward policies were discussed. Those policies had led to serious decline in agricultural production and widespread privation and famine. For his criticism of the Great Leap Forward policies, Peng was dismissed from his posts in the top leadership. During the Cultural Revolution he was jailed and died in prison. After the death of Mao, Peng was posthumously rehabilitated. By the time of the events described in this article, Peng was widely regarded as a popular hero for his resistance to the radical policies of Mao.

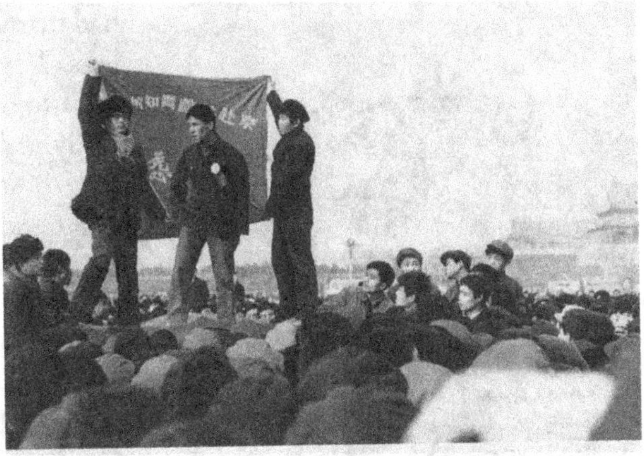

Photo 4: A group of "sent-down" youth from Yunnan Province rallying in Tian'anmen Square, January 8, 1978.

the cities, and in April 1978 the demonstration for Yu Luoke at the Martyrs' Monument.[27]

During those years not many people had a camera—only a few dared to take photos, and those interested in taking photos of these kinds of subjects were even fewer. After people see these photos they say, "These photographers dared to capture those times." But those I know, mentioned above, all took quite a few photos, and of all of them, only a few have been published or exhibited even up to the present. It is hard to get opportunities to see such unusual, nontraditional photos.

In August 1986 the editors of *Popular Photography* (*Dacong Sheying*) received a letter from a reader that said, "I was deeply saddened when I saw Li Xiaobin's *A Petitioner* in *Popular Photography*. I

---

27. University entrance examinations were restored in the spring of 1977 so that selection of university students reverted to the system that had been dropped during the Cultural Revolution. "Democracy Wall" was widely known abroad. Wei Jingsheng became one of the wall's best-known contributors, calling for democracy as a "fifth modernization" to accompany the other "four modernizations" that Deng Xiaoping had called for. Wei Jingsheng was arrested and imprisoned just weeks after Deng's triumphant visit to the United States in January 1979. Yu Luoke was a talented young man repeatedly denied university admission because of his family's capitalist background. He wrote eloquently to refute the idea that class background should be a factor in a individual's credibility as a contributor to revolution. The authorities found Yu's arguments sufficiently compelling as to incriminate him, and he was executed at Workers Stadium in Beijing in March 1970. Like many others similarly condemned in difficult times, Yu was exonerated years later.

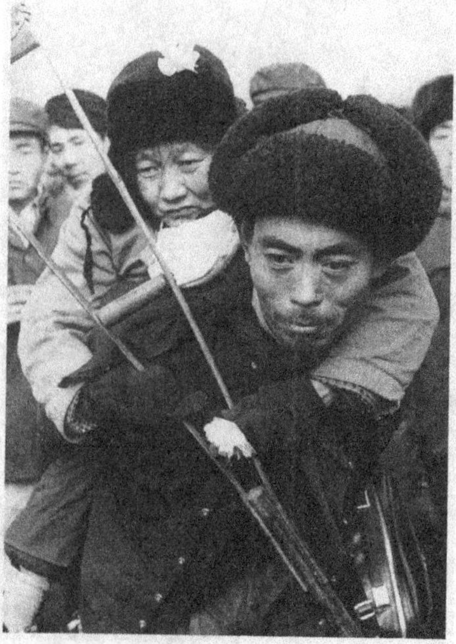

Photo 5: Among petitioners, a man carrying his disabled wife.

had many complicated feelings. I wanted to rip that page to pieces, but as I got ready to do that, I cried and then bowed deeply to this photograph." The letter-writer's father had been in the army and had served in the liberation of Hainan and suffered disabling wounds.[28] Then, in 1957, he had been falsely labeled "a rightist," and got the "double-dismissal" treatment.[29] He petitioned many times but never got any response, and finally he died in anger. The father left behind a leather bag, like the bag the petitioner carries in this photo. The bag was packed full of all the materials for his petitions, and every word of them was written in tears and blood. So the sight of this bag and the writer's memories opened old wounds and brought back the pain of that series of events.

A Petitioner was reprinted several hundred times over the years after I took the photo and was commented on and exhibited countless times. It was the topic of an American graduate thesis in 1991. It appeared in several photo albums both in China and abroad. It was used by CCTV and Beijing TV, and in 1998 it was placed in the collections of the Museum of the Chinese Revolution. It was recognized by many newspapers as "one of the most meaningful historical photographs of the century." It was used many times, even by some very influential newspapers like *Southern Weekend* (*Nanfang Zhoumo*)

28. Hainan is an island in the South China Sea and was part of Guangdong Province before it became a province in 1988.
29. A "double-dismissal" meant that the victim was dismissed from the Party as well as from his workplace.

and *Beijing Youth Daily* (*Beijing Qingnian Bao*), and by some newspapers abroad, but none of these ever sent me any payment for it. It seems that this photo has become public property.

If we only consider intellectual property rights, the question of violating my rights should be pursued.[30] But when you think of the influence this photo has had on China's recent history and the attention people have given it, there is no point in haggling over details. History belongs to the people.

This was written by the photographer, Li Xiaobin. It appeared in Old Photos *no. 15, pp. 126–33.*

30. This comment refers to a larger discussion of intellectual property rights that took place in the late 1990s, prompted by the pressures placed on China by the United States and other Western nations then insisting that China should implement these rights in international business activities.

SIX

Insights on Leaders

Mao Zedong Charms a Liberal Educator-Entrepreneur

Introduction

This brief discussion outlines the 1945 meeting between Mao Zedong and Huang Yanpei along with other figures in China's democracy movement. This meeting became widely known and had great influence. Huang Yanpei was one of the most highly respected liberals among those serving in the Nationalist government. He had a varied career but was best known as an educator with a special interest in vocational education. The timing of this group's visit to Yan'an was important. Although the Anti-Japanese War would end within weeks, that was by no means clear at the time of this meeting.

Huang was most impressed by Mao and the movement he led. He wrote a report on this visit to Yan'an that was published in a newspaper in Chongqing. Some sources report that this became the basis of publications by the Communist press. Authorized or not, these statements became part of Huang's influence on thoughtful people to support Mao and the Communists against Jiang Kaishek and the Nationalists. After Mao became strongly authoritarian himself, many blamed Huang for allowing himself to be duped by Mao's talk of democracy.

Note the contrasts between the diet, salaries, and amenities for members of the Nationalist government, located in Chongqing

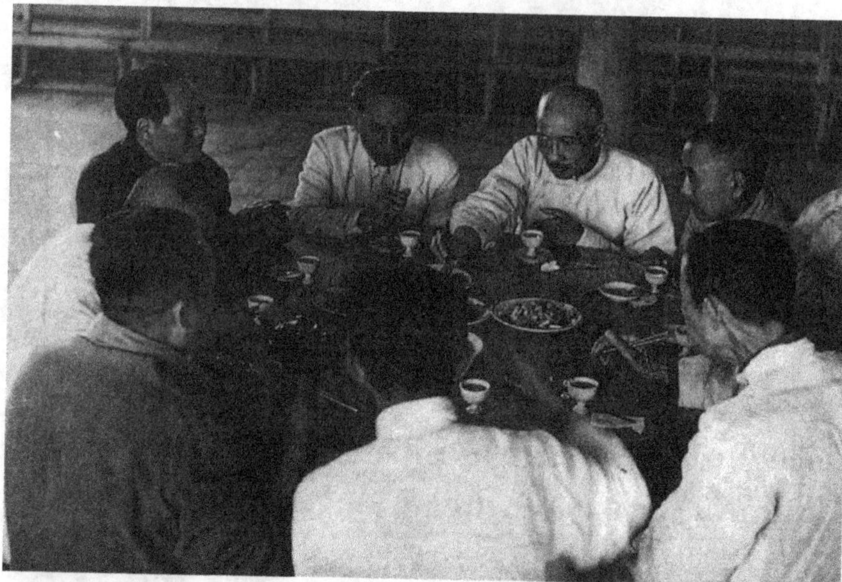

Mao Zedong invited members of the National Advisory Group to dinner at Yan'an. To Mao's left is Zhang Bojun. The second person to Mao's left is Huang Yanpei. The dinner consisted of four dishes and one soup—very simple.

during the war years, and for the Communist leaders at Yan'an after the end of the Long March in 1935.

The Text

Just as the War of Resistance was coming to a close in 1945, Huang Yanpei and Leng Yu, Zhu Fucheng, Zhang Bojun, Zuo Shunsheng, Fu Sinian and six other members of the National Senate flew to Yan'an on July 7 in response to Chairman Mao Zedong's invitation to meet and discuss how to end the single-party government of the Guomindang, establish a united democratic government, and other questions.[1] Mao Zedong, Zhu De, Zhou Enlai, Wang Ruofei and other leaders of the Party Center went to the airport to greet

1. This meeting must have occurred shortly before Mao Zedong went to Chongqing for talks with the Nationalists and also visited the "House of Democracy." Refer again to "The House of Democracy in Wartime Chongqing." Also note that Zhang Bojun appears in the earlier article on Zhang Shenfu, first in a group photo from Berlin, then later in that article in the photo of Democratic League representatives seeing off Zhou Enlai and Deng Yingchao after discussions in Nanjing.

them. As Mao Zedong shook hands with Huang Yanpei, Mao said, "We haven't seen each other in more than twenty years!" Huang was surprised and said this was the first time they had ever met! Mao laughed and said they had met one day in May 1920 as the Jiangsu Education Committee welcomed Dr. Dewey.[2] "You chaired the meeting," Mao said, "and in your remarks you stated that for every one hundred middle school graduates, those who went on to university were only a certain number and those who lost the chance for further education then were another number. In the crowd that heard your comments was one Mao Zedong." Huang praised Mao for his excellent memory. When he got back to Chongqing, every time Huang Yanpei mentioned the trip to Yan'an, he loved to tell about this detail. Who would think that among the masses of our people, there would be such a great hero!

On the afternoon of the next day, the six guests went to Chairman Mao's residence at Yangjialing for a welcoming dinner.

All the Party leaders lived at Yangjialing. It was a short distance out of Yan'an, and the scenery was fine. Well up on the mountainside stood the residences of the Party leaders, all lined up in a row. Amidst these cave-homes stood an impressive auditorium on the hillside, facing the sun. To the right of this auditorium, up the hill and to the rear, was the room where Mao Zedong welcomed guests, as if this were a higher story of the auditorium. This welcome hall was well laid out, with lots of sunshine, a long table in the middle of the room, and an assortment of chairs all around. It could hold twenty people. Various pictures were hung on the four walls, among them a painting by Shen Xuyang with calligraphy by his father Shen

2. The American philosopher and educator John Dewey had great influence in China largely because three influential Chinese leaders studied with him at Columbia University. These were Hu Shi, Jiang Menglin, and Tao Xingzhi. Dewey was traveling in Japan during spring 1919, and in response to these former students, he added China to his itinerary. Dewey and his wife arrived in Shanghai on May 1, 1919, just days before the beginning of the May Fourth student demonstrations that opposed the treatment of China in the Treaty of Versailles that ended World War I. Student protestors in Beijing and other cities specifically resisted the call for the Shandong Peninsula to be turned over to Japan. Impressed by the vitality of the student movement, Dewey eventually extended his stay in China until July 1921, during which time he gave almost two hundred lectures in various settings.

Junru. The subject of the painting was a wine pot, with the characters *maotai* written above, and several cups beside the pot. At the top of the painting was a poem written by Huang Yanpei on the theme of drinking together.

Huang Yanpei said this painting had been done in 1943 at the beginning of the Guomindang's third anti-Communist surge. Xuyang had done the painting for his father's enjoyment. When he asked Huang Yanpei to write the poem, Huang had thought of the rumor that, while they were on the Long March, the Chinese Communist Party people had washed their feet in a pool of *maotai*. Directed at this rumor, the painting and the calligraphy were meant as an allegory. Who would have thought this piece of artwork would appear in the guest room of the Chinese Communist leader!

The dinner was divided into two tables. Sitting with Mao Zedong were Huang Yanpei, Zhang Bojun, and some others. It's not quite correct to call it a dinner since all you saw on the tables was four dishes and a soup. Compared with the banquets for the big figures in Chongqing, this was like the difference between heaven and earth, and yet this was indeed the simple life and arduous struggle of the Communists as compared with the graft and corruption of the Guomindang at the time—a comparison growing worse day by day.

On the afternoon of July 4, Mao invited Huang as a guest in his home, and they talked the whole afternoon. Mao asked Huang what his impressions were after this visit in Yan'an for a few days. Huang spoke frankly:

> I have lived for more than sixty years and have heard more than I can tell. What I have seen with my own eyes matches the saying, "if something rises quickly, it also can decline quickly." Whether it is an individual, a family, an organization, a place, or even an entire country, only rarely does anyone break out of this cycle. At first, everyone stays focused and gives their best efforts. If conditions are difficult, all work hard to survive. But as things get better, people lose focus and get lazy. Then complacency spreads and becomes the norm. At that point even though people are capable, there is no way to turn things around or even to salvage the

> situation. The long arc of history offers many examples: "a lazy ruler enables the eunuchs to control"; "a good ruler dies and his government fails"; or "those who seek glory wind up in disgrace." To sum things up, there isn't anyone who could escape this pattern. Yet I have gradually come to see that all those in the Chinese Communist movement, from the past up to the present, wish to find a new way to break out of this cycle.[3]

Here Huang Yanpei earnestly spoke out, and his statement got everyone's attention.

Mao Zedong happily replied:

> We have already found a new way, and we can make this leap. The new way is democracy. What happens is that the people oversee their government, so that government doesn't get careless. If people take responsibility, government will not fail them.

Mao's reply was also well thought-out; he was looking toward a goal. At the time, "let the people oversee their government" was not just some inspired idea that Mao had, but was really the premise being implemented in the base areas. In the revised draft that Mao made in 1941 to "The Administrative Program for the Shaanxi-Gansu-Ningxia Border Area," rules were set forth for people's democratic rights and honest government.[4] As stated in item 6:

---

3. Huang Yanpei's comments here started what has been called "the cycle conversation." The concept of a dynastic cycle is basic to understanding the pattern of Chinese history. Huang seems to mean that shorter stretches of history also display this same pattern of vitality and decline. In Huang's lifetime the Qing government had periods of reform that ended in frustration. The Republican Revolution failed, and some warlord governments, operating on a provincial scale, set out to implement reform but were not able to sustain their efforts toward that goal.

4. The principles embodied in this program for local government under the Communists support the point that Mao and his associates indeed believed their own propaganda that China needed democracy. A variety of non-Chinese visitors, journalists such as Edgar Snow, and military people like Evans Carlson were also impressed with the Communists' leadership.

> For all those involved in the War of Resistance against Japan, we guarantee human rights, political rights, the right to livelihood, the right of free speech, and to freedom of publication, of public assembly and organization, of belief, of residence, and the right to move one's home. Except for those in the court system and public security organizations acting legally in the performance of their professional duties, no organization, army, or group may seize, interrogate, or punish anyone, and the people have the right to raise a complaint against any illegal action by any public official.

And item 8 of the Program said,

> To strictly enforce honest government, we will seriously punish corrupt behavior by any public official, will prohibit any public official from private gain through his position, and Communist Party members have the duty to correct illegal behavior and punish crime. At the same time, we provide a salary to encourage the principle of honest service and to guarantee that all public service personnel and their families can have a satisfactory material life and a pleasant cultural life.

At the time a Guomindang county head made a monthly salary of 180 *yuan*, while Communist officials at the same level in the border area received a monthly allowance of two *yuan*, and the chair of the border region government received a monthly allowance of five *yuan*. They weren't just a group of old lords becoming officials to get rich but were true public servants seeking to do good for the community. This shows a clear difference between people in government service in two parts of the same country.

This discussion of the "cycle of political fortunes" happened half a century ago. The leaders and prominent figures in this photo are already historical figures who have passed on, but this subject of such a "cycle" is still not outdated. We should say that the democracy Mao Zedong spoke of and the idea that the people should oversee their government in order to avoid the failure of government and deal with the problems of corruption are still a good prescription today. It is just that for some people, this prescription involves some suffering and might be hard to swallow. We know that "good medicine is bitter to the taste before it helps you in your sickness." While it is bitter and you need to take it, you still don't want to take the medicine even though you know you have a fatal disease.

This article is by Wang Zhenyu. It appeared in Old Photos *no. 2, pp. 19–22.*

A University President and His Students

Introduction

Ma Yinchu is a revered figure from the early years of the People's Republic. As this item says, Ma's major role was his service as president of "BeiDa," Beijing University. This vignette suggests why he was such a popular leader. Ma Yinchu is also known for opposing Mao Zedong on population policy for China. Mao saw sheer numbers of people as a resource for the nation and presumably as part of the incentive to develop. Ma noted that excessive population would be a drain on scarce resources and a continuing burden on whatever development might be achieved. Ma said, "The *Ma* of my name isn't the one in Marx (*Makesi*), but rather is the one in Malthus (*Malasi*)."

Respect for Ma increased as years passed. It was never easy to criticize Mao or his policies. Ma had persisted, if gently, and his view was recognized as correct and was implemented in the "one-child policy" initiated in 1978. In 2015, this family planning policy was modified to allow two children per family.

The Text

Beijing University, my alma mater, passed its one-hundred-year anniversary some years ago.[5] In browsing through some photos I took

5. The original text says that Beijing University had its one hundredth anniversary in 1998, when this issue of *Old Photos* was published. The text also says that the date of the

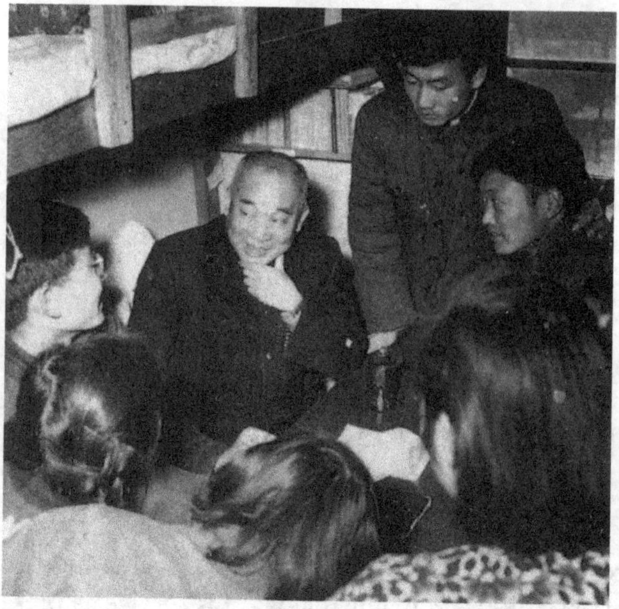

Photo 1: University president Ma Yinchu talks with students in a dormitory to learn about their life at Beijing University.

about forty years ago when I was a student there, I was fortunate to come across two views of Ma Yinchu, who was university president during those years. One shows "Ma Lao" visiting the dormitory where our class lived—the May Fourth class of journalism majors in the Chinese Department. (We called him "Ma Lao" out of respect and because he was a familiar figure to us.[6]) I recall that during autumn of 1956, one day after class, "Ma Lao" suddenly arrived in the dormitory area of the campus and just walked into our room. The students gathered around enthusiastically. He asked kindly about conditions in our studies and living arrangements. I got my camera and took this picture. His pleasant spirit and our attentive mood clearly shows the close relationship between this famous university president and a group of ordinary students.

The second photo shows "Ma Lao" at the officials' platform

---

university's establishment was May 4. The May 4 date is used generically to commemorate the demonstrations that began on that date in 1919, in which students from Beijing University and other schools in the city organized to express their outrage over terms of the Treaty of Versailles regarding Chinese territory.

6. This form, inserting *lao* after the person's surname, is the standard form for such an expression of respect.

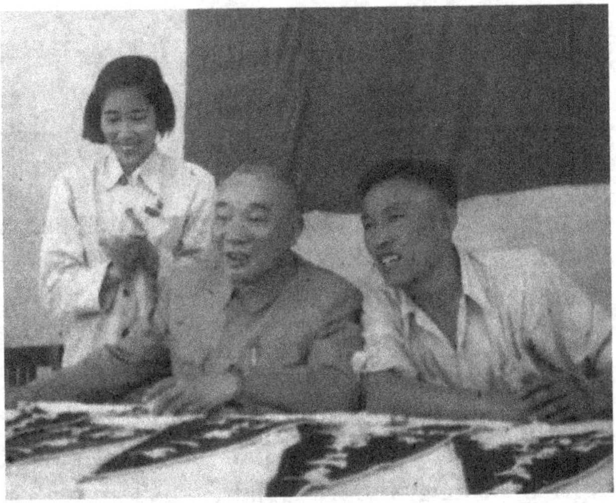

Photo 2: President Ma viewing a student athletic meet.

for the university's athletic meet as he enthusiastically watches the competition while student officials describe the proceedings to their school's president.

Ma Yinchu was a prominent scholar in our country who was known everywhere for his involvement in the struggle against the reactionary forces of Jiang Kaishek. After being appointed president of Beijing University in 1951, he was extremely busy with his work and in public activities (he was also a member of the National Standing Committee at that time), but he still made time to talk seriously with students about their lives and their studies, and to know what was going on in research work at the university. He was good-natured and interacted readily with others. He was revered and beloved by students and teachers alike. These two photos I took then, although the quality is not high (and I no longer have the negatives), are great treasures for me.

This is by Liu Xunhua. It appeared in Old Photos *no. 7, pp. 95–96.*

A National Leader's Local Experience: Hu Yaobang in Nanchong, Sichuan

Introduction

By the end of his career, Hu Yaobang would become one of the most beloved leaders of the Chinese Communist Party, and perhaps its most open-minded leader as well. Born in Hunan in 1915, Hu was still a teenager when he joined the Party at its Jing'gangshan base in the south. He went on the Long March in 1935 and at one point almost died from wounds sustained in a skirmish. While still on the March, Hu was assigned to join the troops of Zhang Guotao, a rival of Mao Zedong who led an advance to areas west of Shaanxi Province, where the main body of the March would wind up at Bao'an and then Yan'an. This Communist force was defeated by troops of the Gansu warlord, Ma Bufang, an ally of Jiang Kaishek. Hu Yaobang and a thousand others in the Communist group were sent as draftees when Jiang Kaishek called on warlord Ma for recruits. But Hu and most of his comrades escaped and rejoined the Communists in Yan'an.

This article jumps ahead to the early years of the People's Republic. As the article notes, Hu Yaobang had served as political commissar of the Communists' Eighteenth Army as they advanced into Sichuan and beyond into the interior parts of China. In this early assignment, he displayed the resourcefulness and human touch that would characterize his later career. Hu's assignment suggests the kinds of challenges the new government faced generally.

In later years, Hu Yaobang was a supporter of Deng Xiaoping and, like Deng, was "sent down" and rehabilitated twice during the Cultural Revolution. With the beginning of reform after 1979, Hu was made the chief of party and government beginning in 1982. He was far more liberal than most others among the top leaders—including Deng Xiaoping himself. When a broad challenge from university students and their leaders developed by the mid-1980s, the Party elders blamed Hu for perceived problems, and he was asked to resign in 1987. He died of a heart attack on April 15, 1989. Thousands of students from Beijing's universities paraded to Tian'anmen

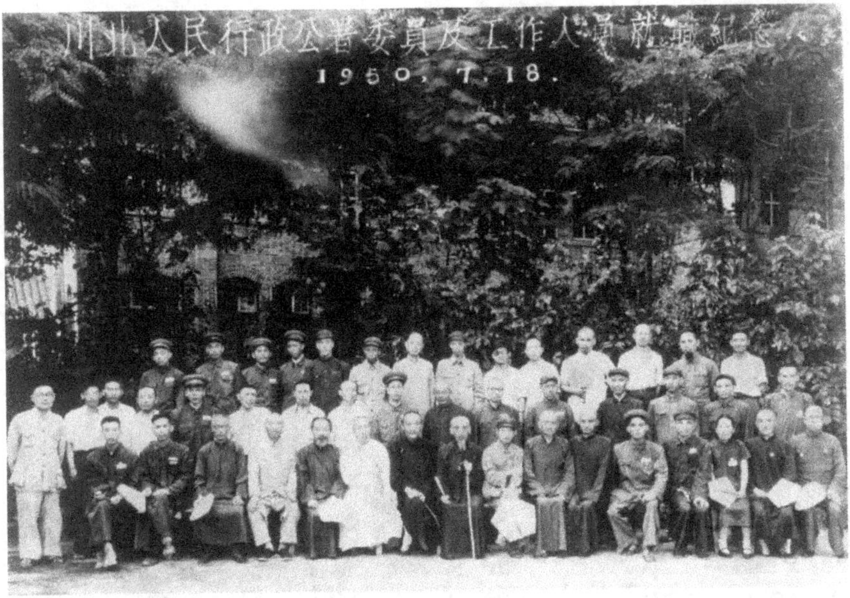

Photo 1: A photo made on July 18, 1950, to record the commencement of duties of Northern Sichuan administrators and officials. Hu Yaobang is eighth from the right in the front row.

Square to express their respect for Hu. Thus began the demonstrations that ended with the tragedy of June 4, when the People's Liberation Army cleared the square, killing hundreds of students there and in nearby areas of Beijing. This was the internal background to the "Tian'anmen Square Incident" of June 4, 1989, although the fact that a visit from Soviet leader Mihail Gorbachev coincided with events already set in motion, brought CNN and other international news organizations onto the scene to create world-wide awareness of these events.

The photos in this item reveal Hu Yaobang's slight physical stature. He shared this feature with his mentor Deng Xiaoping. Here are two men of small physical size who played huge roles in the history of modern China.

The Text

April 15, 2009, is the twentieth anniversary of the death of Comrade Hu Yaobang. The writer has investigated the relevant documents of

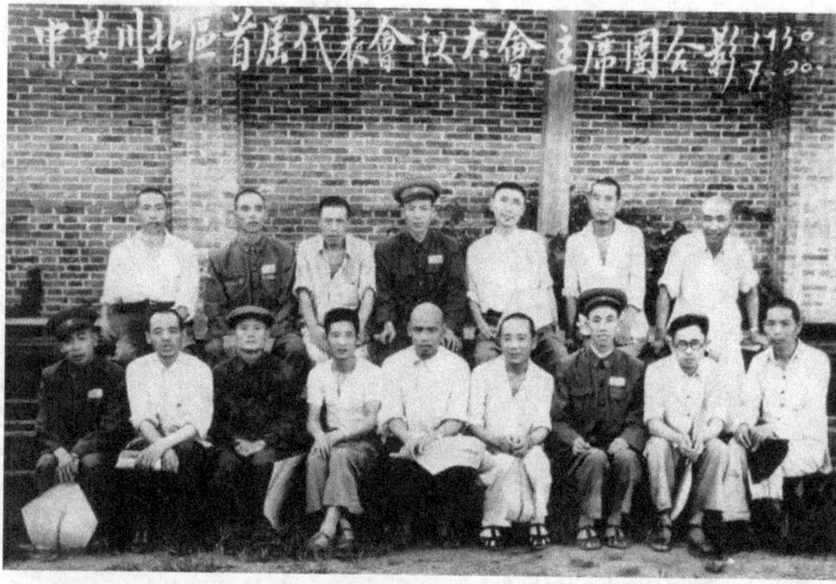

Photo 2: The initial meeting of chairs of representative committees of the Communist Party organization in Northern Sichuan, on July 20, 1950. Hu Yaobang is fourth from the left in the front row.

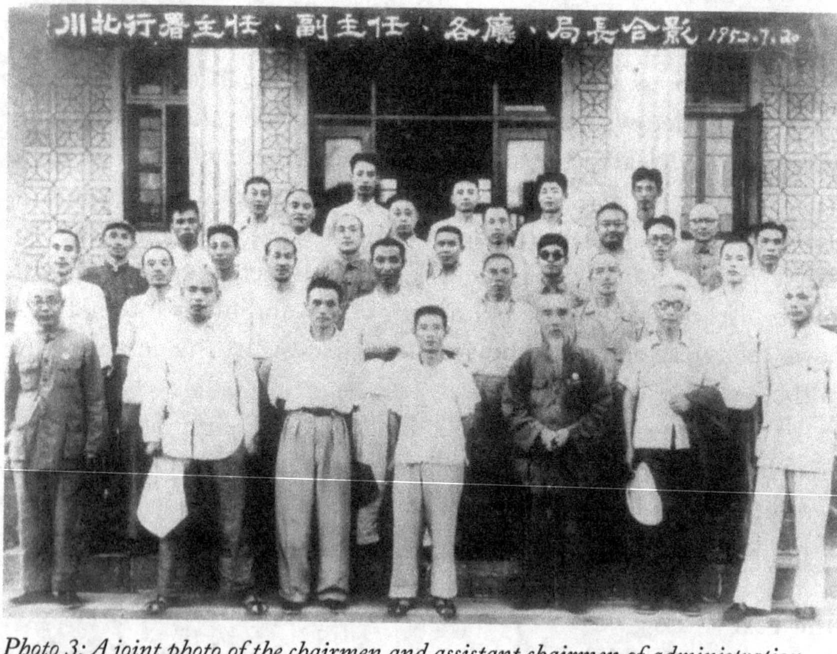

Photo 3: A joint photo of the chairmen and assistant chairmen of administration, offices, and bureaus, also from July 20, 1950. Hu Yaobang is fourth from the right in the front row.

the Northern Sichuan Party Committee's historical materials and has talked with members of that older generation who are in good health. Comrade Hu Yaobang's work and life in Nanchong became vivid to me.

Nanchong is an important small city in northern Sichuan, with beautiful rivers and mountains and excellent human resources. In order to heal the wounds of war and create a new Sichuan, the People's Central Government established four regional administrative districts in Sichuan: North, South, East, and West.

The Northern Sichuan administrative region was ninety thousand square kilometers, with a cultivable area of more than twenty million *mou* (about 6.5 million acres) and a population of seventeen million. It governs Nanchong, Suining, Da County, and Jiange. These four prefectures have thirty-five counties and one city governed directly by the main unit, which is Nanchong. There's an old saying that captures the particular feature of Sichuan society: "If the empire hasn't fallen into disorder, Shu will be chaotic. If the rest of the empire is under control, Shu won't be!"[7]

On January 20, 1950, thirty-five-year-old Hu Yaobang, political commissar of the Eighteenth Army and political committee member for the Sixty-First Army, took office as First Secretary of the Chinese Communist Region of Northern Sichuan, Administrator of Northern Sichuan, and political committee member of the Northern Sichuan Military Region. On February 20, 1950, Hu Yaobang went to Nanchong for formal installation in these offices. The next day he organized a meeting of high-level cadres and prepared "A Preliminary Work Report for the Northern Sichuan District." At the beginning of March, there were reports that General Hu Bozhou of the Northern Sichuan "Guomindang Righteous Army of National Salvation" was planning to bring together dispersing troops with their arms to create a disturbance. Of Communist forces stationed at Nanchong at the time, there was only the militia force at the police station, and there were probably about three thousand bandit agents.

7. "Shu" is a historic name for an extensive area of present-day Sichuan. It originated in the Chengdu plain in early times but extended both north and eastward. Shu was one of the states in the Three Kingdoms period after the fall of the Han Dynasty in early third century CE.

If the new government had needed to call in troops from outside, it would have been difficult to travel and they couldn't have gotten there in time.

On that day the normally peaceful streets of Nanchong were suddenly filled with military vehicles and fully armed troops. They streamed out from the Northern Sichuan Party office. Row on row they came, some carrying machine guns, some carrying mortar shells and all kinds of weapons, and they marched straight toward the Northern Sichuan administrative office. Squadron after squadron of mounted troops were galloping toward the Northern Sichuan Military District. . . .

Where had all the troops come from? Hu Yaobang had released the troops of the security guard to take turns in acting out a great drama of "Massive Forces Gathering to Unite Nanchong." The soldiers came wearing steel helmets and wearing cloth caps. They came marching on foot and mounted on horseback. They came from the east, and they came from the west. The bandit agents thought a great army had come into the city, and they headed for the hills. Disaster was avoided in Nanchong, and the people's resources for their livelihood were safeguarded.

In accordance with the directive of the Central leadership and the Southwest Regional Military and Political Committee, Hu Yaobang made it his major task for 1950 to wipe out various discordant forces in the region.[8] Hu, along with other leaders in northern Sichuan, led the army and the people in carrying out a grand struggle against bandits and Nationalist agents.

Through the cooperation of the people and the PLA of Northern Sichuan, the chief bandit Hu Bozhou and other key figures associated with him were captured. In a public trial held on June 11, Hu Bozhou and other bandit leaders were condemned to be executed. Wang Yimin and others, leaders of the "Anti-Communist Army to Save the Nation," were awed because they were surrounded by antibandit forces and could do nothing but give up their arms and surrender. By the end of June, more than forty-nine thousand bandit

8. Mentions of "Southwestern" offices in this article refer to the perspective of all-China organization, which included Sichuan.

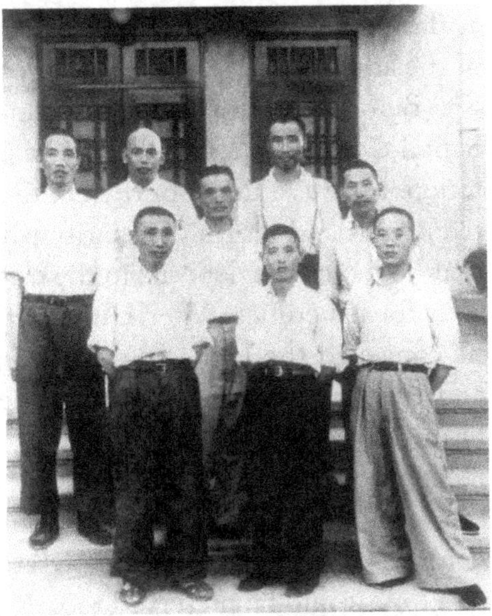

Photo 4: Hu Yaobang (front row, middle) with members of the Standing Committee of the Northern Sichuan District.

forces and some twenty thousand weapons of various kinds were surrendered. The new governmental authority was consolidated.

The Northern Sichuan administrative region is extensive, with a great population. The task of governing here is broad and complex. As a means to strengthen his leadership in the work of the United Front, Hu Yaobang also took on the position of Chief of the United Front Bureau of the Northern Sichuan administrative region.

At the first meeting with individuals of democratic organizations in April 1950, most who spoke were quite reserved. Hu Yaobang said, "We want to make friends with people from outside the Party. We want them to open their minds and hearts and say whatever they think. If we don't speak our minds and express our differences, how can we begin to work together harmoniously? If you have an opinion, please—everybody—state your views openly!" Hu Yaobang spoke openheartedly himself, with good humor, and established a lively air for the meeting. So everyone spoke up and brought out many constructive ideas.

According to Party policy, Hu Yaobang was supposed to appoint some military leaders from the opposition. Pei Changhui, originally Lieutenant-General of the Guomindang Seventh Army Corps, was appointed assistant head administrator of the political region and also director of the Office of Industrial Development. Jia Ziqun, a responsible figure in the Democratic Alliance, was appointed director of the Office of Education. The national capitalist Xi Zhihe was made director of the Office of Commerce. Tan Weigen and Yang Dazhang, also members of the Democratic Alliance, were appointed as director of the Office of Communications and assistant secretary of the administrative region.[9]

Before Liberation, Long Jiesan had served as head of the local Santai County Senate and also had participated in the popular revolution. He was a target of the United Front. During the time of rent reductions and return of security payments, the peasants demanded that he return the deposits to the original households in order to answer for his crimes. After Hu Yaobang learned about this situation, he convened the United Front group and the County Committee, and everyone thought that Long Jiesan should seek help from intermediaries. After some study, it was decided not to hold a struggle meeting but to allow him to go to a meeting of the peasants' representatives and hear the denunciations of the masses. That way, the demands of the masses would be satisfied and the goals of the United Front would be preserved too.

In March 1950, the masses in Bazhong County called for the repression of Zhao Jigang and Li Zilong. But the opinion of the Bazhong County Committee differed on this matter, and they asked for guidance from the Northern Sichuan Party Committee. Hu Yaobang knew that Zhao Jigang was Deputy Commander of the Bazhong County Self-Defense Force, but that in the school he organized he had accepted a number of underground agents for the Party as staff members. When the Liberation Army came into the Southwest Military Area, he made ready to surrender all his organization's weapons and was an effective member of the local uprising to support the Liberation Army's actions. As for Li Zilong, although he had been the County Head under the Guomindang, he had been an upright and honest official. He didn't abuse the masses, he didn't take the soldiers' rations, and he didn't smoke opium. Most of all, on the eve of Liberation, he was very active and undertook many tasks to receive the Liberation Army.

After Liberation, both Zhao and Li were introduced through

9. Concerns about members of democratic organizations occupied much of Hu Yaobang's attention. There were various organizations and one, the Democratic Alliance, that brought together most of the leading figures in the democratic reform movement. See "House of Democracy" in this collection for the Alliance's activities in Chongqing, its headquarters during the wartime years.

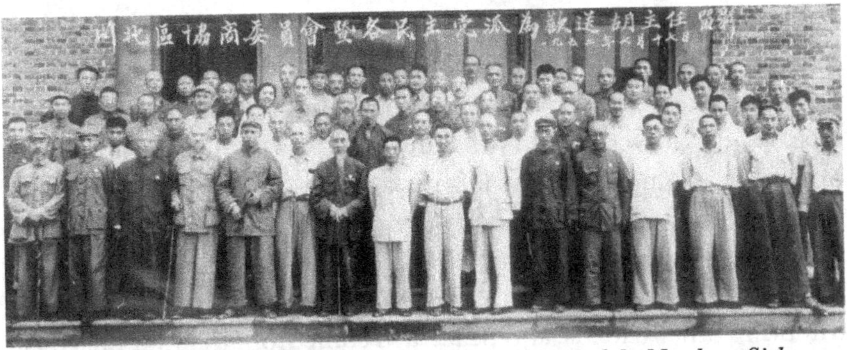

Photo 5: This photo commemorates the farewell gathering of the Northern Sichuan People's Consultative Conference and the democratic groups in the district, on July 17, 1952. Hu Yaobang is eighth from the left in the front row.

the Party underground organization and were already working in the Northern Sichuan administration. Hu Yaobang thought the Bazhong County Committee might make an ill-advised move, so toward evening that day he took a Jeep to Bazhong. He told the County Secretary, "The Communist Party never burns its bridges. Even though Zhao and Li have committed some crimes, they also have done some good things. The masses don't understand this, so you have some work to do!" Later, he also called for the administrators to invite representatives from all groups to a meeting and have Zhao Jigang and Li Zilong talk about their new outlook. After the meeting, Hu Yaobang talked with Zhao and Li and asked them to understand the outlook of the people, and he also said, "With friends like you from the United Front, the Communist Party will treat you fairly."

During his work in Northern Sichuan, Hu Yaobang listened earnestly to the views of the democratic individuals, showed concern about their lives, and tried to take care of their practical problems. By the end of June 1952, the area had more than four hundred people out of the democratic parties doing various jobs in local government. Among these, twenty-six were serving as heads of various offices, thirty-five were department chiefs of specialized offices or assistant county heads, and eighty-six were department heads in the county government.

On June 7, 1952, the Southwest Division received a telegram from Party Central: "Hu Yaobang is to be assigned to work at Party

Central. He is to report to Beijing at the end of July." Before he was to leave for this new assignment, all the different groups in Nanchong planned a big send-off for him. When Hu Yaobang learned about this, he said to everyone in the Political Association, "If you all wish to give me a going-away gift, just make it these words: 'Serve the people'!" The vice-chair of the Northern Sichuan Political Association, Lu Zipeng, gave a deep sigh at this and said, "Now that you are leaving, how can I rest easy again?" Hu Yaobang said in reply, "Heaven will bring other friends, and peace will come from the four quarters."

From February 1950 to July 1952, Hu Yaobang guided the people of Northern Sichuan toward prosperity and did a great job of launching the new political authority and consolidating development. On that day at the end of July 1952, at a farewell gathering at the government administration building, with his characteristic gestures and passionate speaking style, Hu Yaobang made his heartfelt remarks of farewell to all of those in the organizations of government and the body of cadres. Then, in order to evade a clamorous departure, very early on a peaceful morning, Hu Yaobang made a simple start in a modest car, taking only one personal secretary with him.

This was written by Hu Jian. It appeared in Old Photos *no. 64, pp. 121–26.*

Chairman Mao Studies Marxist Economics During the "Three Bad Years"

Introduction

This article offers rare insight into Mao Zedong as leader, and it is also quite ironic. The background to Chairman Mao's decision to undertake study of a Soviet economic text was that the "Great Leap Forward," launched in 1958, was failing disastrously. Peng Dehuai criticized Mao privately at a conference at the resort town of Lushan in July 1959. Mao accused Peng of organizing a rightist clique and dismissed Peng from his post as defense minister. Thus Peng was disgraced for seeking a correct analysis of affairs.

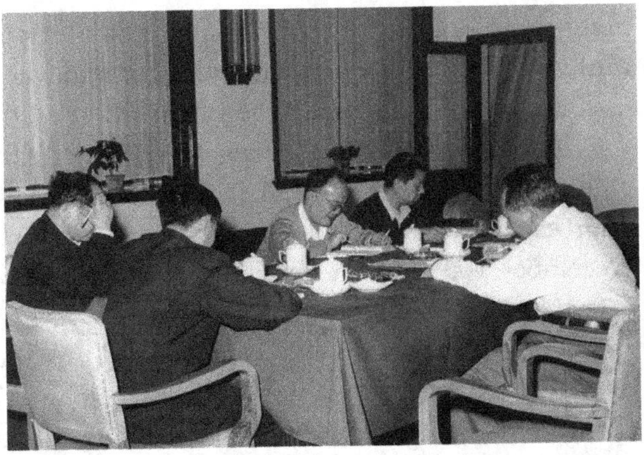

Photo 1: Mao Zedong with the five-member study group in Hangzhou, December 1959. From left, Chen Boda, Tian Jiaying, Hu Sheng, Deng Liqun, and Mao.

Mao's call for study near the end of that same year suggests that he did indeed understand the need to rethink his views on the Great Leap Forward, and he was prepared to learn from the Soviet experience. This too had its background. Mao already had disagreed with Nikita Khrushchev soon after the latter asserted a new path for the USSR following the death of Joseph Stalin in 1956. A telling example of the differences developing between China and the USSR was that when the latter launched a space vehicle named "Sputnik," China gave the same name to the first people's commune, established in 1957.

Thus Mao understood the need to try to correct policies he himself had launched. So he decided to study a text that seemed to offer guidance in his quest for answers. As noted in this article, others in the leadership organized their own study groups, and Mao's staff comprised his group. The writer credits Mao for a number of correct insights, which were easier for Mao to accept in discussion than in real politics during these years. And such was the nature of things that Mao's own comments on these discussions became a text for China's Marxists.

Note the close relationship of this article to "A Shaanxi Peasant's Critique of the Economy During Mao's Great Famine."

The Text

I have kept a photo of Mao and the reading group when they were studying the Soviet book *Political Economy (Textbook Edition)* in

Hangzhou. My father, Tian Jiaying, was a member of the group. Recently my friend Jiang Jie visited Beijing, and when he saw the photo he asked me to write about the story related to it.

In 1990, my mother, Dong Bian, visited Comrade Hu Sheng a number of times. Hu told her what he knew about my father and their experiences in studying this book with Mao Zedong. From December 10, 1959, to February 9, 1960, Mao Zedong organized a study group of five people to study the Soviet textbook chapter by chapter and paragraph by paragraph. The idea was to study and discuss the book as they went along. Their meetings took place in Hangzhou, Shanghai, and Guangzhou, for a full two months.

Why did Mao Zedong want to organize this study group activity in the winter of 1959? This must be traced back to 1958, a very troubled year in China. The "Great Leap Forward" was in full swing, but by 1959 it was difficult to continue. The national economy was in serious trouble. Everywhere you could see abandoned "backyard furnaces." The mass movement to produce iron and steel lost its momentum, and the people's communes, which were experiments in making a transition to full communism, hurt the masses of working people. The community dining halls were dismantled.

Mao Zedong saw quickly that problems were developing. In November 1958, the Central Committee held a meeting in Zhengzhou to correct the leftward trend. In that meeting Mao Zedong issued a letter to the leaders in the central, provincial, prefectural, and county levels of leadership, suggesting that everybody earnestly read Stalin's *Issues in the Socialist Economy in the Soviet Union* and *Marx, Engels, and Lenin on Communism* to begin solving problems of thought and mindset. In his letter he indicated, "When we read these two books, we should relate them to China's socialist revolution and economic construction, so we can take a levelheaded approach in our great economic work." In this letter he also mentioned, "In recent months there has been considerable confusion among some comrades . . . including those who are known as Marxist economists. They are Marxists when they read Marxist political economy, but once they encounter practical problems, their Marxism is discounted."

When Mao suggested we read these books, he also said, "If we have time in the future, we can also read this book edited in the Soviet

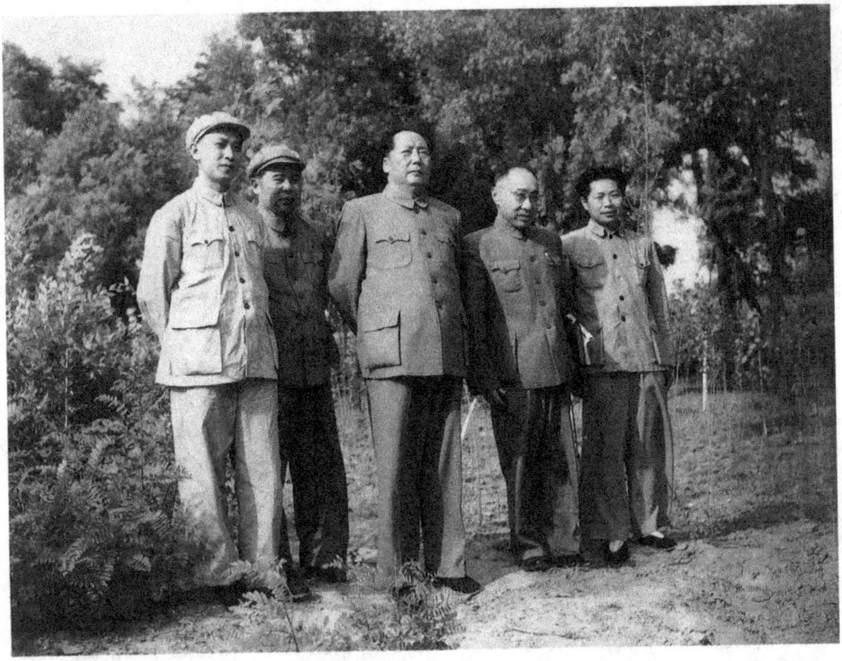

Photo 2: A photo at Hangzhou from 1954. From left, Tian Jiaying, Chen Boda, Mao Zedong, Zhou Zezhao (assistant director of Beijing Hospital), and Luo Guanglu (Mao's confidential secretary).

Union, *Political Economy (Textbook Edition)*." Later at the Wuchang meeting, the Eighth Session of the Sixth Central Committee, he reiterated this request. At the Lushan Conference in 1959, Mao listed nineteen items for the agenda. The first one was to "study," as in making close textual studies. Again, he meant to study *Political Economy (Textbook Edition)*.

Hu recalled that you could see Mao asking everyone to study because there were so many problems emerging from the Great Leap Forward, and in order to solve these problems it was necessary to study Marxist theory and the Soviet experience. Based on Mao's suggestion, the first party secretary of all sections in the central and provincial governments would serve as the chair to organize study groups. Mao Zedong, Liu Shaoqi, and Zhou Enlai all organized their own study groups. Mao's study group included Chen Boda, Hu Sheng, Deng Liqun, and my father, Tian Jiaying. Chen Boda and Tian Jiaying at that time were Mao Zedong's secretaries. Hu Sheng

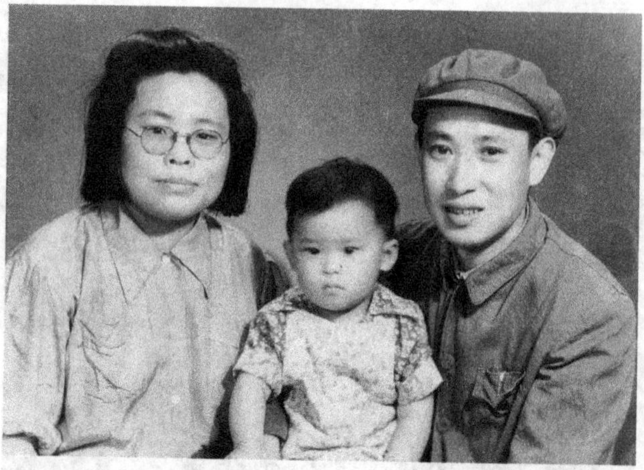

Photo 3: Tian Jiaying with his wife and daughter Zeng Li in Beijing, 1952.

was also secretary of the Central Political Research Office and associate editor of *Red Flag* magazine. Deng Liqun was one of the executive editors of *Red Flag*. The four of them were "flowering talents."

When they reached this point, Hu and my mother recalled that the Lushan Conference was held just before they started the study groups. My mother said at the beginning of the conference, Jiaying, Liqun, and Chen Boda mentioned the problems of the Great Leap Forward when they drafted the minutes of the meeting. Their comments were sharp and critical. They were sympathetic to Peng Dehuai, who wrote a letter to Mao. Mao then said they were a group "only thirty *li* away from being a right-wing clique," especially Hu Qiaomu and Chen Boda.[10] Besides feeling this pressure, Jiaying was also resentful. He thought the conference that began with correcting "left" problems then turned to criticize "right" problems. Some ultra-leftists like Li Jingquan and Ke Qingshi filled Mao's ears with their own words. By the end, the Chairman protected them and made them immune to criticism from the right. After the Lushan Conference, the complexity and the darkness turned Jiaying's inner feelings to apprehension.[11] So when Chairman Mao invited him to

10. Mao used this expression often to indicate his opinion that an individual or group were close to being "rightist."

11. Tian Jiaying would later commit suicide in May 1966, as the Cultural Revolution began. He had been accused of unenthusiastic support for the campaign against Beijing vice mayor Wu Han. Seeing what was developing, Tian opted for suicide. Mao's

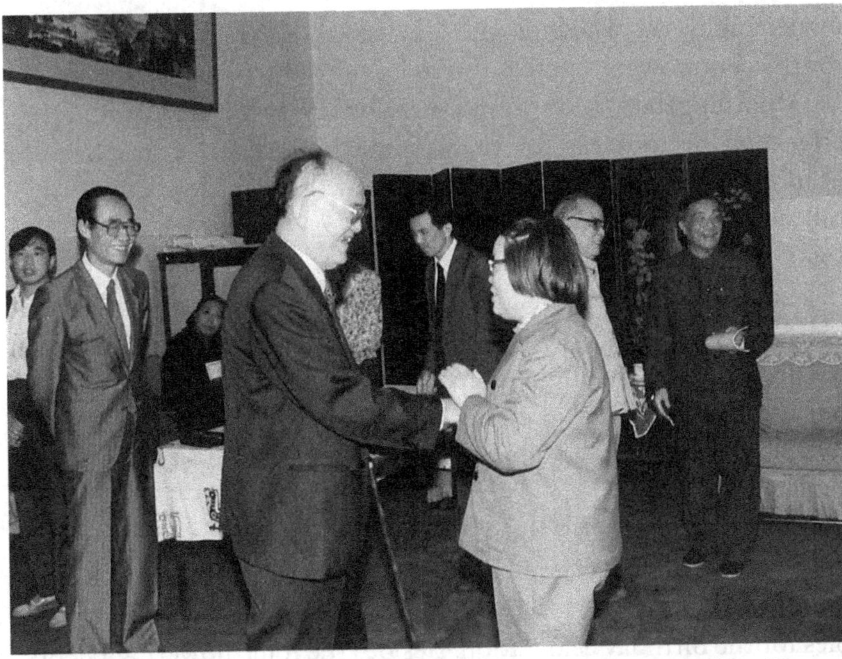

Photo 4: At a 1991 exhibit of Tian Jiaying's collection of writings by Qing Dynasty scholars, held at the China Historical Museum in Beijing. Tian's wife, Dong Bian, greets Hu Sheng in the exhibit lounge.

join this discussion group, Jiaying was happy to think that the leader still regarded him as capable.

Hu Sheng observed that at the Lushan meeting Chairman Mao had said that "these talented people are ours." Then for the discussion group, for him to include Jiaying and Chen Boda indicated that Mao had pardoned them. Hu Qiaomu got sick after the Lushan meeting, or he too would have joined the discussion group. So when Mao talked about how "some comrades think they are Marxist economists," he meant Chen Boda, because in 1958 Chen Boda advocated ending the commune system and collective ownership and returning to general

---

comment on "rightists" included Tian as of 1959. He favored stepping back from collectivization in response to the problems of the Great Leap Forward. Tian was one of a few people who chose suicide during those days. See Roderick MacFarquhar and Michael Schoenhals, *Mao's Last Revolution* (Cambridge: Harvard University Press, 2006), p. 44. Zeng Zi, author of the present article, also wrote on her father's suicide in *Old Photos* no. 87, pp. 1–21.

ownership by the whole people. Chairman Mao had criticized Chen for this, but now he wanted him to join this discussion group.

In Hangzhou, Mao stayed in a cottage at Wangzhuang. Every afternoon he took his car to Nanping to participate in the study group. Most of the activities took place at Dingjia Mountain, a small hill with a few houses. Kang Youwei once stayed there.[12] Mao was fond of this place and liked the house there very much. Sometimes he read and sometimes he just relaxed; sometimes he read beside the swimming pool. The group met from 4 p.m. to about 8 p.m. Dinner would be served about 8 p.m. Mao liked a simple dinner, and everyone ate together. On December 30, we met from 6 p.m. to 10 p.m., four hours, and we read about twenty pages—the most we ever read in one day.

December 26 was the Chairman's birthday, and the reading group met that day as usual. But the dinner was not the same as we usually had. The Zhejiang Party Secretary, Jia Hua, prepared two tables for the birthday celebration. Besides the four "flowering talents," the guests included Mao's secretary Lin Ke, and the party secretary and his wife, Wu Zhonglian. When the dinner was about to begin, Chairman Mao took out copies of four poems that he had recently written and gave them out to everybody.[13]

The food at the dinner was not particularly lavish, but the dishes were well-prepared and delicious. Mao stood up frequently to invite others to eat and drink. He also told various stories and anecdotes from all parts of the country. Every time when others proposed a toast to him, Mao held up his glass and kindly replied, "Thank you, thank you! I wish you all good health." It was a truly pleasant banquet.

According to Hu, Mao suggested that the study group should

---

12. Kang Youwei (1858–1927) was a major leader of the reform movement of 1898 under the Guangxu emperor. Kang remained influential throughout the remainder of his life. West Lake at Hangzhou is one of China's best loved and most visited sites and has always attracted intellectuals in their travels. Many well-known people are buried beside the lake or commemorated in monuments in the area.

13. From November to December 1959, Mao wrote four "poems inspired by reading newspapers" in the classic style of seven characters to a line. None of these poems was published. Mao is known as a poet, and he often expressed himself through poems. But he did not write these as formal poems and did not intend that they be made public or be published.

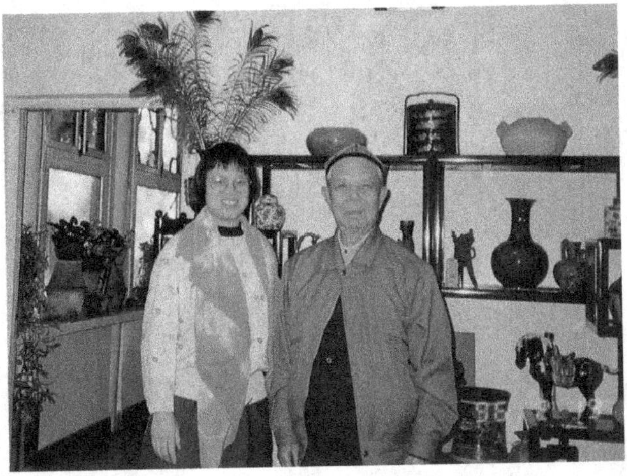

Photo 5: Zeng Zi, the writer of this article, greets her father Tian Jiaying's old friend Deng Liqun at his home (1996).

have three to five people study the book paragraph by paragraph, chapter by chapter. So we all followed Mao's suggestion, and did this seriously. Usually we would sit together, one man reading the original—Tian Jiaying or I [Hu said] would read. After finishing one paragraph or section, we would make comments. Chairman Mao made the most comments, but he also asked everyone else to comment as well. We never read very much, maybe ten pages each day. When we read, the "flowering talents" were willing to listen to what Mao said, but they also managed to write it down. We all knew that Mao didn't like others to make notes as he talked. Elder Hu said that if we take notes as Mao talks, he will scold us. This happened once. Mao saw us all making notes and asked why we didn't talk to him. So we thought of a way. Deng Liqun sat on the side, making notes. We all listened as Mao talked, and sometimes one of us would pipe up when he came to a pause. At that time we didn't have any way to record what he said. He said a lot, and it was impossible to get it all down. The report prepared later, "Mao's Key Comments on the Soviet Book *Political Economy (Textbook Edition)*," was based on Deng Liqun's notes.

At the beginning of January 1960, the reading group went to Shanghai with Mao because there was to be a conference in Shanghai. In Shanghai Mao stayed on the train, so twice the group met on the train, starting late because the conference lasted into the evening. The last part of the book was finished in Guangzhou. My mother

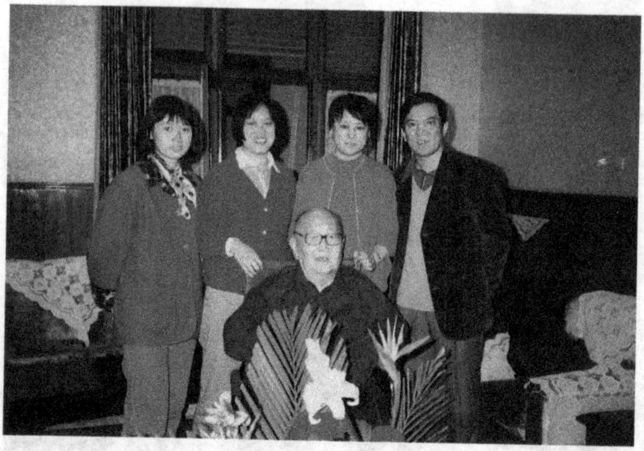

Photo 6: The author and her husband, her sister Zeng Li, and a niece visit Hu Sheng at Jilin Village recovery center (2000).

told Hu Sheng that she found that the journal, *Party Documents* (compiled by the Central Committee Documents Office), published Mao's oral comments in six installments, a total of more than fifty thousand words. The topic of Mao's comments included his world outlook, methodology, democratic revolution, and these elements of the socialist system: (1) The circulation of commerce; (2) Accumulation of capital; (3) A conceptual understanding of these aspects of the socialist system; (4) Reform of the system of ownership of the means of production under the socialist system; (5) Socialist revolution and the proletarian dictatorship; (6) Evaluation and comments on some historical figures and books. The variety of topics and comments were very rich, but because they were spontaneous, they were not very systematic. Hu Sheng, who had experienced all this, spoke on Mao's view of the book, and the most important things left to us from this reading.

Hu replied with a sigh. The frustration of the Great Leap Forward made Mao very worried. He felt deeply that he did not know enough about socialist economy and the rules of socialist development. At that time, we took this book *Political Economy (Textbook Edition)* as a summary of the Soviets' experience with socialist construction. The Soviet Union was the only country that had succeeded to some extent in building a socialist system. What impressed those in the reading group was that the Chairman took a skeptical attitude toward the Soviet experience.

Mao said, "We need to avoid all kinds of blind faith, including that in Soviet experience." One was right, one was not right, and one attitude was wrong. Mao paid particular attention to the Chinese situation and to Chinese practice. Mao also paid attention to Chinese economic construction and tried to relate the reading to that construction. The Great Leap Forward of 1959 in a way was our first experiment in socialist economic construction. As the Chairman read the book, he was also constantly thinking about the experiences we had already had and was already thinking beyond them. In retrospect, many of the issues he raised were correct. For example, we must pay more attention to agriculture and light industry. Mao said that the Soviet Union did not pay enough attention to agriculture, so their agriculture had not developed well. Light industry for everyday items was weak, and this was a great shortcoming. Mao thought these shortcomings were not correct. Clearly Mao was correct on many points. Mao also regarded commerce as an essential part of socialism. Socialism could not abolish exchange of commodities. The market principle of value still had to be allowed to operate. All these comments or suggestions now look quite correct.

Of course, now we cannot put Mao's comments from that time as the final conclusion. We can only say that his thinking about China's socialist construction had a twinkling of insight.

Reading this book with Mao, we got another impression. He repeatedly emphasized, "It is impossible to understand the rules if we only see victory and do not see failure." He repeatedly emphasized that if we are to understand the principles of truth, we have to learn through practice. You accomplish something when you encounter problems or failure. "We must compare success and failure, and gradually make our knowledge accord with reality." This was perhaps Mao's deep and painful reflection following the Lushan Conference.

As I look at these old photos, I still get a bit excited and feel nostalgic. A great deal of time has passed, and most of those people have passed away, including my dear father and Mao Zedong, the revered leader of our youth. From our perspective today, some of their ideas and their efforts to solve problems seem limited. But that generation of leaders and the masses of ordinary people dared to face hardships, to break out of superstition, to shoulder the responsibility for making China a strong and prosperous country, and we will always respect them for this.

This article is by Zeng Zi. It was published in Old Photos *no. 58, pp. 15–25.*

SEVEN

Holding up Half the Sky

The Life of a Woman Scientist

Introduction

This tribute to her mother by a daughter describes the life and career of a talented young woman from a struggling upper-class family in rural Sichuan who was fortunate to have her talent recognized by the uncle who helped raise her. She left home alone as a young teenager to enter a series of schools that had been established as Christian missionary institutions, completing her education at West China Union University in Chengdu, where she studied chemistry. Her abilities led to a career as a university teacher and researcher, a profession still unusual for a woman in the 1930s. Like most other well-educated people, she suffered during the Cultural Revolution. In her case it was not the simple fact that she had been educated in missionary schools that caused her difficulties, but that she supported a foreign teacher who had devoted his career to higher education in China and had stayed in Chengdu into the early 1950s. Her case also re-emphasizes the point that many, perhaps even most, of the excesses of the Cultural Revolution were perpetrated in local situations and not necessarily ordered from above.

The Text

My mother, Xiao Yiju, was born in 1908 to a moderately well-off family in Rongchang County, Sichuan. Her maternal grandfather was a secretary in the county *yamen* and could not only write well but also was known far and wide for his calligraphy. At that time China had many problems—and so did my mother's family. Her grandfather got sick and died. Then Mother's older brother, the pillar of the family who had become a military official, suddenly caught dysentery and died. So the family lost almost all their resources and our grandmother, taking Mother along with her, could only take turns living in the homes of three older sisters who were already married.

From the time she was small, Mother had to work to contribute to the family. The oldest aunt's family opened a shoe shop in the Rongchang county seat, so they began to make a good living. Mother's uncle saw that she was very capable and sent her to the county elementary school to study, on condition that she had to help with household chores. She had to get up early and help prepare breakfast for the family, and then after school she came to the shop to help make shoes. So here was this young girl, less than ten years old, standing in front of a treadle sewing machine making the upper parts of shoes and working until dark every day.[1] When she finished elementary school, all the teachers said she should continue with school, so her uncle raised the money for her to study at the foreign-operated upper elementary school at the Zizhong county seat.

It is about one hundred kilometers from Rongchang to Zizhong, a long, slow road for Mother, who left her family at the age of fourteen. She carried a simple bundle of belongings and was nourished by her hope and some understanding of what lay ahead of her. She used a walking stick to make her way, sometimes slipping or stumbling, along a rugged dirt road, finally reaching Zizhong after five or six days. After the long trip, Mother waited for school to begin and found herself on a difficult path of meeting her expenses. Her uncle had done all he could for her, so Mother had to depend on her own two hands to earn school fees and living expenses.

Jinde School in Zizhong was a Methodist mission school. After a

1. The article on the introduction of sewing machines in China places this change in the middle 1910s (see "Sewing Machines in China" in the first chapter of this collection). Presumably the ten-year-old Xiao Yiju used one of these newly introduced machines. How did her uncle find the money to buy one or more sewing machines? As this article notes, the machines could be purchased in installments, so it could have been this method of payment where everyone won.

simple exam, the school accepted this most special student, my mother. She didn't have anything, not her living expenses and certainly not her school fees. The school agreed to provide Mother's textbooks and arranged the simplest meals for her. They also arranged for her to share a bed with another classmate. In exchange, Mother had to do various kinds of handiwork for the school after class, such as embroidery or knitting. On weekends and during vacation periods, when the other students were happy to get a break from their work and go home, she stayed at school alone and used the needles to meet expenses. During those years, Mother developed the quality of being able to get along with anyone. As she recalled, "There was a very simple logic to this. If I quarreled with my roommate, I couldn't share the quilt that night."

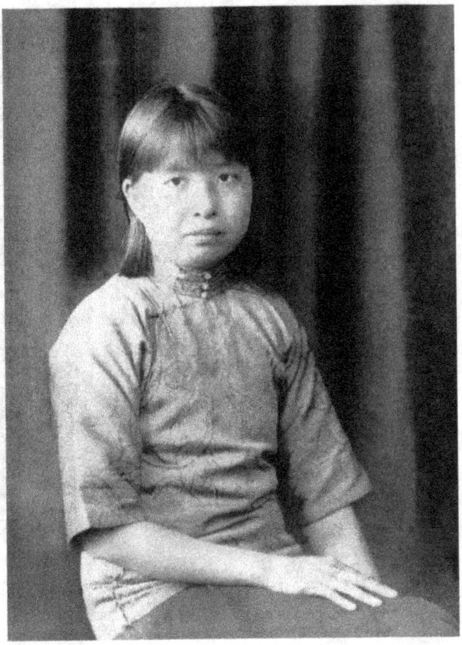

Photo 1: A photo from the 1930s when Mother was a student in the Chemistry Department of West China University.

Two years later, because Mother finished at the top of her class, she received a scholarship to go on to lower middle school, and of course she again needed to work to provide for basic expenses. When she didn't have textbooks, she waited until other students were finished with theirs and then borrowed those books for her own study. When she didn't have a compass to draw circles, she would take a stick from a tree and secure it to her pencil-point, and she'd have a "compass." Despite the difficult conditions, Mother always stood at the top of the list in academic achievement, and she was the first to receive a scholarship.

Amidst the difficulties of her material life, Mother still had some good experiences. She remembered that during the winter vacation

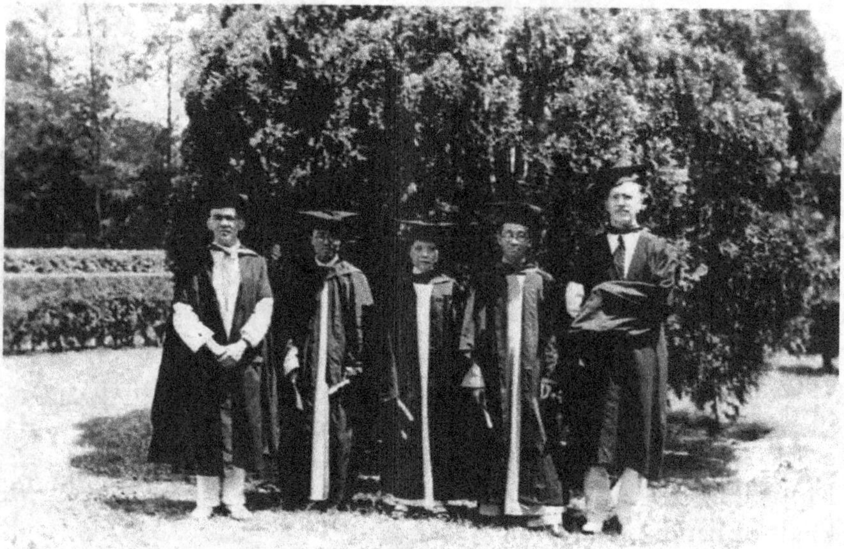

Photo 2: Mother (center) at her graduation in 1935. At left is Prof. Xu Weili, the long-time British professor.

of her third year of middle school, when she stayed at school to work, she knew a "big sister" from the women's normal school next door who had been persecuted under Confucian teaching and took refuge in the Christian church. This older girl sympathized with Mother and her difficulties, and she often invited Mother into her small room and made her a bowl of hot-and-sour noodles on a little wood stove. For Mother, who had not enjoyed such a simple luxury for five years, this became a lifelong memory. More than thirty years later when Mother had become a university teacher, every weekend when we had some good food, Mother would dress up a little and take me with her to Gospel Church on Shaanxi Street in Chengdu to visit her old friend. She helped her out financially sometimes, as long as her old friend lived.

After graduating from middle school, Mother didn't have the funds to continue to study. But because of her achievements, an exception was made and she was recommended to teach at Shude Girls' Middle School in Chongqing. At that time, Shude Middle School was highly regarded in Chongqing. Although Mother had just been in middle school herself, she showed excellent ability as a

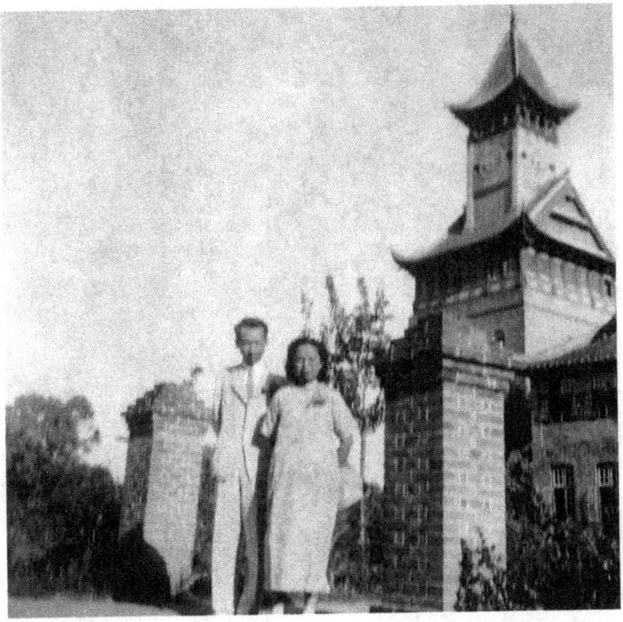

Photo 3: Father and Mother in front of the old bell tower at West China University.

teacher. Even sixty years later some of her students could remember the lively methods Mother used that made them learn to enjoy studying mathematics. Even though Mother had a job that was good enough, she still had her determination to continue her education. She used the money she saved from the two years of teaching to go back to Chengdu and enter Huamei High School.

In 1932 Mother examined into the chemistry department of one of the best-known schools in Chengdu, West China Union University (*Huaxi Daxue*, known as "HuaDa"). With a scholarship and also a loan from the church, Mother completed her four years of study in good order. Again, because she had performed so well, after graduation she was invited to stay on as a teacher, and in another four years she was promoted to lecturer. Then in another four years—this was 1944—she was promoted to assistant professor. In 1947 the Chemistry Association at the university recognized Mother for her eleven years of service by presenting her with a silver cup inscribed "Great Talent in Education." Because of changes in the larger world and Mother's nonaggressive nature, it was almost forty years before she was promoted to full professor. Mother retired in 1986. She had

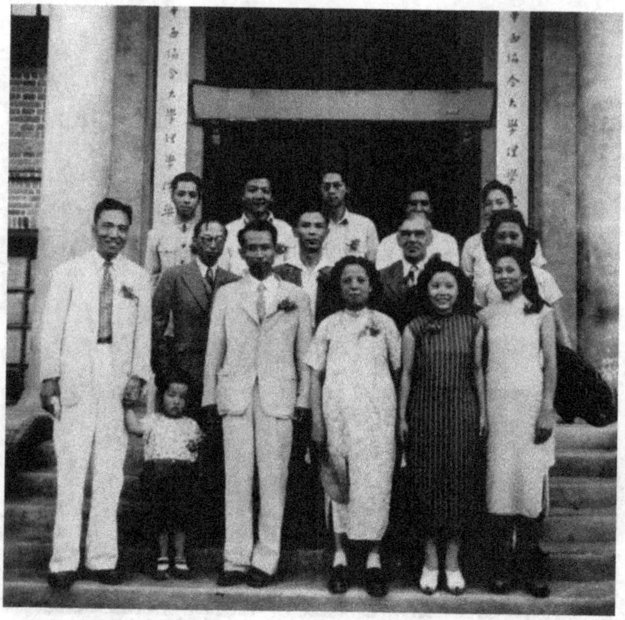

Photo 4: Mother and her Chemistry Department colleagues after the celebration of her eleven years of service at West China University.

taught and done research at Huaxi Medical University for fifty years, and her students could be found everywhere.[2]

When I was studying in England in 1986, I fell into difficulties after my divorce. An English friend of mine thought I might be able to get a scholarship from a church philanthropic foundation, and she brought me two huge catalogs to search through. But all these foundations only supported church members, and I couldn't get any kind of help. My experience led me to ask Mother about her relationship to the church, which had done so much to support her. Why didn't she join the church instead of repaying the loans they made to her for many years after she graduated? Mother told me that while she could not have gotten her education without the church, she had

2. This university has a distinguished history. It was established by Western missionaries and was the only educational institution of its kind in west China. Over time, the university added a medical school. It also became the major center for foreign residents in Chengdu. The university was closed briefly in the mid-1920s during the most intensely nationalist actions of Jiang Kaishek's Republican government. As Ms. Xiao's experiences indicate, it was reopened with its previous name and functioned as before. A number of changes came gradually, not abruptly, after 1949. Today the historic campus is Sichuan Medical University. Sichuan Normal University occupies some of that same older campus space. An agricultural branch was moved to Southwest University in Chongqing.

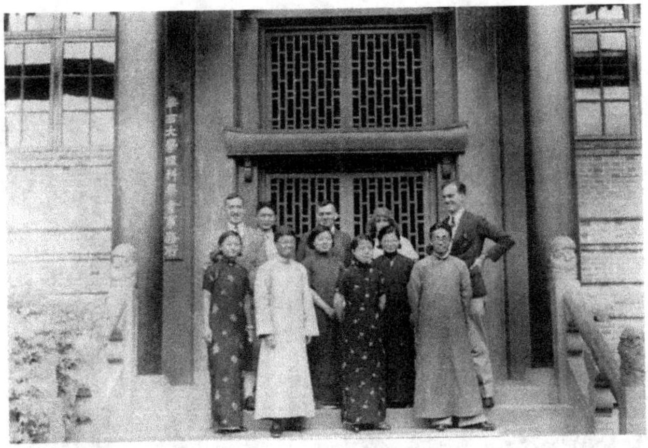

Photo 5: Mother and her colleagues during the 1940s.

repaid their kindness with years of good and faithful service at the university. "I am very grateful to the church, but finally, everything depends on oneself." I realized that the problems I was experiencing—a single mother with a seven-year-old daughter—were much less than those Mother had faced many years ago, and I determined that I would "depend on myself." I found a job and in 1990 I received my PhD degree.

West China Union University was established in 1910 by churches from England, the United States, Canada, and other countries. Many of the foreigners who lived in Chengdu were gathered on the university campus. While Mother had an ongoing relationship with this religious institution, she never fawned over foreigners, but rather understood that some were good and some were bad. There was an Englishman who used the Chinese name Xu Weili. He came to China in 1924 and served as chair of the chemistry department at the university. He was Mother's teacher, colleague, and department head. Professor Xu treated everyone well and was outgoing and helpful to all. He and his wife often sheltered left-wing students who were being pursued by the Guomindang. After Liberation in 1949, Professor Xu was enthusiastic in his support of the new regime, and he joined his Chinese colleagues in discussions on many subjects related to developing a new outlook.

By 1952 Professor Xu was the only foreigner who remained in Chengdu, but the university could no longer employ him. The Chengdu city government wanted to be sure that he and his wife

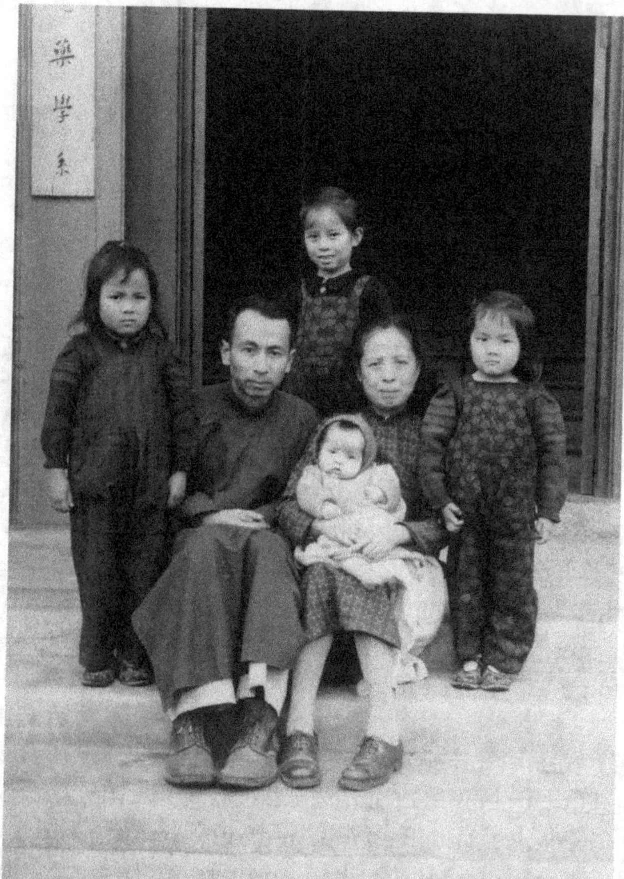

Photo 6: The writer's family. The infant held in her mother's arms is the writer.

had committed no crimes against the Chinese people before they could be allowed to return to England. Almost all of those who had worked alongside him now kept quiet out of fear that they would be associated with "the cloud of imperialism," and the Xus fell into a difficult situation. At this time, Mother and Wei Suxuan, two well-known faculty members who were "timid and retiring" women teachers, boldly stepped forward as guarantors for Professor Xu and his wife. At the time a number of well-intentioned people expressed concern for the two women. Their concerns were well-founded, as Mother did get into trouble over this. During the Cultural Revolution Mother underwent continuous criticism, and people called her a running dog for imperialism. But Mother never had any regret and continued to insist that Professor Xu had never done anything

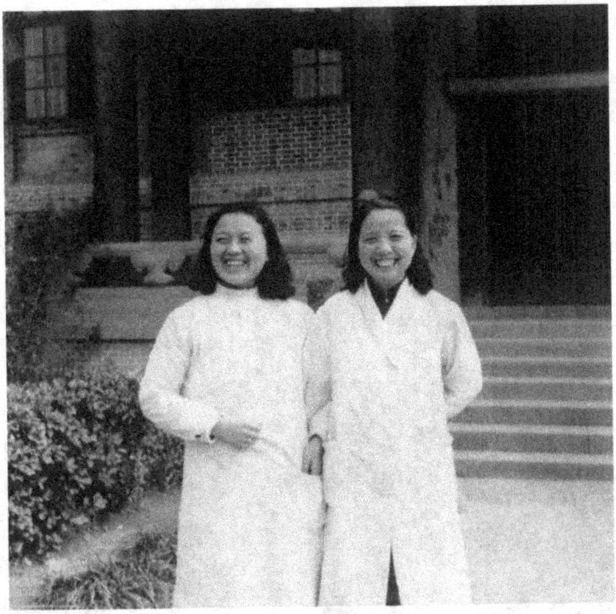

Photo 7: Mother and Wei Suxuan.

wrong. All she and Wei Suxuan did was simply tell the Chengdu city government the truth.

In 1968 Mother, like all the older intellectuals, was accused of political crimes and "sent to the cowshed."[3] Every day they faced accusations. They were exposed and criticized. They had to acknowledge crimes, and they underwent labor reform. One day when I was in line at the dining hall, I suddenly saw Mother being held under guard. She had been beaten and you could see traces of blood all over her face. I was in tremendous agony, but I couldn't say anything. That evening I had a chance to see her because I was to leave the next day, sent down to a village myself, and she had been allowed to come home to see me off. I asked her why she had been beaten so severely and she replied in a soft voice, "They want me to accuse other people, but I don't know anything and I won't make things up."

Even into her nineties, Mother enjoyed talking with me about my own research work and about any scientific subject. On trips home from England, I brought scientific and nature journals, but finally at

3. "Sent to the cowshed" was a phrase widely used to say that someone deemed a rightist was sent to the countryside or reassigned to a difficult kind of work.

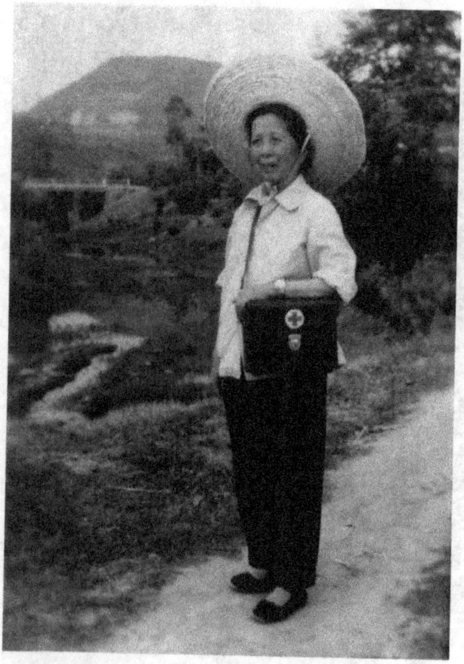

Photo 8: Mother as a "barefoot doctor" at Fenggu in Mianyang (Sichuan) during the 1970s.

age ninety-four, she could no longer see well enough to read them. At that age she also could still get angry at anything that seemed like self-promotion! I had the idea of establishing a scholarship fund named for her and Professor Xu. The scholarships would go to needy students like she had been. But Mother would have nothing to do with the idea that it might be named for her. Actually, she probably wouldn't want me to write these reminiscences about her, but I want people to know what kind of intellectual she was.

Mother died at ninety-four, in May 2003. She was reserved, an ordinary person, a bit shy, kind-hearted and honest. She always told the truth, kept her independence, and firmly upheld her principles of humanity. She was firmly rooted in the Chinese earth, and died as a light cloud slipping silently into the long river of time.

This memoir is by Xiao Bing, from Old Photos *no. 37, pp. 98–108.*

A Taxi Firm Owner and His Two Wives

Introduction

This story is a bit different from others in this collection in showing that the traditional Chinese custom of having a concubine or second wife was still practiced in the middle years of the twentieth century. While the writer focuses almost exclusively on private concerns

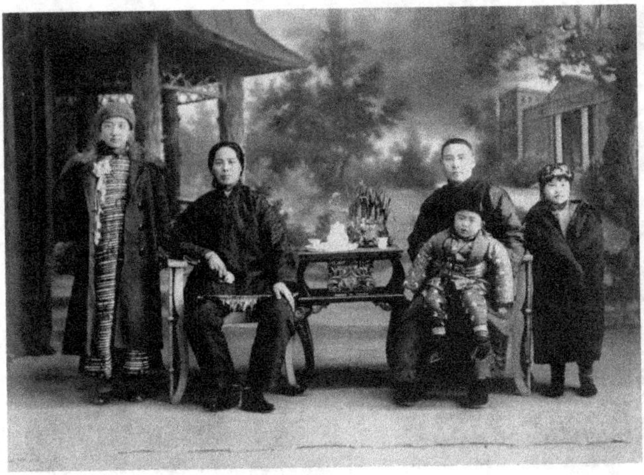

Photo 1: Jiying's family in earlier years.

rather than on the political changes the family lived through, the grandfather's two marriages were not affected by any new social policy, and in his livelihood, he made a transition from private to public management of a taxi company. Like countless millions of people throughout China, the grandfather was powerfully affected by Zhou Enlai's death in 1976, in his case so much affected that his own death followed soon afterward.

The Text

I had two grandmas. Jiying was my grandfather's concubine, whom he married. She liked to take photos, so we have lots of photo albums. Some are scenic photos, and some were taken in a studio or in our home. Unfortunately, some of these were lost during the years of "breaking the Four Olds" in the Cultural Revolution.[4]

My grandpa, Yu Zhenwen, was born in 1909 in Chen'ge Village, Ping'gu County, in suburban Beijing. My great-grandfather was named Yu Huaiqi, and he was the herbal doctor in the village. My great-grandmother was very diligent and thrifty in running the household, and she also did some business in selling medicine, so Grandfather's family was among the wealthiest in the village.

4. "Breaking down the Four Olds" refers to a campaign during the early stage of the Cultural Revolution (1966–1976) that aimed to wipe out Old Customs, Old Culture, Old Habits, and Old Ideas. Countless relics, artifacts, and items of cultural and historical value were destroyed.

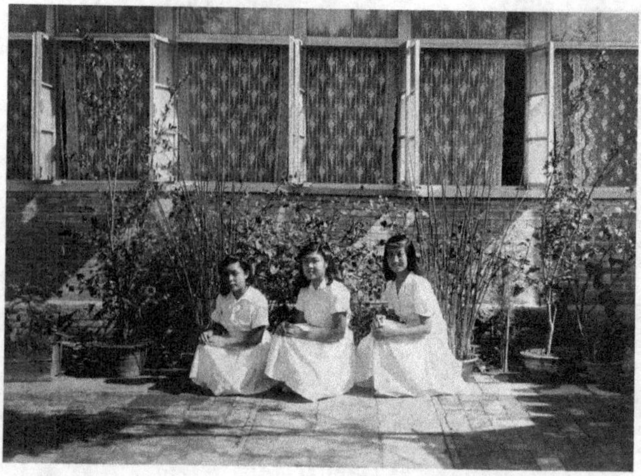

Photo 2: Jiying, still a teenager, with her nurse colleagues at Chuantian Hospital in suburban Beijing.

Grandpa was in a private school for several years, and when he was fifteen he was sent to Beijing to learn a trade. He was very diligent and down-to-earth, so he became an accountant in a rickshaw firm named Fuyou Rickshaw Firm. When he was thirty-two years old, he had saved up enough money to open a taxi company in Beijing—at that time there were fewer than ten in the city. He and Jiying met through the efforts of two friends. Jiying then was tall and slim with an oval face, and she wore her hair short. She knew some Japanese, English, and Latin, and was quite good in social matters and a good dresser. So she quickly caught Grandpa's eye. Photo 1 is a family photo from that time. First on the left is Jiying. She was a native of Shanxi province, and her family was in the winemaking business. Her father was a doctor in Beijing, and his family lived in an elegant home. Photo 2 was taken when Jiying was fifteen or sixteen years old. At that time she was a nurse at Xizhao Hospital and Chuantian Hospital.

In 1943, Grandpa became the manager of the Zhenxing Taxi Firm, which was in an alley called Dongtangzi, or the East Brothel, off Rice Market Boulevard in the Dongdan area. The firm used 1936 Plymouths and 1932 Fords in their fleet. They also used 1941 Ford trucks for long-distance travel. The firm lasted nine years and closed in 1952.

All his life Grandpa had two hobbies. One was travel, and the other was reading. After Jiying and Grandpa started dating, she always accompanied him. She always took lots of photos, among which was Photo 3. When Grandpa's first wife, Jiang Shumin, who stayed in her original village, heard of the relationship, she went to Dongtangzi to see what was going on, intending to demand an apology. When she realized that Grandpa wouldn't be able to leave Jiying, she cooled down and accepted the situation out of fear that she would lose her husband.

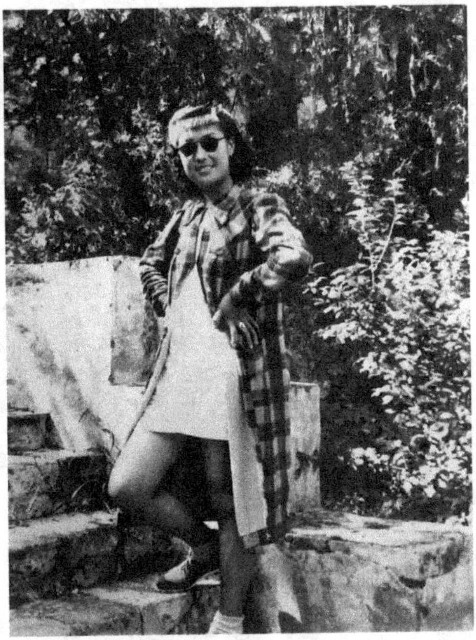

Photo 3: Jiying on an outing with Grandfather before their marriage.

In 1945, Grandpa and Jiying, who was then twenty years old, were formally married in the Hall of Fortune and Longevity. Grandpa was already thirty-six years old at that time (see Photo 4).

During the time when Grandpa was living in Dongtangzi, he became obsessed with mahjong, and my second auntie, Grandma's youngest daughter, was born in 1944 (her mother was the first wife). Second Aunt recalled, "I was already a big child before I saw what my father looked like—every night your grandfather played mahjong all night. And when I saw him in the daytime, he was always sleeping in his pajamas. So I thought that anyone wearing a leather overcoat and driving a car was my father." But it was also true that all his life Grandpa never smoked or drank any alcohol. He never lost his temper or got angry. It seemed to me that he always had a smiling face.

After marriage, Jiying wore her hair in long waves, done by a hairdresser. Photo studios in Wangfujing and Dongdan displayed huge photos of Jiying. When Grandpa went out for social events, for restaurants or movie theatres, or simply for window-shopping, Jiying always accompanied him.

At age twenty a woman is beautiful, but a baby's cry brings a

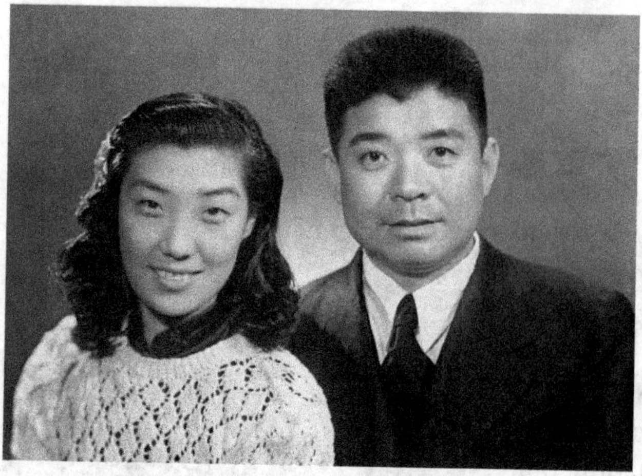

Photo 4: A wedding photo of Grandfather and Jiying, married in 1945.

woman's second life. The Lord did not give Jiying such fortune. She never had a child all her life. During the two years when the family lived in Dongtangzi, my two grandmothers lived under the same roof. They often quarreled for various reasons. Most of the time, Grandma Shumin hit first but Jiying seldom fought back physically. Nor did she talk back. Grandpa knew he was in a difficult situation and decided to separate the household.

In 1947, Grandpa and Jiying purchased a two-story house in East Fourth Archway (present address: No. 3, East Fourth Big Doufu Lane). The house had eight rooms. The second floor was the living area, and the first floor was a garage and an office. Grandpa also rented three rooms near Xidan for Shumin and her daughters so the two women wouldn't have a chance to quarrel. Grandpa regularly sent money to his first wife. The oldest daughter would call and say, "Father, we are out of money!" So Grandpa would take money with Jiying waiting in the car outside. If he stayed there a bit too long, Jiying would honk, which made Grandma Shumin very upset.

Jiying loved Beijing opera and was also an amateur performer, which made her a close friend to a lot of famous opera stars. Photo 5 is Jiying dressed in a Beijing opera costume.

My father, Yu Baoshan, was born in 1930 to Grandmother Shumin. He started out as an apprentice in Grandfather's taxi company at the age of sixteen. He lived with the workers there, eating, repairing cars, and driving with them. My grandfather always taught his children to be independent so they could have a good career when they grew up. My parents' marriage was arranged by my great-grandfather. My Grandpa disapproved of this kind of marriage and intended to find a city girl for my father.

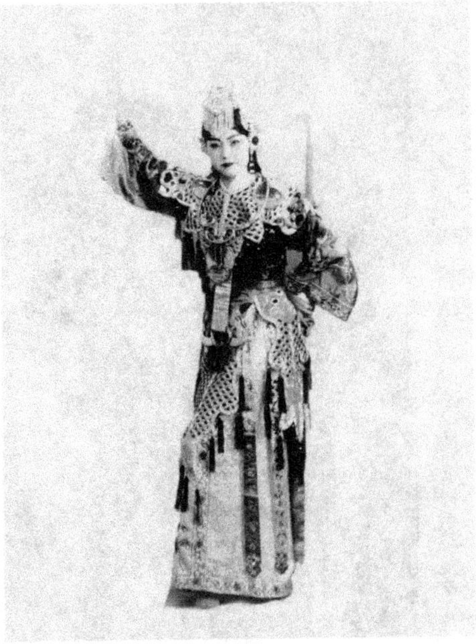

Photo 5: Jiying in costume for Beijing opera.

My mother, Liang Xiuqin, came into my father's family at age twenty. She was expected to do all kinds of household chores. My mother recalled this period of time. "I was pregnant with you for several months, and your Grandma Jiying still asked me to carry a huge basin upstairs to do the laundry. Jiying did not know how to take good care of people. She didn't know how to take good care of money either." In the eyes of Jiying, Mother was more like a maid. Jiying spent a lot of time in front of her mirror. She didn't know how to cook, how to make clothing, or how to do other household chores.

After 1949, in the household registration book, Grandpa's marriage status was listed as bigamy. According to my mother, my birth was a great pleasure for Jiying. She had never had a child of her own, which made her like children. She often called me "sweetheart" or "precious" and always used such terms. She gave me a silver locket and a jade bracelet and clothes. "Come on, baby, let Grandma hold you!" It was as if the family had a new toy who was the center of attention. So Jiying spoiled me from the beginning (Photo 6).

At the beginning of 1953, Grandpa and Jiying purchased a property near Dongzhimen. It had about eight rooms in the front courtyard and was a *siheyuan* or courtyard house. The northern part of the house had two wings. Those rooms had high ceilings and the halls were made of stone. There were glass windows with narrow

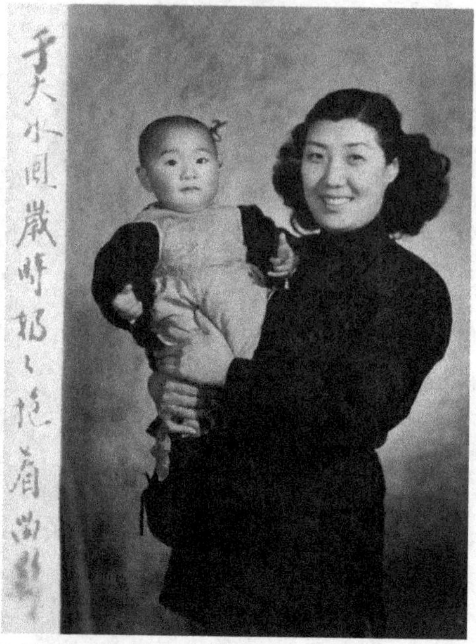

Photo 6: Jiying holding the author as an infant.

wood frames. Both the inside and outside were paved with brick. There were two rectangular raised flower beds with peonies and canna lilies planted in them. In the back yard there was a huge scholar-tree that attracted many magpies who sang nearby.[5] Swallows also came in late spring. These birds suggest the appeal of living in a *siheyuan*. In the early 1950s, three generations of our family lived in this house. Jiying was working in a hospital at that time. She became very kind to everyone then and tried to make peace with Jiang Shumin. But Jiang Shumin's attitude went from resentment to hatred (she is seen in Photo 7). She thought Jiying was the person who disturbed the peace of the family and always found some reason to fight, even physically. Clearly Grandpa felt that his big family might break apart. My second aunt, Father's sister, later commented that they didn't really hate Jiying, but they were not so close to her either. One day Jiang Shumin asked a fortune-teller to tell her future. The fortune-teller said, "Madam, you have a sister who shares a bed with you." Shumin said, "No, I don't!" The fortune-teller said, "Your birthday data shows that you have a sister who shares a bed with you, two people with one rice bowl." So after that, Shumin stopped quarreling and accepted her destiny. So both wives became victims of the old ways in marriage.

5. The Chinese scholar tree is also called "pagoda tree" or "Japanese pagoda tree" (Latin: *Styphnolobium japonicum*). This tree grows to about forty feet high and has a spread of thirty to forty-five feet. It has a rounded canopy with greenish-white to yellow flowers. This tree has been introduced into the United States, and especially in the good growing conditions of the southern states is regarded by some as invasive and undesirable.

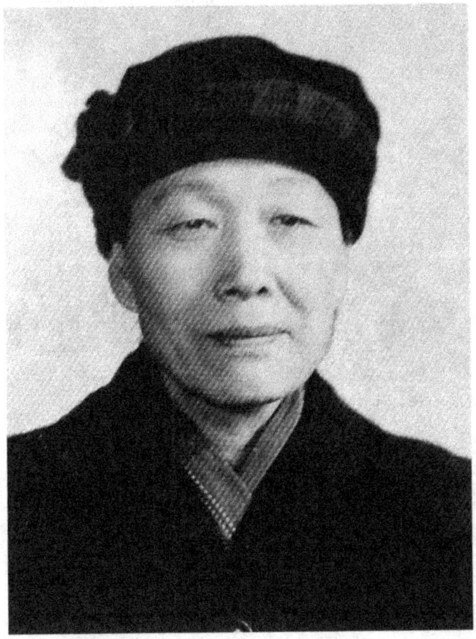

Photo 7: Grandfather's original wife, Jiang Shumin.

The extended family divided in 1955. From then on, the two grandmothers never saw each other. On the day when my first aunt got married, Jiying sold the front part of the house and rented an apartment outside the Guangqu Gate.[6] Grandpa had become used to the larger house and did not like the new place, so he blamed Jiying for the decision. Later, my father's job caused him to relocate to the southern part of Hebei Province, to Shijiazhuang.[7] I spent my grade- and middle-school years in Shijiazhuang, and Jiying sent books and newspapers, especially *Beijing Evening News* (*Beijing Wanbao*). Every summer she invited me back to Beijing. In her diary entry for March 31, 1959, Jiying wrote:

> This morning Lao Yu called me to say that Dashui was coming. I was so excited! I raised this child single-handedly and haven't seen him in more than a year. I didn't have a chance to see him during Spring Festival. I miss him day and night. My love for him is so deep. I asked Lao Yu to pick him up and bring him to the factory. The next day I asked for a day off and took Dashui to the zoo. The day after that, I want to take him to work with me. I can't let him go. . . .

6. Guangqu Gate, built in 1555 and dismantled in 1953, was an eastern gate of the walled city of Beijing.

7. Shijiazhuang is about two hundred miles southwest of Beijing and is the provincial capital of Hebei.

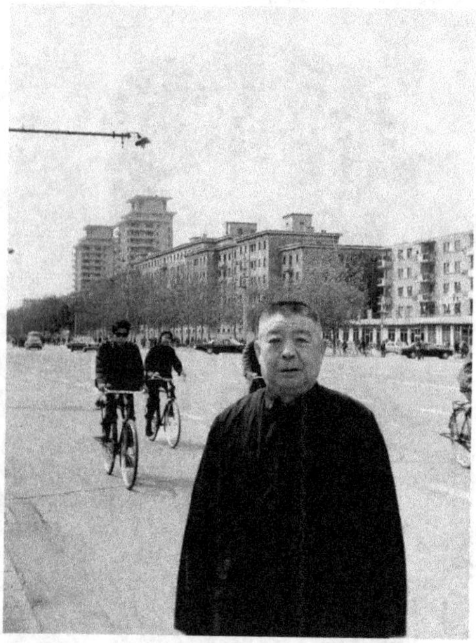

Photo 8: Grandfather in Beijing in 1970, after his retirement.

In 1961, Jiying came to Shijiazhuang in the rain to take me back to Beijing to see a doctor. I saw that she had lost a lot of weight, didn't look good, and was in low spirits. She had been diagnosed with intestinal tuberculosis. With a certificate from the hospital, she was entitled to buy two *liang* [about 6 ounces] of pork and one *jin* [1 pound] of eggs.[8] She made a stir-fried dish of shredded pork and cabbage. I hadn't eaten meat for so long. I ate all the meat when she was in the kitchen. Jiying was unhappy about that and for the first time criticized me: "You should be thoughtful of others!" I remembered this throughout my whole life. Jiying was a nurse, so she was very patient in preparing herbal medicine for me. It usually involved slowly boiling herbal medicine in a large pot until there were only two small teacups of liquid left to serve. She watched me until I drank all the medicine.

At that time, Grandpa moved to an apartment. Grandpa retired in 1970 from the Beijing Transportation Company. He was quite happy, telling me that he had gotten a "red book," indicating that he was a state cadre. As he had state benefits in his retirement, Grandpa could travel to the suburbs for recreation. He enjoyed reciting poetry, but he also enjoyed playing poker with me. When he won, he was as happy as a little boy. (Photo 8 shows him at that time of life.)

On the day of Premier Zhou's memorial service in 1976, Grandpa suddenly had a stroke. I had never seen Grandpa so sad. Every

---

8. One Chinese *jin* is slightly more than an English pound (lb.). There are ten liang per *jin*.

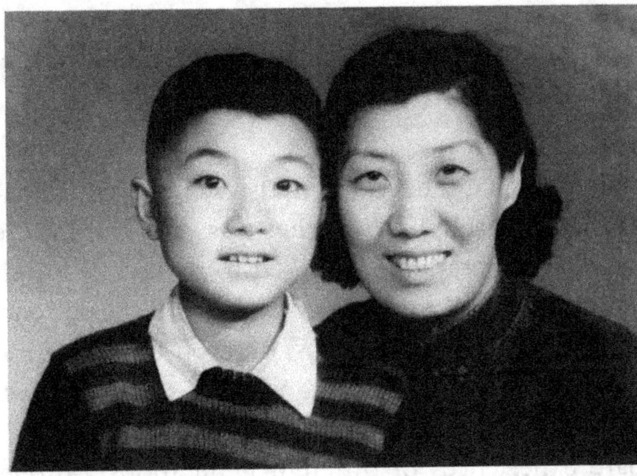

Photo 9: The author with Jiying in 1961.

time he mentioned Premier Zhou, he would weep. After listening to the radio at home during the memorial service, he went to see a neighbor across the alley four or five times. His mood was very unstable, and when he was alone he was always weeping. "Premier Zhou was a good man, so rare. A man like me should die instead of him!"

So Grandpa lay in bed in the emergency room of Fuwai Hospital. The doctor said that after Zhou passed away, many people with a history of heart disease were so sad and disturbed that they had aneurysms. Grandpa lay in bed unconscious for three days and nights in the hospital. Every day I dreamed that he would wake up and recite a beloved poem from the Song period. Ten days later Grandpa passed away. After Grandpa died, Jiying helped me get married. She and my family lived together for nineteen years. She did her very best to take care of the household chores and showed great love to my children. Every night she washed the children's feet, told them bedtime stories, taught them English, and did many other such things. Photo 9 is of Jiying and me in 1961.

My grandpa married two women in his life. Both wives were very loyal to him. Jiang Shumin passed away in 1966 at age fifty-seven. Grandpa died in 1976 at age sixty-seven. Jiying passed away in 1995 at age seventy.

This is by Yu Dashui. It appeared in Old Photos *no. 56, pp. 77–88.*

"Half the Sky" Can Be Heavy: Lin Zhao's Ordeal

Introduction

This item is a reflection on a series of events, not an account of the events themselves. The writer was born during the 1950s and states that his generation was asked to make revolution even as elementary school children. He contrasts their experience with that of the "three skipped-over classes" of 1966 through 1969 who missed out on university entrance examinations and instead were given work assignments in rural areas. Like Americans, Chinese observers compare experiences of different generations. Through the late 1970s, the generations of course shared the experience of various political campaigns and then the Cultural Revolution of 1966–1976.

As indicated here, Lin Zhao was an outspoken student at Beijing University during the mid-1950s, a period when there was still general enthusiasm for the changes then in progress that were creating the "New China." One slogan in the new society was that women would "hold up half the sky." The assumption was that women would share labor and other responsibilities equally with men. Lin Zhao conscientiously undertook an equal share. This is the story of her ordeal. What happened to Lin Zhao perhaps depended on her circumstances in a particular environment. Still, this happened and resembles all too closely the kinds of events that lay ahead for millions of innocent people. The author muses on these conditions and their effects on the lives of people who lived through these years. Like Lin Zhao's piece, this essay is quite outspoken and reflects what is tolerated in publications during the present era.

The Text

Those of us born in the mid-1950s lived through many things during that stormy period of tragic, violent developments in building the nation. As we matured, we came to understand many things. When we reached middle age, that period of insanity seemed to have been put behind us many years ago. Those old events are past, but when you happen to look back on them you feel remorse. Perhaps the greatest

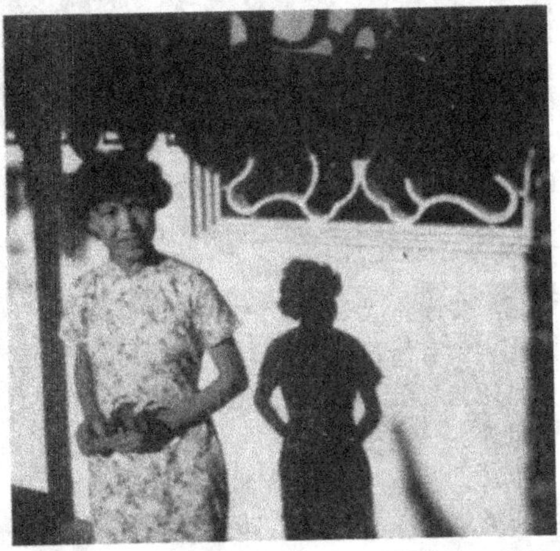

Photo 1: Lin Zhao in Suzhou, 1946.

relief is that you are still alive and you can go onto the Internet and see with a cooler eye how things look in a world that is no longer blocked from us.

I learned about Lin Zhao through the Internet many years after the fact! The earliest I learned about "Lin Zhao's death" was through the article "The Judgment of History," by Mu Qing and others, written a full twenty years after her death and forty years after the situation itself occurred. Zhang Yuanxun lived at that time and was declared a rightist. Forty years later Mu Qing reopened that bitter experience and publicly disclosed the full story of Lin Zhao, a student in the Chinese Department at Beijing University, Class of 1954.[9] Zhang was a classmate of Lin Zhao's then and dared to say she was his "fiancée" when he visited Lin Zhao in prison. This article, written in blood and tears, allows us to reflect again on the disgrace of that time. The article was also published in the press. The response was very strong, and many people were shocked. I was one of those and after I shed many silent tears I became endlessly outraged. Now that I knew everything that had happened, I was incredulous. Sometimes we only gradually come to understand the kinds of things that can happen to people in any kind of society and at any social level. When human nature meets with endless devastation, the greatest tragedies may occur for any living person.

The story begins on a hot night on May 22, 1957. For Lin Zhao, a talented young woman at Beijing University, events that night first pushed her toward the guillotine. Had she known how these events might unfold, Lin Zhao would not have gotten involved in a closely

9. In Chinese higher education at the time, classes were identified by the year the class entered rather the year they graduated.

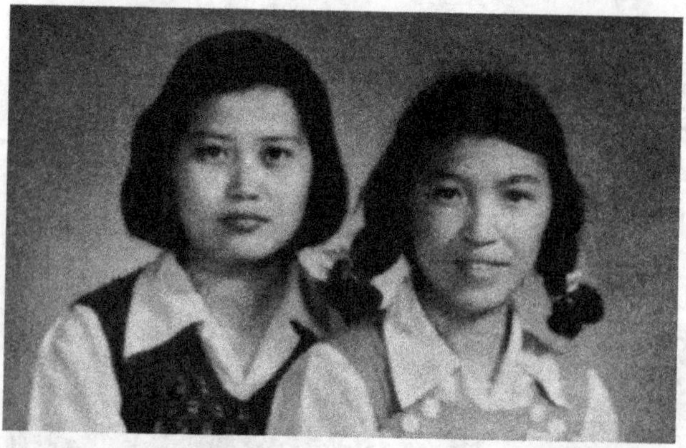

Photo 2: Lin Zhao (right) with a classmate at Beijing University, 1956.

watched "debate" that was taking place on the street outside the University's Building No. 16. Lin Zhao had stood up in a heated defense of Zhang Yuanxun because a small poem he had written was under attack by many people. Facing an armed crowd, Lin Zhao spoke up in her Wu dialect–accented *putonghua* and suddenly the bloodthirsty crowd went quiet.[10] As Lin Zhao saw things, Zhang Yuanxun was not a Party member and not even a member of the Youth League. Why take up arms against him just because he wrote a poem? Lin Zhao asked this question—her conscience asked this question—and she found her voice to speak up. Who would have thought this soft-spoken girl from Suzhou had such courage? Her speech was suddenly interrupted when a stranger's voice roared in the darkness: "Who are you?" Lin Zhao had no way to see clearly who this stranger was, and it turned out to be the voice of an interrogator. Lin Zhao called out angrily, "I am Lin Zhao. Remember that! That's 'Lin' with two trees for thirty-six, and 'Zhao,' a knife over a mouth, with the sun to the left.[11] And who are you—who are all of you? Why don't you

---

10. *Wu* is the dialect spoken in Shanghai and the surrounding area, which includes Suzhou.

11. In China it is common for people to describe characters by the elements of which they are composed. In a conversation in which they are in close proximity, people often write a character in the palm of their hand. The description of her family name, Lin, as "two trees for thirty-six" needs some explanation. The character for "tree" is composed of a cross (or plus sign) which means "ten" and the character for eight, made up of two "legs" that are placed on either side of the lower point where the two lines of the cross intersect. Because of this content, one tree "equals" eighteen, and two trees make thirty-six.

identify yourselves?" There was no reply. The stranger's voice ceased. And in that instant, on that hot, disorderly night, simply because of a "conflict between conscience and organization" in her inner world, this young woman came to the attention of a well-known BeiDa professor, Liu Guo'en, and tumbled into a fate that she would not be able to resist.

That year was a time when anti-rightist activity was expanding and people had to defend themselves. At that time, my generation was still in the cradle, just starting our lives. Although we had the warmth of our mothers' love and the shelter of our homes, and we could temporarily avoid the scene of fishy smells and bloody rain outside the window, our spirits were often melancholy and there was never any way to heal the dark wounds of our spirits. Life does not offer any special favors as to the conditions in which we live. Really, things were worse for our fathers and mothers, who faced ceaseless dread. Many periods in their lives seemed to be dark plots of "great noises and then great stillness," through which they lost their freedom and even sank into captivity.

It has been said, "In any place where people function together, there are those on the left, the center, and the right." Our mothers and fathers faced the necessities of the times, which also were the limits of the times, and they could become only one of several kinds of people. One kind committed crimes through what they said, in spite of themselves, and so neglected the nurturing they should have provided. Another kind barely escaped disaster, and because of this didn't dare speak out. A day passed as a year, and their children had to depend on each other for their needs. Still another kind made a clean break with it all, placed righteousness above family loyalty, and decided to abandon all responsibility for their children. It was as the ancients said: "Those who can take advantage of conditions succeed. Those who can adapt to conditions survive. Those unable to ride the tide of the times will die." As in our modern society, none of these three kinds of fate offer a good choice to anyone. My friend Liu Cui was born in 1958, the year after the Lin Zhao affair. Her father is the well-known poet Gong Liu. From the day she was born, Liu Cui never sucked any milk from her mother. Her mother was a leftist, and she refused to breastfeed any future rightist. In order to nourish her young life, Gong Liu unfailingly provided three pints of fresh milk every day.

In the anti-rightist struggle of the late 1950s, throughout the country more than a half-million people suffered gross injustices. There would be no way to count the number of families affected. Of these, "more than half lost their public employment, a very large number of them were sent to education through labor or prison labor, some of them were set adrift, their families broken and some family members dead." This was disclosed in the memoirs of Li Weihan, United Front Chief at that time.

We all grow up in a particular social context. Our generation had certain distinctive characteristics as compared to other generations. As compared to those born after the 1960s, we had more experience of the world and less drive. Compared to those born in the late 1940s and early 1950s, the "three skipped-over classes," we lacked their depth and enthusiasm.[12] When we were six or seven, we had the three years of natural disasters, so our childhood brought the taste of poverty. Although we didn't understand anything about life, still the lack of food and clothing stayed fresh in our memories.[13] Then the Cultural Revolution followed soon afterward and our education was interrupted, so just at the time we were in primary school we found ourselves amidst the cry to "stop classes and make revolution," and in the space of a single night we became the craziest, least self-confident generation. As shown in later statistics, more than three hundred million children lost their opportunity for the usual form of education. I went out to buy a newspaper for the rebel side, and in the broadcast booth with its canopy of reed mats, the broadcaster urgently read from a leaflet. A classmate's mother had gotten swept up in struggle on the street as she followed all the hubbub. Because we were too small, we had no way to keep up with our fathers and uncles among the "three skipped-over classes," so

12. The members of these classes became "educated youth" and were sent on work assignments "up to the mountains and down to the countryside (*shangshan xiaxiang*)."
13. The "three years of natural disasters" were 1959 to 1962. The economic difficulties of these years resulted from policies introduced during the "Great Leap Forward" of 1957–1958. In agriculture, greatly reduced harvests combined with over-reporting of food supply meant that thousands upon thousands of people simply starved to death.

we would become angry. For a full ten years from childhood to our teenage years, those of our generation were like someone dying from loss of blood, and our bodies had lost the ability to replenish the blood we lost.

Just as we were getting into the dread and "merriment" of the crazy struggle against the rebels, a woman from Beijing University whom we didn't know (she had determinedly stuck by her own views and continued her struggle in jail for many years) was secretly executed on April 29, 1968. (Her family had to turn over a payment of five *fen*, the cost of the bullet that executed her.[14]) Lin Zhao was only thirty-six at the time of her death. Her death fulfilled the prophetic words she blurted out that night: "Lin that is two trees for thirty-six, and Zhao, a knife over a mouth, with the sun to the left." We must understand, bitterly, that there is nothing we may say about such things.

Although we cannot overcome the times in which we live, anyone can be independent in those times, and Lin Zhao was this kind of person. Lin Zhao's death occurred not just because she lived during an unfortunate time. She died because of the way she thought, her beliefs, conscience, and courage, so that ultimately she met with this "disaster resulting from an argument." Without this as her nature, she would not have come to the end of the short life she loved and come to this at the end of the "sole authority's" gun barrel. At first glance, this seems to be a tragic case from a period when reason was lost. Actually it was the inevitable result of an "unfeeling autocracy" in a turbulent society. When it is impossible to destroy the way a person thinks, destroying the person's body is the only choice. Such cases provide insight for people who come later, so that we might hope to see something better happen in our own case. Or at least this might become a key to open our understanding of this painful period of history.

Part of the inhuman treatment she encountered in jail was almost daily beatings by a group of brutal female jailers. The more ruthless their treatment of Lin Zhao, the more they liked to say how "positive" or "meritorious" their work was. When Zhang Yuanxun

---

14. This detail of capital punishment is indeed the practice in China, an ultimate measure of ideas about the state's power.

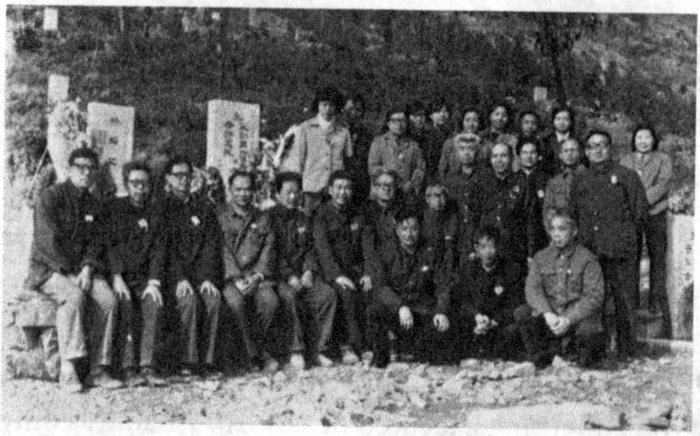

Photo 3: A group of classmates who gathered at Lin Zhao's newly completed grave, on the "May First" holiday, 1982.

called her his "fiancée" and braved danger to visit Lin Zhao in jail, she would point to a group of jailers nearby and say, "They'd like to rape me! All I can do is open a small slit in my clothes when I have to answer nature's call and then sew everything back up when I'm finished." Under such abuse, how could Lin Zhao maintain her human nature? When "class struggle" has been unleashed, how many times has society gone into continuous turmoil with laws trampled and even the leaders of the nation unable to improve the situation? Only under such conditions would anyone trifle with an ordinary person like Lin Zhao. Still, while heaven might change and inhuman behavior never disappear, Lin Zhao never once yielded. She wore a black armband with the word "injustice" embroidered on it in white thread. And at the last moment of her life she used the last few drops of her blood to write the words "history will say I am innocent."

History is a mirror. The death of Lin Zhao shows us how ignorant we can be, and our innocent youth gets a serious after-school lesson. If our generation—and this should include everybody—cannot reach an accurate understanding of the political movements of the first decade after our country was established, we will be unable to have a better understanding of the historical tragedies that occurred after that early period. As the people of a generation grow up, they become acclimated to the passing trends of the times, the powerful storms, including these traumatic memories of a spiritual pain that cannot be put away. Sometimes we keep quiet, but we make this choice in order to survive, for a practical kind of self-preservation in the circumstances. But at the same time this might also be the turning point toward recognizing one's "human nature."

But Lin Zhao didn't have a chance to keep quiet, and became the symbol of nobility, freedom, and commitment for that time.

This is by Fan Hong. It appeared in Old Photos *no. 22, pp. 61–65.*

EIGHT

Education

The Sound Syllabary Displayed on a Wall at Tian'anmen

Introduction

The ongoing effort to attain to universal use of the Chinese language is a good example of a theme in China's ongoing revolution that has been continued by each successive group of revolutionaries. A number of these ongoing themes have been more cultural than political. Efforts at language reform began with late Qing Dynasty reformers, many of whom also joined the revolutionary movement to bring down the Qing government. This article describes efforts to make the language more accessible through a set of symbols indicating how characters should be pronounced.

Another aspect of language reform was to use vernacular Chinese (*baihua*, "plain speech") instead of the more cryptic literary language that had separated educated people from commoners. A *baihua* newspaper was published as early as 1905. This effort continued over the next decade, and by the time of the New Culture Movement in the late 1910s, writers were beginning to write fiction in *baihua*. Another aspect of *baihua* development was to refine the vocabulary of Chinese by creating "compounds," pairs of characters that sharpened terminology.

As indicated here, the vagaries of politics and internal military struggles often stifled progress in language reform. By the late 1920s, the Northern Expedition, an apparently successful military campaign,

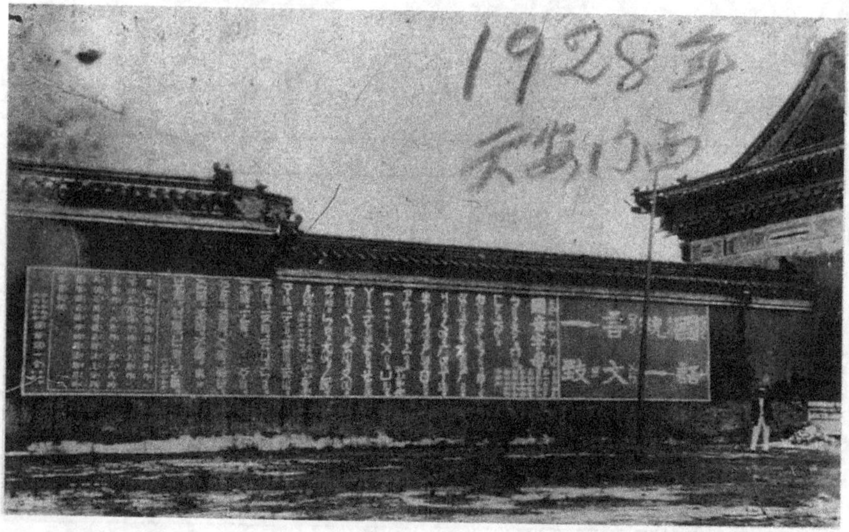

provided the impetus for language reform to make the advance that is described in this article. After another generation—and the defeat of the Nationalists by 1949—the new Communist leadership brought the task of language reform to its fullest implementation by introducing the *pinyin* system, which uses European letters to spell out Chinese pronunciation. And as noted in the article, the syllabary publicized on this wall of Tian'anmen in 1928 departed mainland China with the Nationalists and is still used in Taiwan today, while linguists in the People's Republic still appreciate the role it has played.

The Text

This photo of the west wall of Tian'anmen was taken in 1928. As the note on the photo says, this is a section of the palace wall on the western side. The full content of the writing here makes up a document. The first part on the right says "Unify our national language so that speech and writing will be uniform." This is followed by the heading "Sound Syllabary." Then the forty symbols of the syllabary follow, and on the left end it is signed "Language Unification Committee of the Bureau of Education, The Republic of China."

What is this "Sound Syllabary," and why was it important enough to be displayed on a wall of Tian'anmen? The explanation begins with certain features of the Chinese language. Chinese characters are a form of representational writing. But it is not always clear how they should be pronounced, and in many cases the pronunciation is not apparent from the form of the character. Although the phonetic part of characters that are picto-phonetic provide a ready-made way to suggest pronunciation, they still do not necessarily make this clear. This is even more difficult when a character is not used very often. In their efforts to remedy the weakness in Chinese characters that "have form but not sound," concerned Chinese scholars put in a great deal of effort for many centuries. The main method they used was to "to compensate for internal weaknesses by supplementing from outside"—to use something outside the form of a character to indicate how it should be pronounced.

In order to "unify the national language," a conference was convened in 1913 to "unify the pronunciation of our language." First, the conference spent many months in setting out to establish a "national pronunciation," and agreed on the pronunciation of some 6,500 characters. Then they decided on phonemes and on an alphabet, and made plans on how to proceed. Yet serious problems arose as proponents of each plan upheld their own views and couldn't agree. Lu Xun, who attended the conference, wrote "Comments by an Auditor," in which he noted "Wang Zhao (vice-chair of the pronunciation committee) went to the platform to speak on the question of what to remove and what to keep of the entering tone, and he got into a big fight with Wu Zhihui (the chair of this committee). He punched Mr. Wu in the belly, and his cotton trousers came down!"[1] They discussed these matters for two more months, still with no agreement. Finally they adopted the suggestion of Lu Xun and two associates (Ma Yuzao

1. Lu Xun (1864–1936) later became best known as a writer, but in this early part of his career he was very active in language reform and other cultural concerns. Wang Zhao (1859–1933), like Lu Xun and many other Chinese scholars, had studied in Japan. Wang had been involved in the language reform effort for at least a decade by the time of this conference. Wu Zhihui (1865–1953) had studied in France, where he and several associates were attracted to anarchism. Wu remained concerned with language reform and had enough respect to have been named a chair of this committee. Wu Zhihui and several of his colleagues in France and the early anarchist movement became supporters of Jiang Kaishek by the 1920s because of their ardent opposition to Marxism. Wu went to Taiwan with Jiang Kaishek after 1949 and remained one of Jiang's most influential advisors.

and Xu Shouchang) to use the sound notation syllabary in order to ascertain the pronunciation of characters and to agree on the name "sound notation syllabary" (in 1930 this was changed to "sound notation symbols"). This used the scheme of the famous linguist Zhang Taiyan (Binglin) and also drew on several other formats of the time.[2]

After the syllabary was approved, interest faded until the literary revolution that began in 1918, when the call for reform in Chinese writing arose again.[3] At this point, the issue was turned over to education bureaus in each jurisdiction, who put the system forward as a "sound syllabary" or as the "national language." Under this situation, with those below pushing it on those above, the Ministry of Education formally promulgated the "sound syllabary" on November 23, 1918.

For the next several years, because of the chaos of warlord fighting, the issue of a national language and a syllabary for it faded again. Then in 1928, when the National Government united north and south, everything seemed possible. This task took on new life and quickly reached another high tide.[4] The Ministry of Education set up

---

2. Zhang Taiyan (1868–1936) was a major figure in the anti-Qing movement, a leader among revolutionaries in the lower Yangzi region and a sometime opponent to the overall leadership of Sun Yatsen. Zhang emphasized the importance of tradition in a Chinese revolution and thus was involved in the full range of cultural issues. The scheme he put forward for pronunciation drew on the use of rhyming characters, a method employed in earlier Chinese dictionaries.

3. The literary revolution referred to here was part of the New Culture Movement of the late 1910s in China. This movement drew on many influences from abroad, through mostly younger intellectuals who had sojourned abroad. The major sources for change were several journals that began in 1915 and 1916, chief of which was *New Youth* (*Xin Qingnian*). A major emphasis developed for a new style in literature that would use colloquial Chinese, *baihua* or "plain language," the language of everyday use that also had been used in China's major popular novels such as *Journey to the West* (*Xi you ji*), *Romance of the Three Kingdoms* (*Sanguo yan'yi*), and others. This movement became very successful, so that a succession of excellent Chinese writers put the themes of revolution into understandable everyday language.

4. As mentioned in the introduction to this article, this unification of north and south China was achieved by the "Northern Expedition," the idea of which went back at least a decade. The goal was that forces loyal to Sun Yatsen, and to the Republic established in early 1912 and focused in south China, would bring the northern regions into the realm of the new government. Jiang Kaishek regarded this project as his chief mission when he succeeded to leadership of the Guomindang after Sun's death in 1925. While this campaign succeeded as much through alliances as through military victories, it did provide for the establishment of a new capital at Nanjing and the institutions of a centralized government such as a Ministry of Education.

a "Committee to Prepare for National Language Unification," and called for "every province, city, and county to implement the sound syllabary." This was the goal they put forward: "In the shortest possible period of time, we should call upon all who are literate to put the sound syllabary to use and teach all who are not literate to use this syllabary in learning the language so that we may achieve the goal of having everyone in our country become literate."

They also established one strong measure: "If schoolmasters, responsible people, and those who teach our national language have not become familiar with the syllabary within a suitable period of time, they all should lose their positions." A similar threat was made to officials in all government bureaus and to lower-level government workers as well. Along with all this, various methods would be used to encourage all to learn the system, such as printing handbills of the syllabary, even 100,000 copies per year. In busy areas of cities, the syllabary was to be painted on walls as another means of propagation. So the present photo shows how this was done in one such place.

Although at that time Tian'anmen was not the symbol of national political importance that it has become in more recent times, it was a place that reflected the winds of political change. It is easy to imagine the great influence of having the sound syllabary painted on a wall of Tian'anmen. From this action we can see an impressive part of the movement to modernize the Chinese language.

The sound syllabary deserves great credit for its role in expanding the national language. In Taiwan, popularization of the national language went hand in hand with use of the syllabary. In Taiwan there is a publication called *National Language News* (*Guoyu Ribao*), in which every character used is printed with the appropriate symbol from the syllabary guide alongside as a guide to correct pronunciation.[5] It is said that the number of copies printed gives it the second largest circulation among newspapers in Taiwan.

5. In Taiwan, use of the term *guoyu*, "national language," has been retained, while in mainland China the term used is *putonghua*, "common speech." In Taiwan the Wade-Giles system is still used in transliteration, as in dictionaries and diplomacy. Some in Taiwan also have advocated use of *pinyin* for these purposes. It is understandable but unfortunate that these formats and usages in language have been so thoroughly politicized, but perhaps change will come sometime relatively soon.

In the mainland the "Chinese phonetic system" (*Hanyu pinyin fang'an*), promulgated in 1958, marked the culmination of the Chinese people's experience with such systems over the previous sixty years. *Pinyin* is logical and scientific, and it also has entered into use outside of China. So in mainland China, *pinyin* has been used for pronouncing the sounds of Chinese, building broad use of the standard language (*putonghua*, or "common language"), and in the teaching of Chinese. But as in the *New China Dictionary*, the *Contemporary Dictionary of Chinese*, and other reference works, in presenting symbols to link pronunciation of characters, *pinyin* and the syllabary are both shown alongside the characters. Although the syllabary is still used in Taiwan, the practice of using sound symbols has continued among Chinese in both places.

The words painted on the Tian'anmen wall back then were "Unify our national language; make speech and writing uniform." After almost a century of effort we have accomplished only half of this goal—the uniformity of speech and writing. So "unifying the national language" in today's terms means to expand the use of "common speech" (*putonghua*)—and much remains to be done on this part of the effort.

This is by Fei Jinchang. It appeared in Old Photos *no. 9, pp. 111–13.*

Eliminating Illiteracy in Villages

Introduction

The movement described here was the culmination of a theme in cultural development that began during the early years of the twentieth century. Anti-Manchu revolutionaries understood that any egalitarian goals they envisioned would require language reform to include ordinary Chinese in a nationwide culture and in discussion of and decisions about public questions. Through long centuries the Chinese language had acted as a barrier between the elite and the humble. Competence and the refinement necessary to pass civil service examinations required years to achieve. While extended families commonly pooled their resources to get an education for a promising

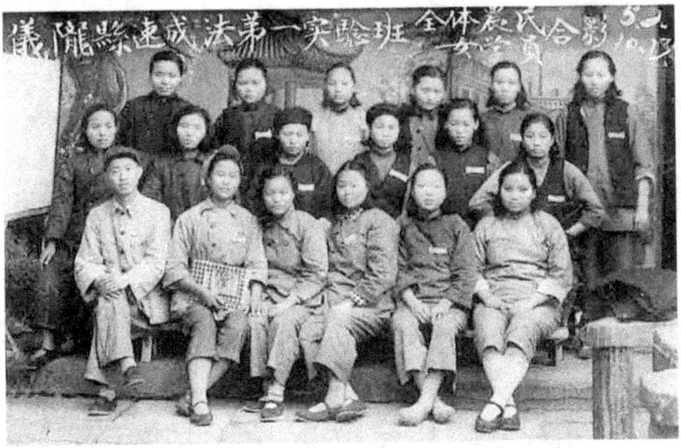

Photo 1: A group picture of peasant women who participated in a class for eliminating illiteracy. Yilong County, Sichuan Province, October 13, 1952.

boy from their numbers, the poorer classes simply could not achieve general literacy and function in the larger culture.

As this article observes, the literacy movement made tremendous steps forward in spreading literacy to the rural masses and began to improve their educational level and the degree of their involvement in the national culture. As the culmination of an ongoing effort in the modernization of Chinese culture to which people of different political views contributed, this writer observes that the literacy campaign ended with the kinds of exaggerations that became all too common in other campaigns during the 1960s and 1970s. Still, this snapshot of language reform efforts during the first decade of the People's Republic suggests the distance that China's rural peasant majority has come in one traditional cycle of sixty years. If positive achievements are emphasized in a survey of PRC history, this one holds a place very high on the list.

The Text

Upon Liberation, 400 million of the nation's total of 600 million people were illiterate. This included more than 80 percent of China's villagers. The great majority of these people couldn't read books or newspapers, couldn't handle accounting, and couldn't write a letter. They couldn't even write their own name, or the names or terms for the farming tools and fertilizers they used every day. When they needed to sign their name, all they could do was use a fingerprint. In keeping records, all they could do was draw circles or lines to

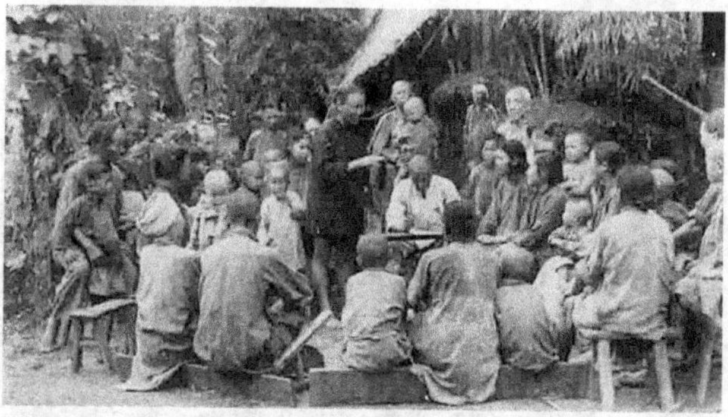

Photo 2: A peasant woman speaking to a crowd about the pain of being illiterate.

represent figures, so that after a time they would have a page full of circles and lines that they couldn't understand and that confused the actual situation.

This lack of cultural development was a hindrance to building a new China. To satisfy the hope of raising the cultural level of the vast peasant population, the state placed major emphasis on its movement to build literacy in the villages. Beginning in 1950, every county (*xian*) established a committee whose sole purpose was to improve literacy. Offices were set up, in every district officials were specifically concerned to address illiteracy, and night schools were started in every village. In some places seasonal classes were started, some at the elementary school level and some at the lower middle school level. Following the principle of "study more when work is slow, study less when work is busy," a variety of active styles were developed for study, always with the goal to build literacy along with current political matters and farming technology. Among textbooks prepared for teaching characters to peasants, one of the most popular was *A Textbook for Peasants' Quick-Study of Characters*, which used the rhyme-form of the *Three Character Classic* (*San zi jing*): "The sun and moon are bright, lighting the four corners; the sky is bright, the earth is light; Chairman Mao and the Communist Party, we and our leaders have great strength. . . ."[6]

6. The *Three-Character Classic* (*Sanzijing*) is a Confucian classic generally attributed to Wang Yinglin (1223–1296) of the Song Dynasty. It became one of the most popular elementary school textbooks for centuries up to the 1940s. All the phrases in the text quoted here end with the rhyme *ang* in Chinese, which cannot be conveyed in the English translation.

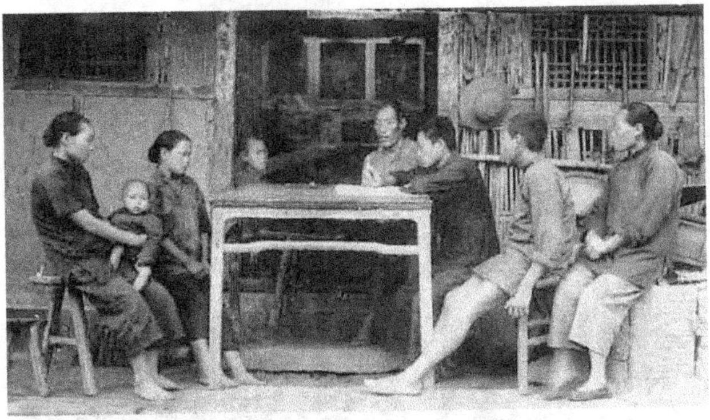

Photo 3: A family-based study group for eliminating illiteracy.

To increase interest in study, many literacy classes began their study by learning the characters for everyone's names. Then they turned to everyday terms and familiarized themselves with the vocabulary of farm life, tools, animals and land. They also learned methods of calculation and forms for accounting and other such subjects. Because the approach was related to daily life, the peasants' enthusiasm for gaining literacy was very high. To avoid "forgetting words if they're not used," most places adopted the goals of "Everyone works to develop good methods. Whatever is convenient is good" and "Just a spare few minutes will do, studying as possible will work for you." In some places people studied on their breaks from work or right out in the fields. They reviewed often in order to strengthen the effectiveness of their study. At that time there was a folk song named "Husband and Wife Study Characters" that was popular everywhere:

> When the sky is pitch-black and you go out, you see the shining stars.
> We write words on the blackboard and, oh, we see how bright things are!
> What words make things bright? Learning the word "study" is alright!

During the 1950s, many peasants sang this song as they entered the classroom to study their language.

This group of old photographs records the conditions in Yilong County, Sichuan, during the early years after Liberation as the literacy

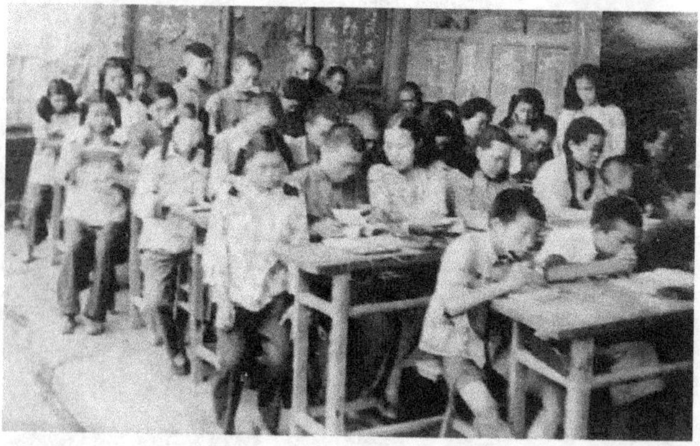

Photo 4: A class of students studying earnestly.

campaign developed there. The *Yilong Local History* reports that villages throughout the county made widespread use of "Qi Jianhua's Quick Method for Studying Characters."[7] In eleven sub-districts and two important districts of this county, a total of thirty-eight Quick Character-Study Classes were launched, with more than 1,800 people enrolled. And throughout the county more than 1,400 night schools were created, with more than 56,000 students involved. By 1956, these totals grew to more than 1,800 schools with more than 90,000 students participating.

The "rapid achievement method" was developed by Qi Jianhua, a teacher in the Southwest Military Region's cultural division. This method enabled students to advance quickly in both reading and writing. Qi's method used 1,200 carefully selected characters with the Chinese phonetic alphabet alongside. These were put on two large charts of new vocabulary items and used as an aid in study. The first step was to master the phonetic alphabet and *pinyin* as tools in the study of characters. The second step was to make a concentrated effort on individual characters and strive to recognize these and be able to use them in speaking. The third step was to use the character textbook in learning and from it to develop various drills for reading, writing, and creation of expressions. This method was adopted by

7. This method was credited to Qi Jianhua (1921–2001), a People's Liberation Army officer. Qi's method was controversial and was not promoted after introduction of the official phonetic system for transcribing the Mandarin (or "common speech"—*putonghua*) pronunciations of Chinese characters into the Latin alphabet, known as *hanyu pinyin*, in 1958.

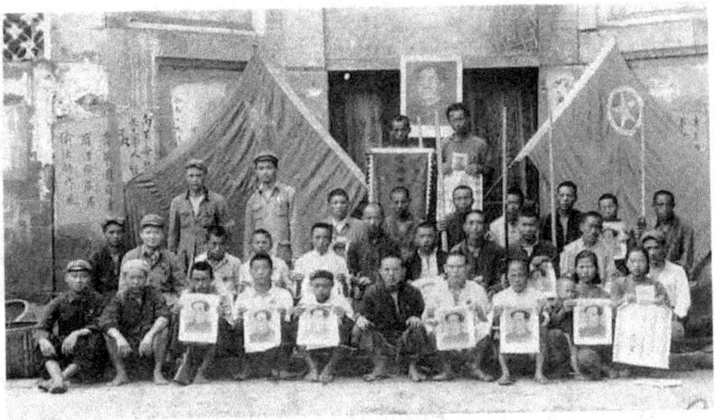

Photo 5: A group of successful students with rewards for their achievements.

the army in 1951 and later was gradually expanded throughout the entire country.

In May 1952, the Education Bureau announced the "Notice on Developing the 'Rapid Study Method for Learning Characters' Throughout the Country as a Task in Pedagogical Experiment" and designated Hebei Province as an experimental area for this method. *People's Daily* also published a special editorial on "Comprehensive Implementation of the 'Rapid Achievement Method for Studying Characters.'" Then the Education Bureau put out a bulletin telling how Shanxi, the Northeast, Tianjin, and other places had used the "rapid achievement method" in their programs to build literacy. Since the "rapid achievement method" had been implemented, the cities and rural areas participating in the literacy movement had increased greatly, and the workers and peasants involved during 1953 totaled almost twenty million. Of these, almost seven million were using the "rapid achievement method."

In December 1955, the central leadership of the Chinese Youth League announced their "Decisions on Eliminating Illiterate Youth Throughout the Nation Within Seven Years." A basic decision here was to use the 30 million literate young people in the villages to wipe out illiteracy among the other 70 million village young people who were still illiterate. They sought to enable 80 percent of illiterate village youth to overcome their illiteracy and envisioned that each would develop a vocabulary of 1,500 characters. At the same time the Youth League also set out its "Methods to Reward Positive Youth in the Movement to Eliminate Illiteracy." In order to foster unity and organization among educated young people in the

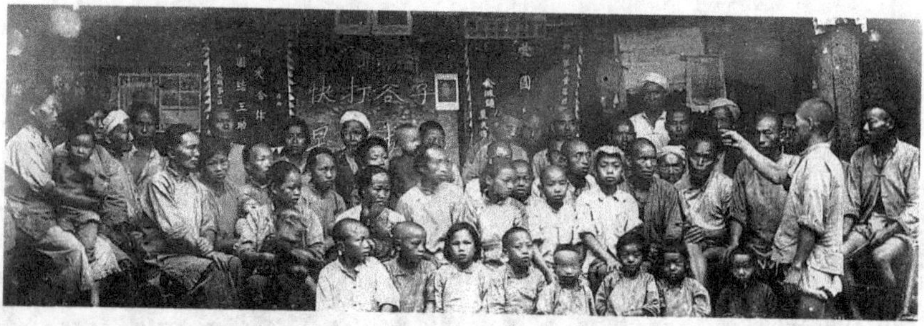

Photo 6: The leader of a cooperative group guiding the students in learning new words and sentences: "Thrash the crop quickly to get a good harvest."

villages, make better use of them, and work out teacher shortages, the Youth League leadership in January 1956 issued a "Notice on Regularizing the Establishment of Youth Teams to Eliminate Illiteracy." This called on village units to create standard organizations of youth teams to eliminate illiteracy, to organize educated youth in the villages to be responsible for village schools, to record work points in the literacy classes, and to serve as teachers and assistants in the small-group classes. There followed a high tide in the movement to eliminate illiteracy on a nationwide scale. In a single year after the autumn of 1955, villages throughout China achieved basic literacy for some 70 million people.

The vast numbers of women in China's villages had never enjoyed learning or cultural awareness. As the levels of their education improved, not only was the productive strength of villages improved, so too was their social position.

As the literacy campaign continued, because the central government called on local leaders to work with maximum effort to attack illiteracy, some lower-level officials exaggerated their political achievements. In their efforts not to fall behind, they also made the campaign into a race to be fastest to achieve the goals, so that false, boastful reports began to increase. According to some such reports, "It only took three days to completely overcome illiteracy"; the "fastest" village claimed to have accomplished the goal in five hours! In 1958, Yilong County in Sichuan, whose activities have been described above, claimed that "ten thousand have taught, and everyone has studied," and "with a month of bitter struggle, we have broken through the barrier of illiteracy." They made other such claims in the style of Great Leap Forward slogans of that time. At the end of that year, Yilong County made an overall report from communes and other local reports that more than 110,000 people had become literate. Given that those figures could not be verified, because the report claimed that the county had succeeded in eliminating illiteracy, the State Council still gave the county an award for "overcoming illiteracy." Soon afterward, because of the mistakes of the Great Leap Forward and the difficulties of everyone's life, this county's movement to overcome illiteracy came to an end.

It should be acknowledged that the movement to overcome illiteracy opened the way to intellectual culture for broad numbers of peasants and served as a first step in changing the basic conditions of backwardness in China's villages.

This article is by Yi Ge. It appears in Old Photos *no. 74, pp. 82–87.*

Spring Fieldtrip for the Students of a Village School

Introduction

This charming description of a local elementary school's spring outing reveals much about life in a remote part of China in 1945. Heyang County lies about one hundred miles northeast of Xi'an, the provincial capital of Shaanxi. Its eastern boundary is the Yellow River as the river flows southward between Shaanxi and Shanxi.

The recollections of Mr. Qiao, the long-retired teacher who organized this school's spring outing, provide several details on popular feeling and local style. Some of the children wear military-style caps that reflect their patriotism as the Anti-Japanese War nears its end, and their appreciation of American support of China. Young male teachers sport canes—in this case made by a carpenter hired by the school. Entertainment includes amateur singing of Shaanxi regional opera and local folk songs.

The Text

This photograph is as old as I am. It was preserved by my middle school language teacher, Lei Lei. Teacher Lei is already quite old, and fortunately I saw this photo at the home of his son, Lei Chunlin. Chunlin is younger than I am, so he was not clear about the content of this photo. I thought the photo was interesting, and I took it with me once to the county seat. From some of the words written on the photo, "Wufu Middle and Primary," I was able to find someone who knew the circumstances. It was fortunate that in continuing to pursue this, I finally located Qiao Zhenhu, a teacher in my primary school, who was able to tell me about the photo in great detail. Teacher Qiao is eighty-three years old, but he is hale and mentally very clear. He said he also has this photo, but the one he has is in color. "Teacher Lei liked art, and he did the color work himself."

The teacher toward the front and left is the principal, Cui Chengcai. Sitting nearby with his legs crossed is Lei Lei, who was the head teacher. The two of them and Qiao Zhenhu had been classmates at Tongzhou Teachers College. In the education system at that time, the principal advertised for teachers, so it was natural that Cui as principal would first consider people he knew. Seated still further to the right is Teacher Liu Haicheng, who played the *banhu* very beautifully.[8] All the way to the right is Teacher Kang Langting, who liked to sing local opera. At that time a district's central primary school would be likely to have teachers with such capabilities. Wufu District lay in the northwestern part of Heyang County, close to Mt. Liang.[9] The district government seat was Huangfu, near the throat of Yellow Dragon Mountain, but the school was located about twenty *li* (seven miles) south of Huangfu, at Fangluzhai, because this place was more or less central and the students could get to school more easily.

Every year after Qingming Festival, the primary schools in the area liked to organize a spring outing for the students. At the

8. The *banhu* is a stringed instrument with a thin wooden sounding board, played with a bow. This is one of a variety of such viol-style folk instruments found in various regions of China.

9. The term translated as "district" is *xiang*. This is a subdivision of a *xian*, "county."

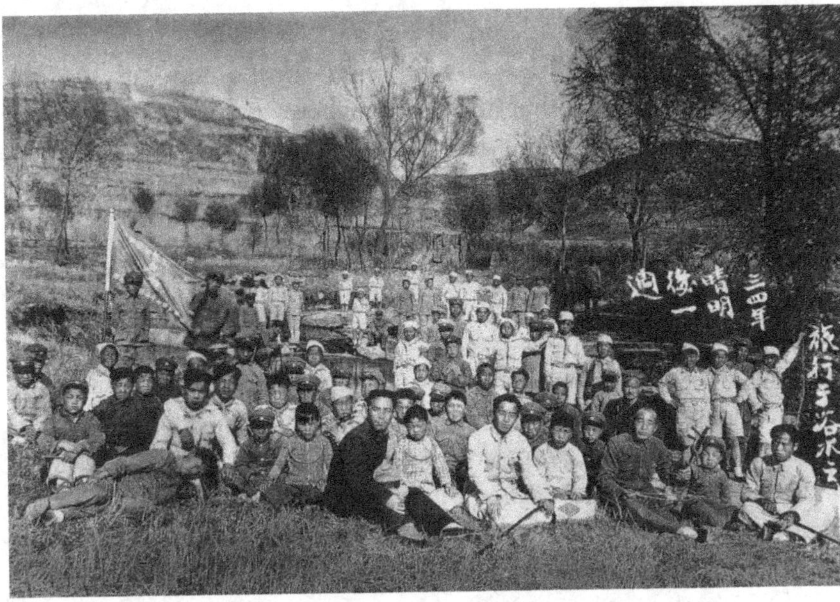

time this was called a "hike" or "tour."[10] Teacher Qiao said he was twenty-four then and was teaching physical education, so the school decided he should lead this spring outing. The destination was to be the Dayu River valley, the river that formed the boundary between Dengcheng and Heyang counties. The place where this photo was taken is called Zhujia (the Zhu Family) River, the only place where a village lay on both banks of the river. It is a beautiful place with a rich history.[11] During Tang times the high official Erzhu Kui chose this place to live as a hermit. Later generations dropped the character *er* from the name and used only *Zhu* as the family name; thus this place came to be known as "Zhu Family River." The village backs onto the hill there and faces the river. It was well-situated in the landscape, with sturdy structures of brick and stone and old scholar trees shading the area. The scenery is most pleasing.

10. Qingming Festival is celebrated in early April, a traditional day for honoring the dead and showing respect for previous generations. Grave cleaning is commonly done during this festival.

11. The historical and geographical information here belongs to local lore. While the people of every society have such lore, people in China seem unusually devoted to it. Perhaps in some places tradition gives weight to fact. On some such questions, there is ongoing questioning and study.

There is a high memorial archway in the village, and a "double-dragon spring" that flows throughout the year with streamlets running close together and a spot where "the cypress embraces the scholar tree." In a line of elm trees in Yubu Garden, there is a gap between the trees that is different from other places. In the river outside the village there's a "Fossil Pool," where pebbles fall into the pool without a trace, and the "Scorpion's Tomb," a hill shaped like a huge scorpion. Set in the midst of yellow earth so dry and treeless, this spot was indeed a fine place for their spring outing. When they finished their walk, everyone sat down on the grass for a while, sang songs, and told jokes. It was all so satisfying![12] As for the students who made the outing from school that day, they were like birds who had been let out of their cages and were happy beyond words. In the photo there is one boy lying down (front left). The teacher chose not to scold him for this behavior.

During those times some students wore children's military outfits, shorts, and "boat-shaped" caps, a good many of them in American military style. This isn't surprising, because the Allied Forces were helping China resist Japan at that time. The thirty-fourth year of the Republic was 1945, and in August it was announced that the Japanese had made an unconditional surrender. Still other students were wearing school uniforms, their flat-top hats decorated with the "blue sky–white sun" emblem of the Republic of China.[13] Teacher Qiao said that students then were very poor, and their uniforms were homemade from coarse cloth sewed after dyeing. Still other children in this photo are quite small. Teacher Qiao said these might be beginning primary school students, or maybe children of the teachers,

---

12. Here the original says Qinqiang, which means "Shaanxi tunes." This term is a strong provincial reference suggesting the local opera and the high-pitched nasal approach to song that is beloved by many Shaanxi people. The renowned contemporary writer Jia Pingwa used this term as the title for a novel published in 2005.

13. This description of the flag of the Republic of China refers to the blue field and white circle with multiple regular points around the circle. This flag is sometimes identified more graphically as the "gear-wheel flag" because this is what the sun feature resembles. Articles in *Old Photos* concerning earlier periods of the Republican era display other flags, often one with multiple stripes representing the major ethnic groupings in China's population. The "blue sky–white sun" flag became official when Nanjing was made the capital in 1928.

who never imagined they would be allowed to go on this outing. To the left of Cui Chengcai sits a small boy, and to his right a small girl—and it's likely this girl was his daughter, because at that time it was extremely rare for a girl to go to school. Both children wear a silver ring around their necks; this was often seen in villages as an emblem of "a beautiful child," and usually only the children of relatively prosperous families would wear these.

Teacher Lei and Teacher Kang both have canes resting beside them. Mr. Qiao said this was a fashion at the time, and he also had a cane. They all were in their early twenties at the time, strong young men with some flair—and you wouldn't really use the cane as you walked along. Back then such a cane was called a "culture staff"; whether you used it or not, it should be in your hand as a symbol of a certain social level. But these were all poor teachers in a village and didn't really have the money to afford a "culture staff." It happened that the school had hired a carpenter, and they had arranged with him to make these canes, so that they looked very good.

Teacher Qiao said the teachers and students stayed at Zhujia River most of the day, and everything had been very relaxed. Finally they decided to take a photograph at a spot with a nice scenic background. He was the organizer of this outing, and he ran around trying to get the teachers and students well arranged for the photo. Finally he found a big rock beside the river where he could squat with the camera and take the picture. He hadn't realized that the layout wasn't quite as good as it might have been. For example, to the right of the school flag there are three of the students wearing white military-style outfits that distract from the flag a bit, and also the angle on the flag is not so good. A couple of the students apparently decided to show off a bit. One right behind Principal Cui looks ready to play his flute, and one farther back and to the right has extended his arms as if he's ready to sing a song for the group. This photo is a good record of a pleasant school outing those many years ago.

This piece is by Shi Yaozeng. It appeared in Old Photos *no. 25, pp. 136–38.*

Is It Worthwhile for Village Kids to Go to School?

Introduction

This reminiscence of school life in a village gives a picture of educational conditions in much of rural China. As the writer himself gradually realized, not every young person who was of school age could actually go to school. A family's economic situation determined which child would get an education and which would simply have to miss out on that opportunity. As is true in most agricultural societies, going to school yields to the need for hands to plant or harvest crops.

The writer almost seems to be musing, "Were we all small because we were from a small place?" Not even the teachers were physically large. Further comments on the teachers indicate another common situation in village schools. Few teachers were fully certified, and the use of "community teachers" was common. It was a continuing goal to get "community teachers" into teachers' colleges, which are the equivalent of two-year institutions in the United States.

Finally, the writer's New Year visit with his uncle reveals contemporary concerns among those who aspire to higher education. The uncle had "done the math" and concluded that the return on education, especially university education, was not worth the investment. For most urban families this calculation is more about status than family finances. Still, the uncle's calculated decision also suggests that higher education remains an issue in China today. Do the high costs of higher education ultimately help a family prosper? Some American families face similar concerns regarding higher education costs today.

The Text

This photo was taken in 1979 when I graduated from primary school, and shows the students and teachers together. It is also the last photograph of my childhood years. It was taken at the photo studio in Daigang Town in neighboring Mengyin County. In order to have this photo taken, we got up very early, and three teachers led us along

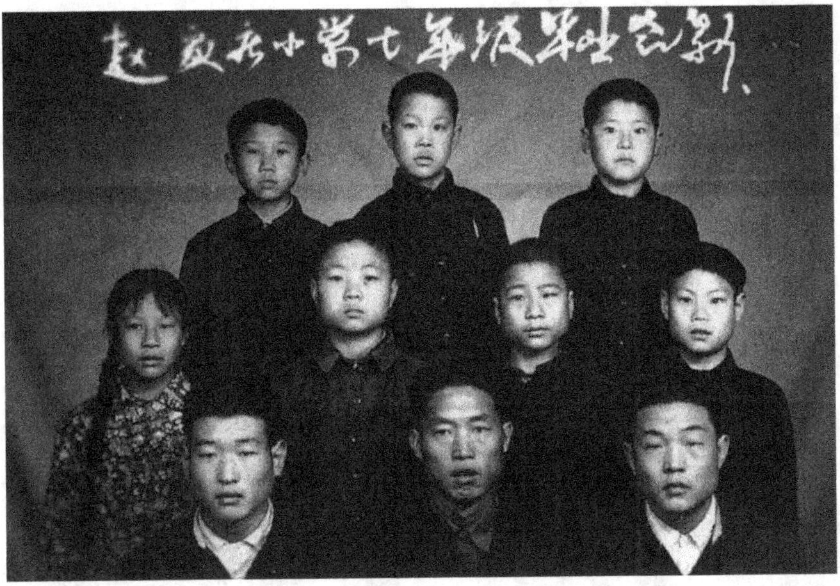

a mountain path and through the hills and the valleys. It was about forty *li* [thirteen miles], and it took us the whole day.

The last two rows in the photo are the seven students from the fifth grade in our school (the writing on the photo has it wrong—it says seventh grade). These were all the students in our grade! The three people seated in the front row are all the teachers in the school. This photo has three points of interest. First, the students' ages cover quite a range. I am the one in the middle of the back row. At that time I had just turned twelve, and I was the youngest in the class. The oldest person in the class was named Qi Jingxing (second from the right in the middle row). He was already seventeen! The second interesting thing is that most of us were surnamed Qi. Aside from two in the back row, one named Yang and the other named Dong, all were named Qi. Although our village was called Zhao Family Village, people with the surname Qi made up 80 percent of those who lived there. The third point of interest in the photo is that the students were small, and the teachers too were small.

Of the three teachers in the photo, the one sitting on the left is Yang Qinghong, and the one on the right is Gan Shaorun. Later, they both examined into teachers' college and went from getting subsidies as community school teachers to being formally qualified teachers who received a government salary.[14] The one in the middle, the oldest, was named Guo Chenghou. Because he was older and his educational level was low, he took the exam several times but never passed. After serving twenty years as a community school teacher, he finally gained recognition as a fully qualified teacher just a few years ago. Later when I graduated from middle school, I examined into teachers' college. After graduating, I returned to my village and taught classes with those who had taught me.

As I look at this photo now, some questions occur to me. Why should there be a class with students this small? Was it because our village was small? Although our village was indeed small, with less than a hundred households, as I recall there were no fewer than ten children born the same year I was born (1968). Why were there only seven students in our class, and not the entire group of children born that year?

Recently my father came to visit me, and I asked him some of these questions. My father said I had recalled correctly that there were more than ten children born that year, but of those who went to school, all were enrolled because the teachers went to people's homes and got the children enrolled. If there were several children in a family, they would let the teacher take one, who then would go to school for a year-and-a-half. Then the older ones would go and join the production brigade to earn a wage. Many of the children were young teenagers when they got their opportunity to go to school for first grade. Only a few were able to stay long enough to graduate from primary school. No wonder some of them never got a chance

14. These teachers came from their local communities and lacked training. They were allowed to serve as teachers in order to fill out the staff in their communities' schools. In the late 1970s, there were several million of these teachers throughout China (a reliable source puts the total at 4,910,000 such teachers for 1977). As early as 1977, policy change set out to reduce the number of these teachers. As life in general improved through the 1990s, the relevant government agencies looked to reduce the number of these teachers to a minimum, so that as of 2000 there were only 250,000 of them still serving.

Note that the Chinese term for these teachers is *minban jiaoshi*, literally, "teachers managed by the people." The term used here is meant to indicate community effort to provide decent schooling in small places such as villages.

to go to school! So it was not strange that our class had only seven students. At that time it was unusual for someone to go to lower middle school, and it was even rarer to go to higher middle school.

When I started the second year of lower middle school, we launched the responsibility system in farming.[15] All the teachers at our school were community school teachers from nearby villages. Before the responsibility system was put into effect, they could draw a salary without participating in burdensome physical labor. It was a job that people envied. But once the responsibility system was started, the situation changed greatly—from getting their food without being involved in the labor system, to receiving a subsidy because they didn't have any knowledge about farming and couldn't farm well. Family receipts didn't match up to the number of people to be fed. Teachers in community schools soon came to be seen as being in a difficult situation.

As things developed to raise the social position of teachers, this also gave impetus to the task of education. In order to improve teachers' salaries, many teachers' colleges offered community school teachers opportunities to examine into college, and in some areas new mid-level teachers' colleges were launched for community school teachers. At one time there was quite a surge in community school teachers studying so they could take these exams. They had lots to do then. They taught their classes as always, they tried to decide the best way to review, and they also had to keep up with their work on the land. Teachers really were busy during that time!

Sometimes the entire school would make it a project to help an older teacher whose educational level was low to make it into one of these colleges. If that person didn't have to teach every day, didn't have to live up to his contract, his study for the exam could be more relaxed. We weren't thinking about the future, which seemed far off. We were just going along day to day, living free and easy, with no idea we were passing through an unusual time that we should

---

15. This new policy in agriculture developed gradually during the late 1970s and was generally promulgated by 1981. This was the beginning of the market system in agriculture. The essential change moved from centrally assigned quotas throughout the country to contracts for each farming family to meet less ambitious quotas than before, with permission to sell any produce exceeding the quota on the open market.

treasure. During that time, everyone did what they enjoyed doing. Success and enjoyment came easily. When I traced over characters in a copybook as I taught the children to write, I "invented" pictures of people in ancient dress. The students would use their writing tools or toys and exchange their artwork with me. As we teachers saw it, nothing could compare with such things to make us happy. School was simply a paradise.

In the seasons when crops were planted or harvested, the teachers would take a few days off, and the students would take turns helping them with farm work. Many families would have the children leave school and return home to help plant. So the group who helped the teachers would dwindle as the days ran out! At school, we had more than one way to skin a cat in such matters.

After a few years more, a good many teachers examined into teachers' college. People finally began to see that for village children, staying in school could be a good way out of poverty and the grind of village life. Many of those who thought it was better to let the children return home and work in the fields now began to see that this was not a good plan. Before this, my parents took me to a key middle school in a town about twenty *li* (about seven miles) from home so I could study. Many of my schoolmates had already wasted precious time when they might have been studying—including my friends in this photo.

This year I returned to my old home shortly before Spring Festival. In talking with my Third Uncle, I learned that his son Anjun had graduated from lower middle school and then gone off to Jinan to study motorcycle repair. Anjun was the only child of this uncle. Although I thought I was prepared for whatever news he might have about Anjun, I still was amazed. Why didn't he continue his studies? They reasoned that school fees were just too much. They couldn't get that much money together. My uncle said, "If a family has two children they simply quit school and don't continue, even though they think it's good when a family does as yours did and graduates from school."

My uncle sighed and gradually laid out the figures for me. "To continue through three years of higher middle school, the tuition, cost of books, and living expenses would add up to three times the substantial amount of three thousand *yuan*, which would be the large figure of ten thousand *yuan*. Then if you examined into a university, this would be another three years at least, and if you add up all those expenses, the yearly average would be another large figure and three times that for three years! Then after graduation you would be responsible for finding a job on your own, buying a house, getting married—and it would be several more large figures, several times ten thousand!

"If your aunt and I work hard for a year, we can't make three thousand *yuan*, so would you say it's worth it to keep studying? Really, it isn't worth it! It's a complicated thing. School is good and students strive to be at their best. But it remains true that teachers don't even get paid every month, and they have to struggle to make ends meet, so do you still want to be a teacher? In schools like this, it just doesn't accomplish anything for you! And it does make sense to spend a few thousand *yuan* to learn a trade. You make enough in a year to take care of everything, and can still help me with the farm work."

Would you say that my uncle was short-sighted? My classmates from that time are now farming in the village. They have orchards, they go off to drive a truck, and they're prosperous. Some have been making ten thousand *yuan* for a long time. So what might I say? Really, there don't seem to be any good responses to my uncle's argument.

This article is by Qi Xinjun. It appeared in Old Photos *no. 12, pp. 97–100.*

Going to School During the Cultural Revolution

Introduction

The writer's reminiscences of his school years, which spanned those of the Cultural Revolution, reveal much about that period at an everyday level in an important aspect of life. As noted even in the opening section, it was difficult or even impossible to tell where education ended and politics began. These were the conditions of that turbulent period.

In recalling his experiences then, the writer adopts a casual approach and sees the humor—and the irony—in many of the situations of his school years. From the very beginning, however, note the importance of being identified as "poor peasant." To fail to get this classification could mean problems for a family at any time.

This selection also provides a sense of how the popular culture of the Cultural Revolution reached China's younger generation. In homes and public places throughout the country, radio was broadcast for the consumption of all and difficult to tune out. Often, a "contemporary Beijing opera" would be a big part of the program. Jiang Qing, Mao Zedong's wife and the major figure in the "Gang of Four," fostered the updating of this important feature of Chinese culture. Many would regard this as a positive contribution, as many of the tunes and characters from these stories remain popular up to the present.

The Text

Donglin Primary School Takes on a "Red" Name

In 1966, I started to school in the first grade. At that time I was living with my maternal grandmother in Wuxi, Jiangsu. The place where I was studying was the Donglin Primary School. This school was on the site of the Donglin Academy of Ming times, the center of an influential group of scholars. I remember that on a wall in a certain place at the school four characters were carved about a meter high: "Donglin Academy."[16] School had not been going very long when introduction of new revolutionary names began, and Donglin Primary School became Little Red Soldier Primary School. In language class we studied the phonetic symbol system known as *pinyin*, but later we only studied "*bo, po, mo, fo,*" and the rest of the phonetic system. Then we switched to study the quotations of Chairman Mao.[17]

---

16. The Donglin Academy arose in the early seventeenth century as a philosophical movement opposed to the influence of Wang Yangming (1472–1529), a free-thinking Confucian scholar whose influence the Donglin group sought to combat. The Donglin group inevitably became involved in Ming court politics, where they never developed great influence. Their center was indeed in Wuxi, about eighty miles west of Shanghai.

17. The "*bo, po, mo, fo* system," as it is often referred to, arose out of the efforts to develop a usable sound syllabary earlier in the Republican period. See "The Sound Syllabary" in this collection.

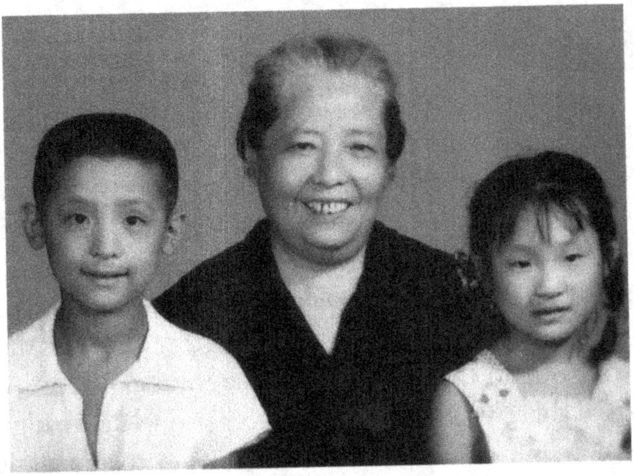

Photo 1: The author and his sister with their maternal grandmother in Wuxi, 1966.

It happened quite a few times that we would arrive at school in the morning to hear the fifth-grade students tell us, "Today the teachers will have a struggle meeting and we won't have class." So we first-graders were very happy to go back home and play.

Quotations from Chairman Mao

My grandmother had a small copy of *Quotations from Chairman Mao,* its red plastic cover embossed with those words. My parents sent me an edition twice as large, with the characters for *Quotations from Chairman Mao* in gold print and an image of Chairman Mao on the front cover. Grandmother was covetous and we talked about it several times. She wanted to borrow my large edition to take to the Neighborhood Committee's study sessions, but I never said okay. She brought it up one more time and I said, "It was Father who gave it to me." Grandmother got very angry. She pointed to the padded cotton jacket I was wearing and said, "I bought this for you!" I took the jacket off, and she immediately put it back on me—and she never brought this up again. I sometimes went with Grandmother to these neighborhood study sessions, and a group of old women would sing songs using the quotations. One was "We support all that our enemies oppose, and we oppose all that our enemies support." Over and over they would sing "*diening*" for *diren*, "enemy," "*yongwu*" for *yonghu*, "support," and *"woli"* for *women*, "we"—all this is Wuxi dialect.[18]

Chairman Mao Badges

Grandmother and I sometimes went to visit a cousin of hers. Once when the cousin took me to the vegetable market, I had a nice round badge of Chairman Mao, gold in color, pinned to my chest, and another kid about my age came along and snatched it. The market was crowded. I spent a while looking for him but never found him. I was so mad that I cried, but there was nothing I could do! In those days there was a saying that to steal a Mao badge was not a crime, which must have meant this was a way to show one's fervent love for Chairman Mao.

The cousin had a nice Mao badge too, a square one with an image of his face on the left and a book on the right, on which was written "Serve the people." It was painted with bright red lacquer and was even nicer than my badge that was pilfered. The cousin's daughter had given her this great badge, and normally she wouldn't have been wearing it. The cousin gave me the badge and I finally stopped crying. We went on home together.[19]

"Chocolate" and "*Doufu* Flowers"

Primary school students didn't have class very often, so I often went out for a stroll. Once I found a couple of dollars that my grandmother had put under the mattress. I went out and bought a small piece of chocolate and a bowl of jellied bean curd [*doufu nao*], which is called "bean curd flowers" [*doufu hua*] in Wuxi dialect. I also bought a few small snacks. There was more than a dollar left, and I went home and

18. Regional dialects in Chinese feature such differences in pronunciation or usage as shown here. Native Chinese people adapt to these differences in a matter of weeks. This adjustment is more difficult for foreigners learning the language, who must focus on learning standard pronunciation. Today the dialects are diminishing somewhat through the influence of nationwide television programming and other aspects of popular culture, while at the same time individuals might choose to watch only a regional television station mostly using dialect.

19. Mao badges were produced in a vast range of forms, done by a great variety of institutions and groups. Collectors of these badges in China and abroad are still engaged in sorting out the variety of this activity during the Cultural Revolution years.

gave it back to my grandmother, but she still paddled my behind. So the neighbors made a joke of this, saying that I had founded a "chocolate squad" or "bean curd flower advance group."

Checking Your Background When Registering for School

When the time came for my second year of primary school, my parents were afraid that my maternal grandmother wouldn't be able to control me. So they took me to the countryside near Changzhou (still in Jiangsu Province), where I lived with my paternal grandfather and grandmother. My school at this time would be in Wujin County at the Ben'niu Commune, Yellow Lane School in the Western Woods Brigade.

The school was located in a temple that no longer had any Buddhist images. My cousin took me to school to register, and when the teacher asked, "What is your class background?" my cousin broke in and said, "Poor peasant, poor peasant." Only later did I learn that my grandfather actually was a middle-class peasant. There were no landlords in the village, and there was one rich peasant, who suffered extreme discrimination.

At school we were taught a song named "I'm a Good Descendant of a Poor Peasant Family." Regardless of their actual background, all of the students sang these words: "I'm a good descendant of a poor peasant family. I'll take to heart all the Party's teachings, follow in the footsteps of earlier generations, and hold high the Red Flag of revolution. . . ."

The Report Card

I studied at Yellow Lane Primary School through the fourth year. I have kept my report card from the first semester of that year, a rather small sheet of paper printed by stencil. At the top of the left side are the words "Highest Directives," which means a quotation of Chairman Mao: "Our educational policy is to achieve development in all of these aspects: moral, intellectual, and physical education, and to make students knowledgeable workers with socialist consciousness."

Then, there is a letter to parents:

> The Proletarian Cultural Revolution rolls along in a great tide, thoroughly smashing the revisionist educational system. With the unbroken successes of Mao Zedong Thought brightening the cultural educational front, a brand new proletarian education system is now taking shape. Under the management of our poor and middle-class peasants, a generation of new people is growing strong and healthy through the nourishment of Mao Zedong Thought. . . .

The seal of "The Revolutionary Education Committee of Western Woods Brigade" is printed, and the date shown is June 30, 1970. On the right half of the page is the teacher's evaluation:

> During this term he has held high the great red banner of Mao Zedong Thought, and during after-class time he has been busy in revolution and has progressed rather quickly. He has been able to lead the propaganda activity of the Little Red Guards in a positive way. He has also made positive progress in his studies and has participated in criticism in a positive way. . . .
>
> Language arts: Outstanding. Mathematics: Excellent.
>
> Things to pay attention to during the summer vacation: (1) Resolutely study the works of Chairman Mao. (2) Participate in Little Red Guard activities in a positive way. (3) Work hard in physical exercise and play in appropriate recreation.
>
> Opinion of the head of the household (filled in by my grandfather): At home, Zhang Xiaoliang is not very modest. He is still proud and impetuous. He is not very diligent in his tasks at home.

Along with my cousins and friends in the village, I carried a signboard with a quotation from Chairman Mao as I went to school. Grandfather made straw sandals for my uncle, and I begged him to make some for me too. So Grandfather wound strips of cloth together and made me a pair of "straw sandals." When school was out each day, my classmates lined up based on village and shouted out a slogan before going home. Seeing that I was wearing "straw sandals," the teacher praised me for living simply. This must have been the source of the comment in my report card that ". . . he has been able to lead the propaganda activity of the Little Red Guards in a positive way." On the comment that I had "participated in criticism in a positive way," I recall that I wrote a couple of small criticisms commenting on ideas that Liu Shaoqi had allegedly put out in his articles entitled "Reading Books to Become an Official" and "Reading Books Is Useless." If you can become an official by reading books, how can you say that reading books is of no use?[20] At that stage of my life I didn't give this much more thought. Ultimately, everything I wrote was from what our teachers taught us and what was passed on from the adults in my life.

Studying Singing and Performing

Every home in the village was wired for radio, and broadcasts were run three times each day—morning, noon, and evening. The texts and songs were from the eight model plays and were kept on file to be used again.[21] I remember at noon one day the broadcast was a play named *Sha Family Creek*. Grandma Sha would sing a line, and all those listening could repeat it for practice.[22] I thought this was new

20. Liu Shaoqi (1898–1969) was Mao Zedong's assumed successor as the Cultural Revolution began. He was known for writing a pamphlet entitled "How to Be a Good Communist," a likely source for the comment mentioned here. Liu fell out of favor with Mao. His background had been in labor work and urban revolution, which led him to prefer "expert as well as red" rather than "red over expert" as appropriate for China's ongoing revolution. Liu died in 1969, having been purged from the Party and hounded to death.

21. The eight model plays were a project of Mao Zedong's wife, Jiang Qing. The *Sha Family Creek*, *Red Lantern*, and *Taking Tiger Mountain* mentioned in the article were among these eight model plays. At the height of the Cultural Revolution, these were virtually the only plays allowed to be performed in China.

22. A translation of this play appears in Lois Wheeler Snow's *China on Stage* (New York: Vintage, 1973) as "Shachiapang," pp. 115–90 [pinyin *Shajiabang*]. This and the operas, ballets, and movies mentioned in the text were among the works authorized for public performance during the Cultural Revolution. A popular saying at the time was "eight operas for eight hundred million people." A discussion of the appeal of these artistic works, even to the present, is Barbara Mittler's *A Continuous Revolution: Making Sense of Cultural Revolution Culture* (Cambridge, MA: Harvard University East Asian Monographs, 2012). Also see Xiaomei Chen, *Staging Chinese Revolution: Theater, Film, and the Afterlives of Propaganda* (Columbia U. Press, 2016)

and different. I waited for the broadcast to end, went directly and climbed up on a stool, and facing the radio cabinet I continued to run through a few lines from the play. I thought about this later and wondered if the radio had another head that heard me practicing in this way.

One New Year, my grandfather gave me five *mao* (fifty cents) as a New Year gift. I went to the New China Bookstore in Ben'niu Town and bought the comic-book versions of *Red Lantern* and *Taking Tiger Mountain*. When I finished reading them, I loaned them to my cousins to read, and then to other buddies in the village. In feeding pigs and other farm activities I wasn't really one of them, but these two comic books linked us in good feelings.

Father Joined a "Rebellion" and Suffered

Father graduated from higher middle school in 1948. He joined the People's Liberation Army in 1949, and over the years he served in the Southwestern District Air Force Command's Office of Military Tactics, the Wuhan District Air Force Command's Office of Military Tactics, and as a staff officer in the Wuhan Military District Command's Military Tactics Division. During the Cultural Revolution, following along with a second-level leader of the Wuhan Military District, Father joined in some "rebellious" activity. The result was that he was jailed by the Wuhan Military District commander, Chen Zaidao, for half a month and then was sent to the District's "May Seventh" cadre school for labor. Father sent Grandfather a letter saying he had been jailed and he sent a bag of peanuts—his product in the cadre school farm. This was during 1969 and 1970.

In 1961, 1962, and 1963, Father was honored three times with a third-class award of merit, which is not easy to get during peacetime. But after he was involved in "rebellion" during the Cultural Revolution, Father's career began to decline. He became terribly depressed and died at the age of fifty-two.

Self-Criticism Summarized as Eight Flaws and Six Strengths

By the fifth year of primary school, I went to the primary school attached to the Wuhan Hydroelectric Institute to study. I have kept my report card from that year. The "Highest Directive" at the beginning was as before: "Our plan for education is to cause those being educated to achieve development in moral education, intellectual education, and physical education, and to become cultured workers with a socialist consciousness."

> Achievement: Politics (see the appraisal); Language: outstanding; mathematics: good; everyday knowledge: outstanding; labor (see the appraisal); military and physical education: good; revolutionary arts—music, good; art, outstanding.
>
> Appraisal of the class committee and teachers: he is able to actively study and apply the works of Chairman Mao, and positively "recite and apply." At times he needs to struggle with bad people and bad behavior. He is relatively good at labor, and he joins physical discipline in a positive way. He respects the rules of the classroom, and he conscientiously completes assignments in all his classes. During the first evaluation period for this term he was praised in his company (note: at that time the school began militarizing its organization, the company was your grade level, the platoon was your class, and your team was the small group). We hope he will continue his active study and application of the works of Chairman Mao, will be a good person and engage in good activities, will be open-minded in listening to the opinions of others, will strengthen revolutionary unity, and will move ahead in elevating self-consciousness for carrying out revolutionary discipline.

In the individual and small group parts, I applied three of the "highest directives": "the need to carry out self-struggle, criticism, and correction." "Study well and advance daily." "Conscientiously summarize your experiences." Here are some of my eight shortcomings. "If a classmate hits me, I hit him back. I'm not always kind in my speech. Sometimes I arrive late, and don't turn in my assignments on time. . . ." And then there were six good points. "Every day I persistently read from three important things (Mao badges, *Quotations from Chairman Mao*, and *Chapters from the Heart of a Little Red*

Guard). I do quite a few good things, and I try to bring to light bad people and bad events. I study and apply (the Works of Chairman Mao) actively. . . ."

Performing Selections from *The White-Haired Girl*

Our fifth-grade textbook included selections from the ballet *The White-Haired Girl*, and our teacher organized the classmates to perform this as a way to improve the effectiveness of teaching and learning. Dachun and other major roles didn't go to a newcomer like me, just in from the countryside. The roles of Huang Shiren and Mu Renzhi went to a couple of mischievous classmates who asserted themselves, and I was given a supporting role as a militiaman. I saw that the classmate who played Huang Shiren wore his father's wool overcoat to show that Huang was wealthy. And I got a piece of black cloth from the drawer of my mother's sewing machine and sewed it to a white shirt, to reflect the hard labors of a liberated peasant. These details weren't suggested by our teacher but depended on our own enthusiasm as the performers. What we performed were excerpts. We didn't do the part of the story in which Yang Bailao gives Xi'er a length of red cord to tie up her hair. The classmate who played Yang Bailao actually didn't have anything to say. He just came onstage and died.

"How many hired hands were oppressed, how many Xi'ers were enslaved . . . there is no way to express our hatred! Let the rivers and oceans flood together into mighty billows reaching to the sky! They can't resist our flames of fury. We must destroy the darkness of the old society!" We performers went up on the two wings of the stage, and the body of Yang Bailao, just now dead, is seen. Xi'er comes on stage, sobbing: "Daddy, Daddy! What a tragic death. . . ." The Eighth Route Army comes, and the top landlord with another classmate and I as two militiamen come onstage, taking Huang Shiren and Mu Renzhi into custody. These two bad eggs kneel on the floor, trembling but also smiling furtively. Wearing my "patched" clothes, I look majestic with my red-tasseled spear, standing behind these two.

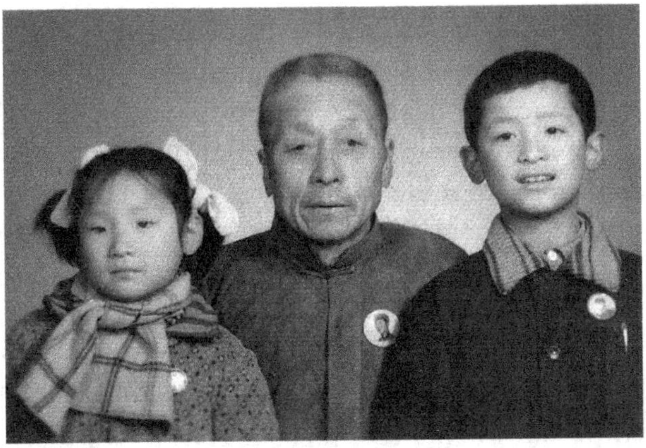

Photo 2: The author and his sister with their grandfather in Changzhou, Jiangsu, 1968.

"Speaking What Is Useful" and New Expressions

In fifth grade we often had meetings to "study and apply." The full title for this was "Meeting to report on studying and applying the works of Chairman Mao." Some students would say that after studying Chairman Mao's works, on the way home from school they had helped someone push their cart or helped an elderly person cross the street. Everybody spoke this way, and after all that talk, such statements became worn-out sayings without any new meaning.

So what should we do? I asked a neighbor girl who was three years older than I, and she enlightened me. This was done to stimulate struggle in thought. If you see someone pulling their cart up a hill, you could envision two kids who would disagree about what to do. One would say, "Let's go right now to help push," and the other would say, "Let's go home and do our homework." At that time, a strong voice beside me would say, "Serve the people with all your heart and all your will." Or, "It's not hard to do a good deed for someone." So I went right away to help push the cart.

Jadeite, Night Soil, and Straw Rope

Our school had a paddy field, and the three classes of fifth-graders took turns working there. I wrote a composition on the silky green paddy field called "A Jadeite Mosaic on the Great Earth of our Country."

Once another classmate and I were sent to fertilize the paddy field. We took the bucket and the gourd ladle and went looking for night-soil behind Classroom Building B. We lifted the cover of the tank that the toilet system drained into but didn't find anything. Just as we decided to leave, we heard the sound of a toilet flushing in the building. We went straight back to the pipe with our gourd ladle and came up with some "goods." We went back to the road and made up a ditty, "without the stench of night-soil, there can be no aroma in the paddy-field," to encourage ourselves as we went along.

The crop in the paddy-field ripened in fall, and after it was cut we needed a way to move it. During my three years in the village I had learned how to twist straw into rope, so I volunteered to make straw rope to bind the crop. The teacher saw that I was a "technological talent" and asked me on the spot to teach my classmates how to make straw rope.

The Influence of Movies

In the three years in the village I only saw a movie twice. Once was a documentary film of the National Day celebration in 1969. It showed the military attaché from the Albanian embassy endlessly brandishing a red-covered copy of *Quotations from Chairman Mao*. All of us kids were fascinated with his military hat and his big nose. The next day we heard that the neighboring brigade would show a movie. All of us went over to see it, and it was the same documentary.[23]

After I went to the city, I got opportunities to see a movie every week. The Hydroelectric Institute had an open-air theater. It was five *fen* for admission, and you took your own little stool to sit on. In order to have more of an audience, a silver curtain was put up in the middle of the floor, so the audience could sit on either side to see the movie. Before the feature film there had to be a "News Briefing." In these there were lots of images of Chairman Mao, usually receiving guests from abroad. All it took was for Chairman Mao to appear on the screen, and the audience would begin to applaud. In the cartoon *Cockcrow in the Night*, in order to hurry the hired hands

23. By the time of the Cultural Revolution, Albania was China's only ally among Eastern bloc nations. The long-standing relationship between China and the Soviet Union ended during 1959–1960 over disagreement about how to continue the Marxist revolution. Albania chose to follow China's example rather than that of the Soviets, and thus held a special place in Chinese esteem.

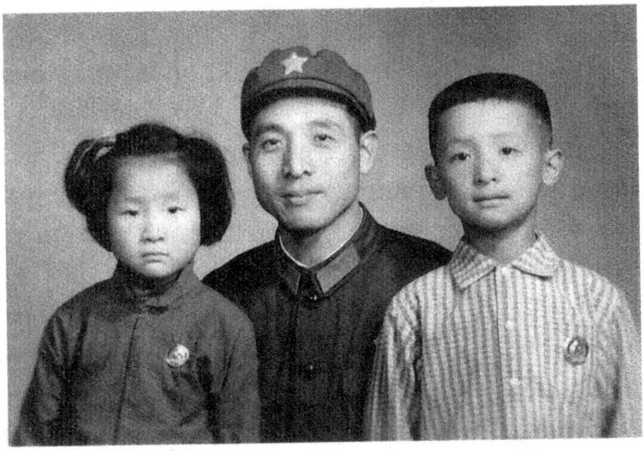

Photo 3: The author and his sister with their father in Ben'niu Town, Changzhou, 1969.

to start work earlier, the landlord Zhou Papi took note as to when the rooster crowed in the middle of the night, and then Gao Yubao figured out how to give the landlord a painful beating. Not long after this an English-language version of *Cockcrow in the Night* was made, probably in order to help university students in their study of English. Up to today, I remember Zhou Papi, big stick in hand, going into the workers' room and beating on their quilts, saying, "Lazy bones, lazy bones."

The Poor and Lower Middle-Class Peasant Lights a Lamp

During my first and second years I attended the middle school attached to Wuhan Hydroelectric Institute. The school organized students to go to a village in the suburbs for labor. We took our own bedding, and the group of five or six students slept in the main room of a peasant's home. At the time, the way of categorizing peasants still wasn't very specific, and according to "class viewpoint," these were called "poor and lower middle-class peasants."

One of our boys got up during the night to go to the bathroom. He couldn't locate the chamber pot, and he couldn't find the light switch either, so he yelled, "Poor and lower middle-class peasant, poor and lower middle-class peasant, turn on the light!"

Dancing "The Waterwheel"

Again in the first year of middle school, the teachers chose four boys and four girls from each class group, planning to do the dance performance called *The Waterwheel.* I don't know why I was chosen. This dance had four male Liberation army soldiers and four female commune members, dancing and singing: "The waterwheel is turning, *yo-wei.* The waterwheel is turning, *yo-wei.* Tread the wheel, tread the wheel. I feel happy, happy!" Our pant legs were rolled up to our knees, and with white kerchiefs around our necks, we moved our bare feet as if we were treading a waterwheel.[24]

When we gave a repeat performance, the parents came to see it. My mother said the way I danced looked like a wooden puppet. At that time, boys and girls didn't talk to each other much at all. This was called "keeping a clear line of demarcation between boys and girls." In the performance there is an action expressing friendship between Liberation army soldiers and commune members by shaking hands. We did this by brushing past each other, but we weren't sure about this. At that time the school had an army propaganda team with troops stationed there. We did this performance to show our support for the troops, and before the show they loaned us four genuine military hats to wear. One of our male classmates said to everyone, "This performance today is for the soldiers. We want to do the handshake well and really shake hands."

"Class Education" by a Brigade Secretary

During my second year of lower middle school, the school sent several students from each class to a village in the suburbs to make investigations of society and help with the writing of a history of the village. Our supervising teacher recommended me because I could "both write and draw."

The Party Branch Secretary at Mo Mountain, Secretary Cai, spoke to us. Before Liberation, a poor peasant was leading his cow past the landlord's gate, and the cow let loose a pile of manure. Secretary Cai said this represented the rebellious spirit of the peasant. To show this I made a picture that I titled "Releasing a Load of Manure When Passing the Landlord's Gate." I made the landlord's gate red

---

24. The kind of waterwheel referred to here is a treadle about the size of a bicycle wheel that moves a line of relatively small horizontal containers filled with water. Other kinds of waterwheels used in China might have similar pedaled mechanisms.

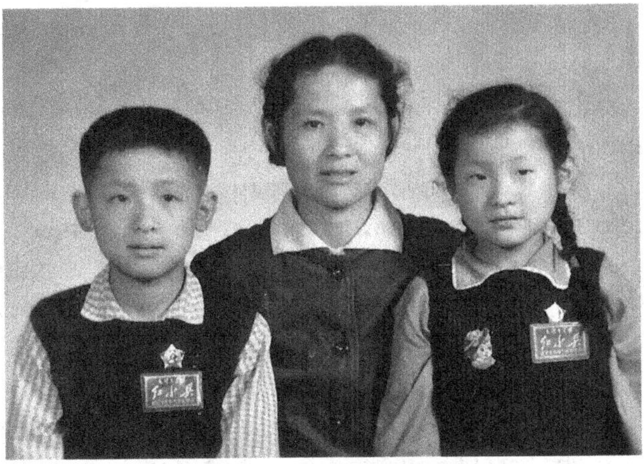

Photo 4: The author and his sister with their mother in Wuhan, 1971.

to correspond with a line in a famous Tang Dynasty poem, "Wine and meat rot behind vermilion gates."[25] Secretary Cai also told us how, after Liberation when a landlord had lost his tiled roof to poor and middle-class peasants, his house then got a straw roof. So the landlord wrote a New Year couplet for his door: "When a great rain came, we had the bottles, the jars, and the pots. When the sun came out, we had eggs from chickens, ducks, and geese." The first line says that the straw roof was not waterproof. One had to use all these containers to hold the water on a rainy day. The second line says that when sun came out, one could see through the broken roof with holes ["eggs"] in a variety of shapes. Secretary Cai pointed out that this is the new direction in class struggle. The old landlord used this to criticize the new situation of villages under socialism.

Studying Workers, Peasants, and Soldiers

On May 7, 1966, Mao Zedong wrote a letter to Lin Biao, which later became known as "The May 7 Directive." An important part of it said, "Students are also this way. They must make study the main thing, but they also need to study other kinds of things. That is, they shouldn't just study the arts and humanities, they also need to study workers, peasants, and soldiers, and they need to learn how to criticize the capitalist class." Many times during my elementary and

25. This refers to a poem written by Du Fu (712–770). The full line reads "Wine and meat rot behind vermilion gates while at the roadside people freeze to death."

middle school years, I joined in the study of workers, peasants, and soldiers.[26]

In studying soldiers, I participated in firing live ammunition and fired one shot.

In studying workers, I spent three months in the technological reform group of the workshop in the 401 Boiler Factory in Wuhan.

In studying peasants, the longest stint was during the first year of higher middle school, when I spent a full term. For language class we interviewed poor peasants and wrote essays on their situation. In math class we surveyed small reassessment situations. We didn't have physics or chemistry classes. These were replaced by work that brought basic knowledge in industry and agriculture.

When the term was over, all the students in our class sang a rousing rendition of the major song in the movie *A Decisive Break* called the "Communist Labor University Loyalty Song." "The flowers of our university open into red petals, the revolution in education sets a red wind in motion. Work and study hard, join the vanguard of the Revolution for Three Great Goals."[27]

Attend Class, Take Exams—and Wield the Silver Needle

"The thousand-year iron tree has put forth blossoms, the withered bamboo has put out new shoots. For the first time ever, mute children have been able to speak. . . ." This song, "The Thousand-Year Iron Tree Has Blossomed," supported the effort during the Cultural Revolution to treat deaf-mute children with acupuncture so that they would be able to speak. At one time, it seemed that acupuncture could deal with almost any medical problem. During higher middle school, physiology was started as a health class: the breathing system, the digestive system, and others, all done superficially. The reproductive system was done through self-study. The teacher only lectured on acupuncture, and did this well, showing how you actually use the needle. For the examination you could get a maximum score

26. Workers, peasants, and soldiers were designated the leading classes in the Chinese revolution, and thus models. See below for another aspect of their role.

27. The "Three Great [Goals of] Revolution" were set forth early during the Cultural Revolution years and were "class struggle, struggle for production, and scientific experiment."

for hitting ten acupuncture points. I was a member of the study committee and thus took the lead in doing acupuncture. I worked with another classmate and got the point correctly on the first try. When we got one of the points, we raised our hand to report. The teacher came to inspect, and he would say, "Swollen! Numb!"

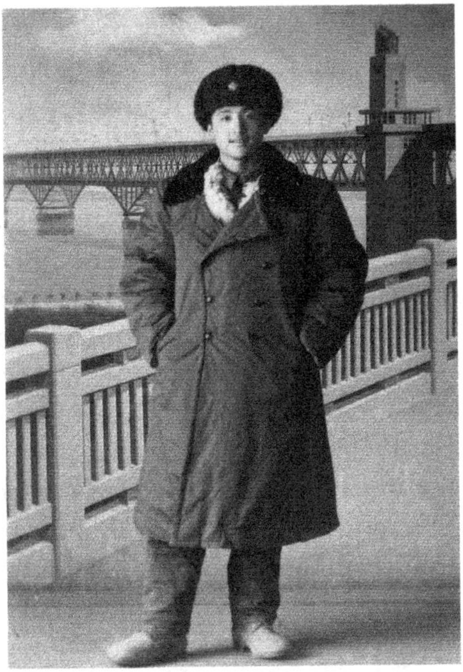

Photo 5: The author in a photo taken at the photo studio on his army base in Qinghai, 1979.

Imitating the Criticisms of "The Gang of Four"

In 1976 the "Gang of Four" was brought down. Our school carried out parades and activities as set forth by higher levels. Every class lined up, carried placards, and shouted slogans. Our first-year class decided to do something different, so we sought out classmates to take the roles of the "Gang of Four" and two others who became Liberation Army soldiers who escorted them to the front of our classroom. Zhang Chunqiao was easy to do, wearing a Sun Yatsen suit and a pair of glasses. Wang Hongwen and Yao Wenyuan were distinguished in our choice of a tall, thin classmate and a short fat one. No one wanted to be Jiang Qing, but a boy classmate finally volunteered. He borrowed a black skirt, used black yarn on his head for her hair, donned a pair of glasses, and wore a pair of high galoshes as leather boots. But his outfit didn't cover his entire body, and his young man's hairy legs were readily seen. This brought lots of laughter from the audience.[28]

28. By the end of the Cultural Revolution, these members of the "Gang of Four" were well known and ridiculed in private, if not publicly. Even coming close to presenting

Worker-Peasant-Soldier Students Didn't Know "Yang Guifei"

In 1977 I did my second year of higher middle school at Hongshan Middle School in Wuhan. A group of worker-peasant-soldier students visited our school to observe.[29] At the time, the fifth volume of *The Selected Works of Mao Zedong* was published, and everybody had a copy. The school called on us to read the whole thing and to read certain parts intensively. One of the worker-peasant-soldier students read before the group from Mao's article, "Criticizing Liang Souming's Reactionary Thought," the following: "Liang Souming, what have you achieved? In your entire life, what have you done for the people? There is not one little bit, not the smallest thing. Yet in your writings you have made it out that you are the most beautiful person under heaven, more beautiful than Xi Shi, more beautiful than Wang Zhaozhun, and even more beautiful than Yang Gui*ji*." [This last character should be read *fei*, not *ji*. The reading error indicated that the student did not know Yang Guifei (719–756), an imperial consort whose name has been a synonym for "beauty" in China for centuries. Ed. note][30]

Jiang Qing, Mao Zedong's wife and the gang's chief leader, was sufficient to draw laughter as well as great relief. Many such demonstrations were held throughout China.

29. University entrance examinations had been suspended in 1966, and no new students were admitted. Instead, about eighteen million university students and aspiring middle school students were sent to the countryside to be "reeducated by poor and lower-middle class peasants." By the early 1970s, authorities looked for a way to reopen universities, and in 1972 several of China's major universities accepted their first classes of "worker-peasant-soldier study personnel" (*gongnongbing xueyuan*). This policy for recruiting incoming university students continued until 1977, and university entrance examinations were reinstated in 1978. The lesser level of preparation is reflected in the writer's observation that the student at his school mistook the last character in the commonly known name of Yang Guifei, beloved concubine of the Tang Dynasty Emperor Xuanzong (reigned 712–756), who was forced to commit suicide after the emperor's attachment to her made possible the An Lushan Rebellion of the middle eighth century.

30. This fifth volume of Mao's works was published after Mao's death in 1976. It had been many years since the previous volume was published. Liang Souming (also pronounced "Shuming") had built a reputation as a Confucian traditionalist over a long career. In many ways Liang was an ideological opposite of Mao, as Mao's castigation of him suggests. But Liang had spent decades working in the rural reconstruction movement, so he shared basic concerns of Mao and the Communists. The major provocation for Mao's break with and attacks against Liang began in 1953, when Liang spoke at a meeting of the People's Political Consultative Conference to oppose Mao's plan to adopt the Soviet model in dealing with China's rural development. Mao's attack on Liang was as emotional as it was rational. This background is a major theme in the

"Watchdog for the Revolution"

At the end of 1978, I put on a soldier's uniform and rode a freight train out to the Northwest. I had three years of military life in the sands of the Gobi Desert. During that period, I heard an anecdote relevant to the Cultural Revolution.

During the Cultural Revolution the words "cow" and "dog" took on new political meanings. "Cow" was a positive image—to bow your head and willingly become a cow to serve children, an extremely hard-working old yellow cow.[31] "Dog" was a negative image—you were a lackey of imperialism, or capitalism, or the landlords, and you might be beaten as a dog that fell into the water.

In the army we had meetings to discuss and use terms from the works of Chairman Mao. The next speaker picked up from the last of the previous speaker's words to elaborate further. "Be a good soldier for Chairman Mao, be an old yellow cow (i.e., a diligent worker) for the revolution." On one occasion, someone decided to change this a bit, so he said, "Let's be running dogs for the revolution." Those listening off to the side looked puzzled, but because he put "for the revolution" in his analogy, they gave him a good round of applause.

This is by Zhang Xiaoliang. It appeared in Old Photos *no. 85, pp. 135–50.*

excellent biography of Liang Souming by Guy Alitto, *The Last Confucian: Liang Shuming and the Chinese Dilemma of Modernity* (Berkeley: University of California Press, 1979).

31. This comes from a line in a poem written by Lu Xun (1881–1936). The whole poem line reads, "Fierce-browed, I coolly defy a thousand pointing fingers. Head bowed, like a willing ox I serve the children."

NINE

Culture

Early Progressives in Chengdu

Introduction

Chengdu is the capital of Sichuan Province, in the western part of "China Proper." This is a historic city, its people more closely attuned to national affairs than the province and city's distance from Beijing would suggest. This photo is from 1925, almost a decade after *New Youth* and other periodicals began publishing and after the New Culture movement began in Beijing. The interests and activities of these progressive young people reflect the influence of these currents of thought and action in Chengdu.

The name Ba Jin appears here, as a friend of one of these early progressives. Like some in this Chengdu group, Ba Jin was greatly influenced by anarchism and other progressive thought. He began with his own experiences in a traditionalistic family and the larger setting of this provincial city. Readers may gain deeper background on family and city life in Chengdu in the recent fine book by Kristin Stapleton, *Fact in Fiction: 1920s China and Ba Jin's* Family (Stanford University Press, 2016).

Like many other of the stories translated here, this one demonstrates the tragic historical truth that being "progressive" in one period of time did not mean a person would be seen as progressive at a later time. Someone who was a "progressive" in the teens or twenties was often seen as "bourgeois" and "right-wing" during the Cultural Revolution.

The Text

This photograph was taken in Chengdu in 1925. After the May Fourth Movement, the six people in this photo all made contributions to culture or education in Chengdu. Among them, four are mentioned in the "Story of Hair-Cutting" that I saw in *Old Photos*, Number 5. Two of the young women from that story are shown here with their friends. My comments here might be of assistance to those studying women's liberation and other topics.

Here is more about each of the people in this photo, starting from the left:

First on the left is Chen Zhuying, born in Chengdu in 1903. Her father, Chen Huanpu, a *juren* of late Qing times, served as education administrator of Luojiang County, Sichuan (the same position as head of the education bureau today).[1] Zhuying was his second daughter, most beloved by her parents. When the influence of the "New Culture" reached Chengdu, an expert started a class in the "sound syllabary" at the former regent's palace in Chengdu, and Chen Zhuying joined this class. She met a number of progressive young people there. They responded to the appeal of the various new periodicals such as *New Tide* (*Xin chao*), *New Youth* (*Xin Qingnian*), *Young China* (*Shaonian Zhongguo*), and *Liberation and Construction* (*Jiefang yu gaizao*).[2] They became very enthusiastic for these new ideas and opposed feudalism and all forms of persecution. They advocated colloquial literature, women's liberation, and open social relations.[3]

1. This degree, *juren*, was the second highest of three official degrees during the Qing period and sufficient qualification for the appointment mentioned here.

2. These periodicals were among the most influential during the "New Culture" and "May Fourth" years. They were launched by various individuals or groups as editors and received contributions from these same progressive intellectuals. The earliest of these journals were begun in Beijing and developed a broader circulation. Such journals were later launched in other cities, also using progressive-sounding titles. The cumulative influence of these journals was tremendous. Most of a new generation of educated Chinese youth found the "new culture" extremely appealing and helped to make it a national movement.

3. The cluster of progressive ideas listed here were major themes in "new thought" throughout these years. "Feudalism" is a catch-all term still used today (sometimes too conveniently). For the young progressives in this article, "feudalism" represented the old order in which the family and elders dominated the thinking, goals, and life-decisions of

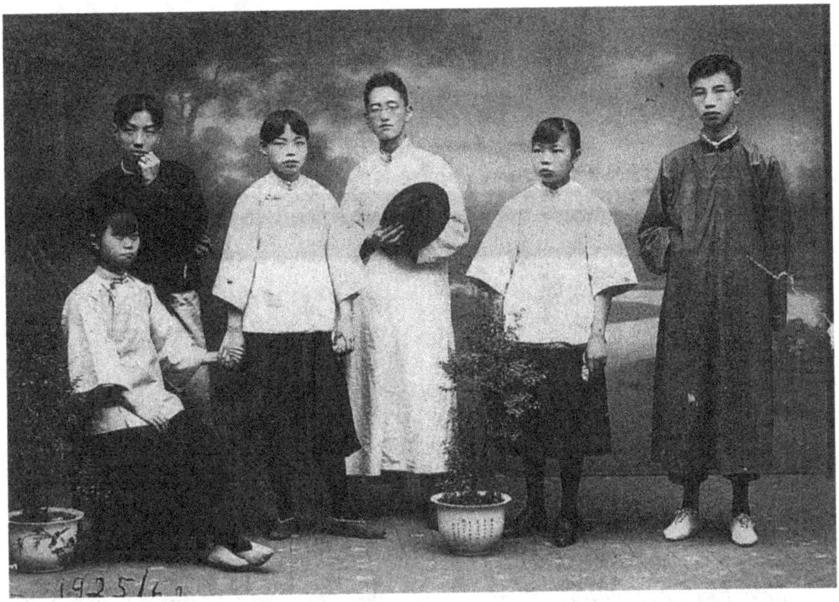

In about 1922, Chen Zhuying went to Shanghai where she joined a "Patriotic Women Students" group and became friends with Chen Xuezhao. After returning to Chengdu she became a teacher at Pengshan Girls Middle School, and she started publishing *Women's Monthly* (*Funu Yuekan*).[4] On the twenty-fifth anniversary of West China Cooperative Middle School, Chen, Wang Yi'an, and others performed *The Aroma of a Plum Basket*, a four-act play, which was quite a success.[5]

Before Liberation, Chen Zhuying lost her position once and

---

their children. This was the order they set out to change—and by the 1920s had begun to have some success in this project.

4. The title that Ms. Chen chose for her journal is a good example of the pattern referred to in note 2. There were a number of journals with this title, or others such as *Women's Liberation* (*Funü jiefang*).

5. Also during these years Chinese students and many urban adults became interested in Western-style drama, introduced by foreign teachers and also by Chinese intellectuals who had studied abroad. One of the latter was Hu Shi, China's leading liberal thinker during the first half of the twentieth century. Hu was an editor and contributor to *New Youth* and other progressive journals mentioned above. Hu attempted to launch a movement to write modern Chinese drama in *baihua*, with a play entitled *Life's Great Event* on the conflict between the traditional arranged marriage system and marriage by choice of the partners.

then became a cadre in the provincial Party Women's organization. After Liberation, she volunteered to go to the Northeast and became a teacher in Qiqihar.[6] When she returned to Chengdu the next year, she became a teacher of Chinese at the Nationalities School.[7] Then during the Cultural Revolution, she was sent off to labor reform. She left the school in anger and went to live in secret in a women's refuge for a few months.[8] Her oldest child, Ma Yiwen, was sent to do labor in Yunnan during the Cultural Revolution and injured her leg. Chen Zhuying became outraged at the injustices of that time. In 1973, she began vomiting blood and died.

Second from left is Ma Jingchen—Chen Zhuying's husband. From Qingchuan County, he graduated from the middle school associated with the Teachers' Academy. He was a stellar student and interacted well with people. With his classmate Liu Xianliang, he joined the "Direct Consciousness Society" (*Zhijue she*).[9] He edited *Sichuan Daily* (*Sichuan Rikan*) and *New Sichuan News* (*Xin Chuan bao*). In later years he worked in the postal service. After Liberation he taught Chinese in the Chengdu County Middle School and later taught at Lanting Middle School. He died after the Cultural Revolution.

---

6. Qiqihar is a smaller city in Inner Mongolia, relatively close to Harbin and a long, long way from Chengdu.

7. "Nationalities" schools and universities have been established for the broad project of acculturating the many ethnic groups in China. The teaching of Chinese is a basic part of their curriculum. For some ethnic groups, learning Chinese has been critical to building a better life in the larger Chinese society. For other groups, especially the Uighurs in Xinjiang, while they must learn Chinese, Uighur children are no longer allowed to study their own language in formal schooling. The analogy to mainstream treatment of Native Americans in the United States is all too clear.

Sichuan has a large population of Tibetans, most of whom live in the mountain ranges that run along the western frontiers of the province (ethnic Tibetan settlements continue into Yunnan). The Qiang people are another relatively large ethnic group in Sichuan, often interspersed among the Tibetans. Some other ethnic groups found in Yunnan and Guizhou also live in southern Sichuan.

8. The Chinese here is *baomeiqu*, a home for women without families to assist them in times of difficulty.

9. The name of this small group is also typical of similar groups throughout urban China during the New Culture years, again reflecting the idea that a new consciousness was available, a new way of looking at the world. The "Fortnightly Society" and its journal is another example. Its name perhaps suggests a concern to appeal to others besides progressive young people.

Third from left is Shen Dujie, who graduated from Yizhou Girls Middle School. She married Wu Xianyou and served as a health worker at Chengdu Number Thirteen Middle School.

Fourth from left is Wu Xianyou, who was from Chengdu. He did foreign studies and then moved to the education department at West China Union University, where he graduated. With several others he started the "Fortnightly Society" (*Banyue she*) and published the journal *Fortnightly* (*Banyue*). His colleagues in this endeavor were Shen Ruoxian, Zhang Shiyi, Shi Rucong, Shi Jufu, Lu Jianbo, Mao Yibo, and others. Their journal showed the influence of the May Fourth Movement. He also began to study Esperanto with a teacher of Korean background, Gao Zixing, who was living in Chengdu. They organized an Esperanto Study Society and held classes at Chengdu's Popular Education Hall. They prepared *An Esperanto Reader* (*Shijieyu Duben*) for their teaching materials.[10]

Wu Xianyou was very hardworking. He had been influenced by Kropotkin, Bakunin, Tolstoy, and the thinking of other such people. He advocated vegetarianism and the principle that one's actions should match one's words.[11] In his autobiography, Ba Jin said that, among his friends in Chengdu, Wu was the one he held in highest esteem.[12] Ba Jin said in his autobiography that he "had been influ-

10. Esperanto was another progressive cause during these years. Chinese anarchists were among the most enthusiastic in efforts to propagate Esperanto. This theme reflects the idealism of Ludwig Zamenhof, who developed Esperanto in an effort to overcome the lack of language unity, and thus of understanding, among the varied peoples of Eastern Europe. Anarchists and other progressive younger Chinese hoped to overcome the difficulty of communication with people of other cultures through Esperanto. This movement has lasted more than a century in China; although numbers are not large, Esperanto groups can be found today in a number of Chinese cities.

11. The Russian thinkers mentioned here are a thinly veiled reference to the fact that Wu Xianyou was influenced by anarchism. Again, many educated young Chinese had been introduced to these and other Western anarchists through anarchist journals during the decade before the New Culture Movement. Anarchism was a major influence in the birth of the New Culture movement. Leo Tolstoy was the specific source for vegetarianism, reflecting the influence of his own late-life advocacy of a pure, direct Christianity.

12. Ba Jin was also an anarchist. He studied in France during the late 1920s, when a number of French anarchists were still active. These individuals had helped to attract Wu Zhihui and other early Chinese intellectuals to anarchism during sojourns in France. Ba Jin was probably influenced even more by Shifu (Liu Sifu), one of the most admired Chinese anarchists, who had encountered anarchism while in Japan during 1907–1908 and also pursued the more science-oriented themes of European anarchism. During the

enced by Wu's moral sense." On a trip back to Chengdu in 1937, Ba Jin often went with Wu to teahouses in the Xiwanglong Street area.

Wu enjoyed music and drama, and he wrote a tribute to Beethoven, praising the composer for his spirit in opposing oppression. He enjoyed hearing the blind singer of Chengdu, Jia, chant folk tales using his bamboo fiddle. Wu Xianyou taught Esperanto at Sichuan University. Later, he served as principal of Chengdu Number Thirteen Middle School. He died of a sickness in 1956.

Fifth from left is Li Qingyun, of Zigong. As a child she moved to Chengdu with her father and brother. Her father Li Yunxian, a doctor of Chinese medicine, worked as a pioneer in underground activity. He was very progressive in his thinking, so Li Qingyun didn't have to be concerned about pressure at home. She joined the "Direct Consciousness Society," and all went well in her marriage to the like-minded Wang Yi'an (who is not in this photo). Wang Yi'an was a multi-talented social activist. He joined various groups of amateur Beijing opera performers.[13] Wang also edited a newspaper called *The Report* (*Baobao*) modeled on the Shanghai paper *Brilliance* (*Jing bao*). After a few issues, people began to think well of this paper.

The last person from left here is Li Xiaoyun, Qingyun's oldest brother. At that time he had just examined into the English department of Chengdu University. He too joined the "Direct Consciousness Society." I was a classmate of his, and I know that he refused to

---

early part of his writing career, Ba Jin said that his pen name came from the first syllable of Bakunin's name as transliterated into Chinese, Ba, and the second part from the last syllable in Kropotkin's name in Chinese, Jin. Ba Jin suffered greatly during the Cultural Revolution, if not directly for his anarchist principles, then for his forthright devotion to honesty, which was much influenced by his anarchist principles. He was made to walk on his knees on broken glass, in a public humiliation at Workers' Stadium in Beijing. See the biography of Ba Jin by Olga Lang, *Pa Chin and His Writings: Chinese Youth Between Two Revolutions* (Harvard University Press, 1967).

13. In Sichuan and other provinces there were many such local groups of amateur opera performers. Many of these enthusiasts became polished performers, and certainly their interest helped to perpetuate one of China's long-standing art forms, which more recently has fallen on harder times despite the existence of TV stations that present Beijing and other styles of opera all day long. For the kind of popular group that Wang Yi'an belonged to, see "'Play-Friends,' a Grass-Roots Sichuan Opera Group," elsewhere in this volume.

join the Guomindang. Sadly, on the eve of the "February Sixteenth Tragedy" in 1928, he died after bleeding from the mouth.[14]

These events from the lives of these six people are things I myself know through my friendships with them. I too joined in some of their activities, and I have kept this photo for many, many years. I thought it would be a good idea to lend it the editors of *Old Photos* to help with recollections of these people's lives and to express memories that I find I can't let go of!

This article is by Min Zhentao. It was published in Old Photos *no. 9, pp. 121–23.*

A Period of Hero Worship

Introduction

This brief item describes popular culture in China shortly before the Cultural Revolution. Heyang County is approximately one hundred miles northeast of Xi'an. Also see "Spring Fieldtrip for the Students of a Village School," which concerns local life in Heyang some fifteen years earlier. Perhaps in any generation of a modern society, teenage youth readily take up the symbols of popular culture. In China during the early 1960s that culture was enthusiastically revolutionary. As a rural county distant from Xi'an or other major urban centers, Heyang reflected this kind of involvement in the national culture. As the school head noted, middle school students were supposed to be aware of the larger culture, and they certainly were.

14. This event was unique to Chengdu. It resulted from student protests to a fee increase—the occurrence of which in itself seems powerful testimony to the appeal of "new thought" in this provincial capital. A local official, Yang Yanquan, sent police into the university area of the city in efforts to make the students desist. Modest violence in this first attack increased the students' fervor. A second and even more violent action on February 16, 1928, resulted in the killing of fourteen students. Li Xiaoyun might have died after injury during the first police action or become so exercised that his death was from induced "natural" causes.

The Text

This is the graduation photo of Higher Middle School class 64-A in Heyang, Shaanxi, my class. This year, many in our class went to the funeral of Yang Tianxiang. When we saw this photo on the wall at the Yangs' home, Lei Yingxian (third from the left on the front row) pointed to the female classmates and asked me, "Do you remember why we girl students wore white scarves that year?" Because of family difficulties, I'd had to drop out of school, so I'm not in this graduation photo. This photo was given to me as a memento by my favorite classmate Wang Chang'an (whose pen name is Yanwen—this name is written on the back of the photo). Anyway, I still know the answer to her question. This was a symbol of hero worship characteristic of that generation of young people.

During the 1960s, there were a dozen novels that became fashionable throughout the country. These included *Red Crag, Keep the Red Flag Flying, The Red Sun,* and *The Builders* (these four were called "the three reds and a great undertaking"); also included were *Tracks in the Snowy Forest* and *Song of Youth.*[15] Whenever

15. The novels lauded here belong to a chapter of Chinese literature of the early years of the People's Republic. *Red Crag,* written by Luo Guangbin (1924–1967) and Yang Yiyan (1925–) and published in 1961, takes its name from the Communist Party headquarters in suburban Chongqing. This was the place that Zhou Enlai worked from as he interacted with the Nationalists and fostered those who might sympathize with the Communists (see "House of Democracy" elsewhere in this collection). The most identifiable figures in this novel were Communist prisoners at Zhazedong, a Nationalist operation. The Nationalists set out to execute these prisoners, a last act of spite as Communist armies approached Chongqing in November 1949.

In *Red Crag,* US military personnel working with the Sino-American Cooperative Organization (SACO) share blame with Jiang Kaishek's Nationalist operatives in carrying out the executions. More recent study of SACO operations has shown that the US side was no longer involved by 1949. On this, see especially work by Xujun Eberlein as pursued through Internet sites.

One of the characters in *Red Crag* is a woman, Jiang Zhuyun (1920-1949), who was the inspiration for the modern opera *Sister Jiang.*

Keep the Red Flag Flying was published in 1957. Its author is Liang Bin (1914–1996). Liang grew up in Hebei in a difficult situation. The novel is built around peasant-landlord struggles during the early decades of the century, episodes that Liang experienced personally.

The Red Sun pursues the period of the late 1940s when the Communist challenge to the Nationalists turned into full-scale civil war. Written by Wu Qiang (1910–1990), this novel was published in the early 1960s.

The Builders, by Liu Qing (1916–1978), appeared in 1960. Its major theme is land

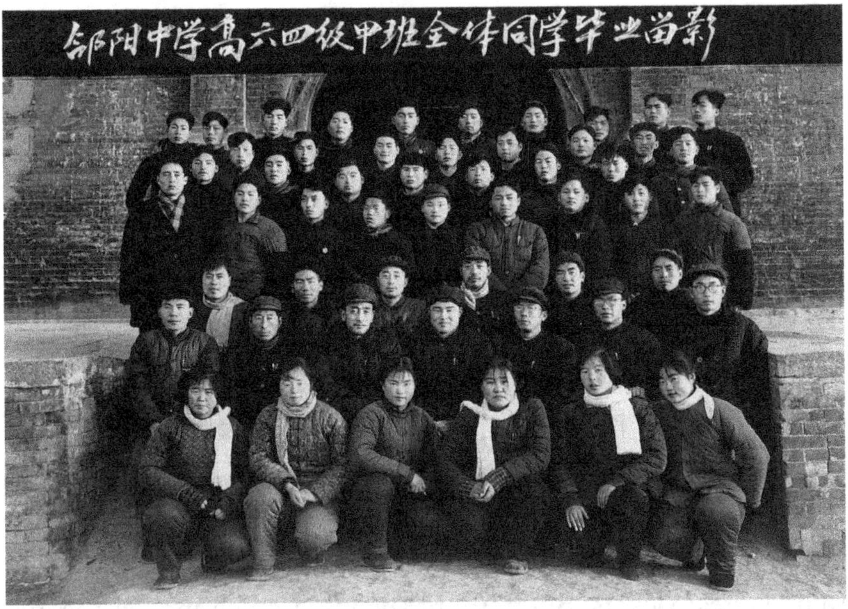

a group of young people talked among themselves, it was embarrassing if you had not read these novels. To use the words of our school's head, Han Chaonong, "Heyang Middle School is the highest educational institution in Heyang County," so of course we were first to know about such things. Even if they hadn't read

---

reform in a north China village. Two generations of the Liang family reflect differing attitudes toward change though neither opposes it. Liu Qing is credited with true creativity in his portrayal of Grandfather Liang San.

Tracks in the Snowy Forest also concerns the civil war period in the Northeast. Written by Qu Bo (1923–2002), this novel was published in 1957. The novel follows the work of a group of special troops as they pursue bandits in mountain areas. All of these novels were made into movies and some became modern Beijing opera. This one reemerged as one of the favorite new-style operas, *The Taking of Tiger Mountain*.

Song of Youth, published in 1958, whose main character Lin Daojing the girls found so appealing, was written by Yang Mo (1914–1995), a woman writer who shared the same background as her subject. Yang herself had "seen the light" of the socialist analysis of society and the cause of revolution. As the writer of this article indicates, their classmate from a similar background was seen as not making that transition.

Originating as fiction that drew on life experience from the years of the Communist revolution, these stories became films, modern Beijing opera, and more recently television series. While this ongoing evolution in form perhaps reflects present-day tastes among television programmers, the stories come from a truly impressive period in China's evolution during the twentieth century.

these novels to the end, the students still idolized the heroic characters in them. The most respected figure for the boys was Yang Zirong in *Tracks in the Snowy Forest*, who was always prepared to put himself into the most difficult situation. For the girls, it was Lin Daojing from *Song of Youth*. They would fall at the feet of this Lin Daojing, who wore white from head to toe. But their families didn't have enough money to dress them that way because we all were children of peasant families.

Class 64 at Heyang Middle School had 165 students, but by graduation, more than 100 of us were not in school. Many families were having money problems, so how could they provide their children with nice clothes? From the photo you can see that the girls are wearing cotton oversleeves made at home (this is an accessory worn by Shaanxi women to stave off the cold). These cotton items made it difficult to move freely, yet no one could afford to make real jackets. But without spending much money, the girl classmates could pull a couple of feet of white knit cloth around their neck, lay it around in back and have the front part hanging at their bosom. So as you can see, with such a simple "prop," the girls could add a touch of class. It made the boys really envious and helped to keep the girls warm in the winter.

The girl all the way to the right end in the front row was named Lu Qiqi. She was a special student in the class, so she didn't know why the other girls were wearing these scarves and didn't wear hers that way. Then you can see that Lei Yingxian, the one who asked me about this, is not wearing a scarf herself! Maybe she forgot it in her haste, or it might not have gotten dry after she washed it. Let's just say she didn't think of it. The girl second from left in the front row is He Xialin, whose class status was both landlord and capitalist. So she was like Lin Daojing in the novel, and because of this she went through many difficulties during the Cultural Revolution. She was graceful and intelligent, and that white scarf about her lovely fair face would have made her as beautiful as a clear spring day. So why didn't she wear the revered white scarf for the graduation photo?[16]

16. In a black-and-white photo it isn't clear how far off-white He Xialin's scarf is.

That a small white scarf that would express the hero worship of that generation of young people would, I'm afraid, be taken as a joke by the youth of today. It was so pure and naïve. But there was nothing at all phony about it. For that was an era of worship for revolutionary heroes.

This is by Shi Yaozeng; it appeared in Old Photos *no. 29, pp. 124–25.*

"Play-Friends," a Grass-Roots Sichuan Opera Group

Introduction

This article presents a kind of amateur organization found in most areas of China. Its members love traditional Chinese opera, and they help to keep this form of old-style popular culture alive. It also describes a variety of regional and local styles, which would be difficult for newcomers to opera to differentiate. While the two examples of traditional opera included here demonstrate how widespread its appeal was in its heyday, in recent decades opera has been in decline. Still, foreign travelers in China regularly encounter individuals or small groups playing traditional instruments in public parks or at tourist attractions. Some retirees are happy to receive tips for their amateur performances, but they also enjoy sharing this aspect of traditional culture with visitors.

Traditional opera was much more than a performance. It communicated history, lore, and values in dramatic form, reinforcing both the educated and commoners in these aspects of Chinese tradition. The "revolutionary operas" fostered by Mao Zedong's wife Jiang Qing during the Cultural Revolution modified this traditional art form quite successfully as it instilled transformed values in younger generations of modern Chinese.

Perhaps she wished to share her classmates' enthusiasms but because of her background, wasn't sure they would appreciate her displaying that style.

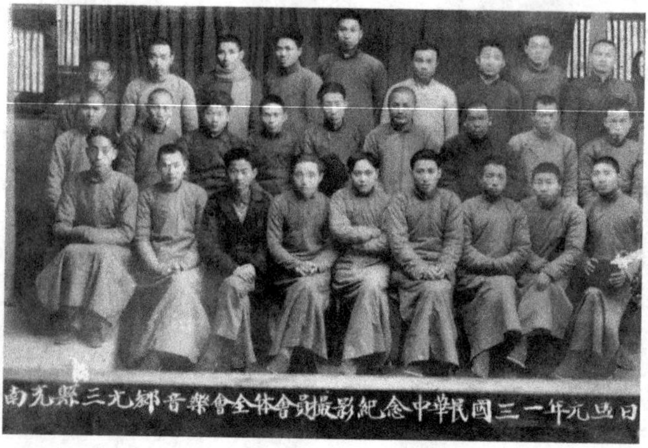

Photo 1: The entire group of friends in Sanyuan district, Nanchong, in 1942.

The Text

These photos show a music group from Sanyuan District, Nanchong County, Sichuan. Most of the photos are from the 1940s. This was not the kind of music group or club that we have now. According to the *Annals of Culture and the Arts in the Nanchong Area*, a "music group" was popularly called a "play-friends" group. This was one kind of popular organization, self-organized by people who made a hobby of Sichuan opera to put on performances.

The various regional opera companies that specialized in Beijing opera, Yunnan opera, and others, often called themselves "ticket friends."[17] But for Sichuan opera the term was "play-friends." These popular organizations in Sichuan opera are self-organized. They manage their own instruments and get together the funds to buy the drums and the *erhu* (two-string Chinese viol) and other instruments for their performances. They entertain themselves humming and singing as in an opera performance. And because this is an enterprise of friends, it is called "play-friends." This activity is also sometimes called "listening to play-friends" or "beating the drums."

Sichuan opera is a style of local opera that is popular throughout Sichuan and in parts of Yunnan and Guizhou. It includes folk operas such as lantern-plays, Kunshan opera (from Jiangsu), high-pitched singing, and story-telling as done in several places in southern China. These various forms of the folk opera art have been performed

17. Typically translated as "amateur players."

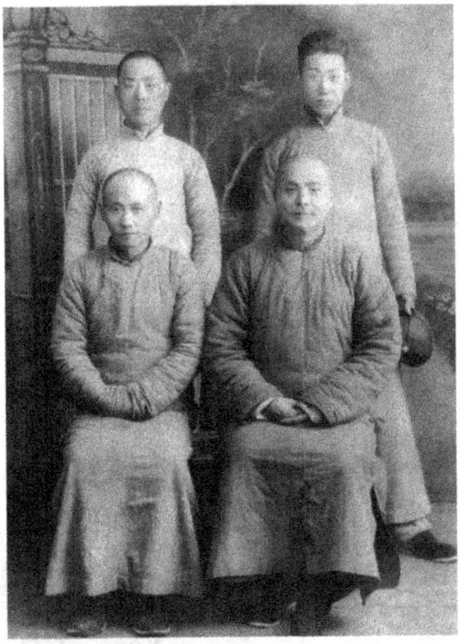

Photo 2: The four organizers of the Nanchong group.

at different times in Sichuan, but since the Qianlong era of the Qing period (1736–1796), for reasons already suggested, they have tended to have mutual influences and to merge, so that they gradually took on a common quality. By late Qing times they had taken on the name of "Sichuan plays" or "Sichuan opera." Sichuan opera is divided into four schools named after four rivers in Sichuan. These are "Northern River style," "Ziyang River style," "Luzhou River style," and "Suining River style." Nanchong is in the northern part of Sichuan, so the style there belongs to the "Northern River school," with the story-telling approach the major form. It has also been somewhat influenced by *Qinqiang*, the opera of Shaanxi Province to the north.

As reported in the *Nanchong County Gazetteer* for the Daoguang period of Qing (1821–1851) until the Republican period (1912–1949), the groups formed to present Sichuan opera were usually itinerant groups.[18] They had interesting names like "Great Tide," "Longevity," "Perpetual Peace Professionals," "Golden Dragon Professionals," or "Egalitarians." Many factories, schools, and teahouses had "play-friend" groups for Sichuan opera. The group would perform while seated informally for the purpose of exchanging views on their technique. These performances were also given to welcome or see off guests, entertain groups of people, or participate in happy occasions for either "red" or "white" (wedding and funeral) events, or for religious celebrations or other kinds of rituals in society. Through all these kinds of activities, some members of these groups gradually developed into professional performing artists in Sichuan opera.

18. A gazetteer is a local history. This is a standard form in Chinese historiography and often a great source for research.

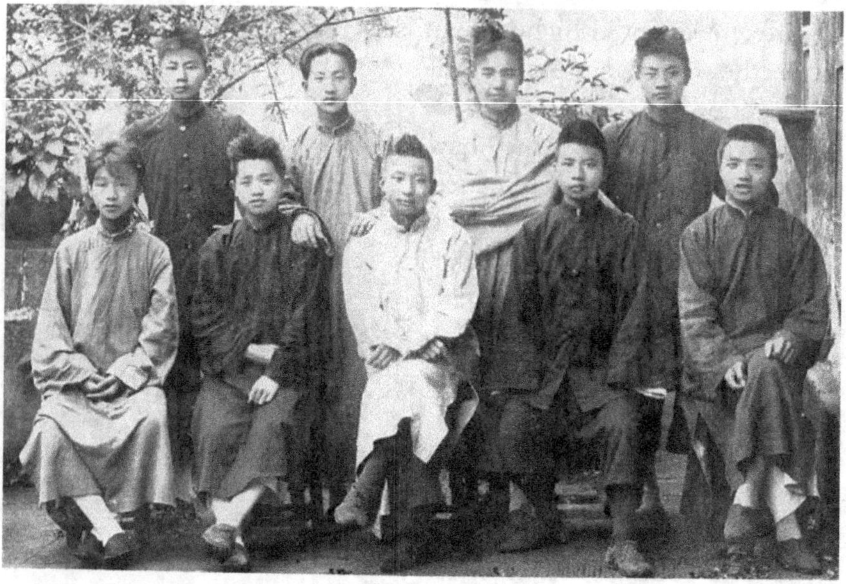

Photo 3: During the Republican period, "play-friends" groups had many such enthusiastic young members.

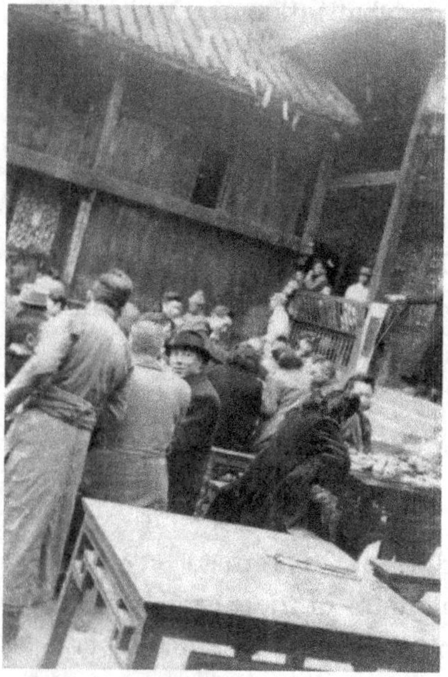

Photo 4: A crowd gathers, waiting for the "play-friends" group to arrive at a teahouse.

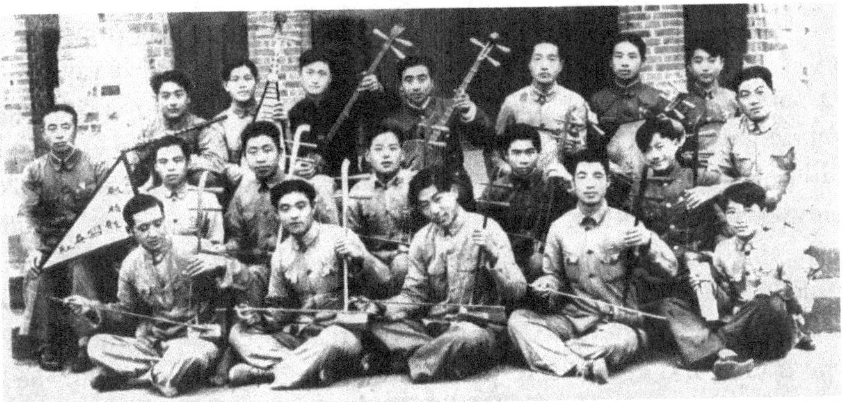

Photo 5: The members of a group called "Rows of Dragons" (1946).

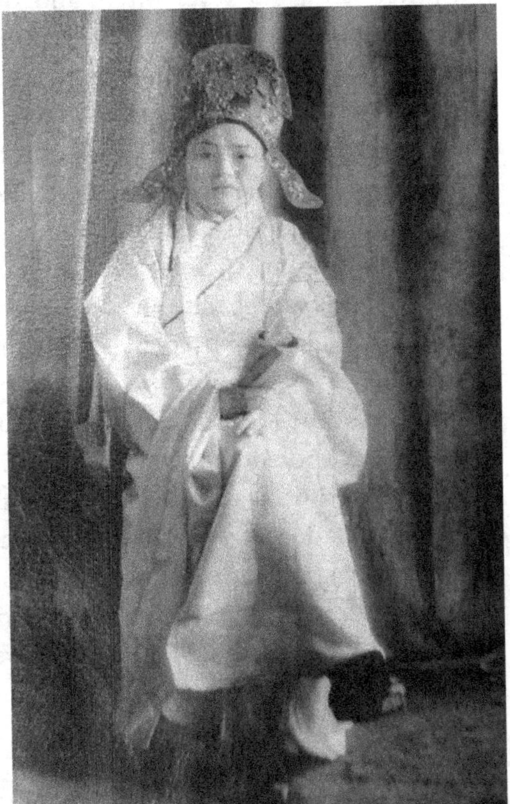

Photo 6: A "play-friend" made up for a female role. Until very recent times, female parts were also played by men.

Back in those days, tea houses were one of the most important places of amusement for ordinary Sichuan people, and many older people and others who enjoyed listening to or performing opera music gathered at teahouses to drink tea and pass the time of day. "Play-friends" made tea houses their meeting places. Through this simple form of sitting to play their music, the "play-friends" could get contributions large and small to support their performances, and usually there was no need to get into costume or find a place to stage an opera. They just gathered at a teahouse and began to beat the drums and cymbals, to play the *erhu,* and start to sing—so began a performance. And when the excitement of the performance reached its peak, with the drums and cymbals beating away, the *erhu* and other strings at fever pitch, and the singers intoning in their high nasal way, a great crowd of listeners would gather at the teahouse. I have heard older people say that, especially during festival periods, at every tea house where these groups performed Sichuan opera in this informal way, everyone in the crowd would wave their heads in time and the place would be filled with excitement. During the Republican period, among the towns with the largest number of "play-friend" groups such as the Nanchong county seat, Dongguan, Longmen, Changle, Lidu, and other such places, these groups were extremely popular and particularly distinctive.

After 1949, some of those who had participated in "play-friend" groups joined the local professional theatrical troupe to perform Sichuan opera, and some participated in literary and arts groups organized at various local levels, so that the "play-friends" groups of earlier years faded away. What happened was that Sichuan opera, this grass-roots art form, which earlier had been performed in local places, found its way to more professional opera stages. The "play-friends" who performed the opera so well simply joined the more formal organizations. By the 1980s, though, you could once again see "play-friend" groups emerging and performing as before, especially in county seat towns and other larger places. These groups are the heart and soul of Sichuan opera. And now in many places there are contests among these groups. The performers are judged on the extensiveness of their selections, the richness of their repertoire, and the entertainment level of their performance style. All in all, these contests display the enchantment of Sichuan opera and the love that the great mass of people have for it.

In short, without the contributions of "play-friends," Sichuan opera would not have endured from one generation to the next.

This is by Yi Ge. It appeared in Old Photos *no. 71, pp. 171–85.*

Old Theaters in Weihai

Introduction

This writer outlines the background and construction of theaters in Weihai, on the northern coast of Shandong. Opera performances were a regular feature during festivals and market days throughout China. As is suggested here, it was an achievement to develop proper facilities for performances, and local benefactors often provided support. Note the care and preciseness applied in building a theater.

As in the preceding article on opera in Sichuan, the styles of opera presented in Weihai theaters also receive attention here. Besides Beijing opera, which had broad influence, there was a variety of regional styles. In Weihai and many other places, different styles were performed. The photos here that show audiences further attest to the great appeal of this aspect of traditional culture.

As already noted, traumatic change in China's modern century undermined both theaters and traditional opera, a major form of popular entertainment even well into the twentieth century. For comment on the influence of the revival of opera during the Cultural Revolution, see "Going to School During the Cultural Revolution."

The Text

In earlier times the most common leisure activity of Weihai[19] (Shandong) people was to go see Chinese opera. Because this was such a common activity, there were quite a number of theaters both in the city itself and in the suburban areas. From the time they were built,

---

19. Weihaiwei is another name for this city. It became a British concession in 1898, so despite its relatively small size, it became an international city.

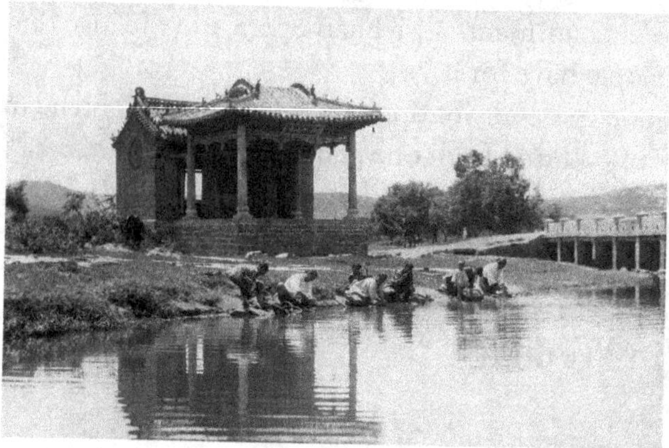

Photo 1: The old Weihai District opera theater.

these theaters were closely linked to the lives of ordinary people and were part of what they enjoyed in life. From old times until quite recently, an important part of folk life in Weihai was the performance of operas in these theaters. At the New Year, on market days, and during all the other festival periods, people came in droves to the theaters and there was a rush of activity.

Weihai's theaters were like those found throughout China, built in every time period and in sizes large and small. The usual form was "a stage in front and a room in back." As all the relevant local histories recorded, the Heavenly Mother's Palace, the Temple of the Three Old Men, Guandi's Temple, the Temple of the Fire God, the Temple of Li the Dragon King, and other such temples all had theaters. During the reign of the Qing Emperor Daoguang (r. 1820–1850) a Weihai man, Wang Zhezheng, who had served as the magistrate of Wu County in Sichuan, returned home and built a fine house with a theater that everyone regarded as an excellent example of a private theater.

Nowadays, aside from the two theaters at the Temple of Li the Dragon King on Liu Gong Island and the Admiral's Academy, all the theaters have been destroyed in one way or another over the years. The theater on Liu Gong Island has been preserved up to the present, probably because the island was controlled by the military and maintained as a restricted area. An investigation of the relevant historical materials showed that this theater at the Temple of Li the Dragon King was built at the same time as the headquarters for the

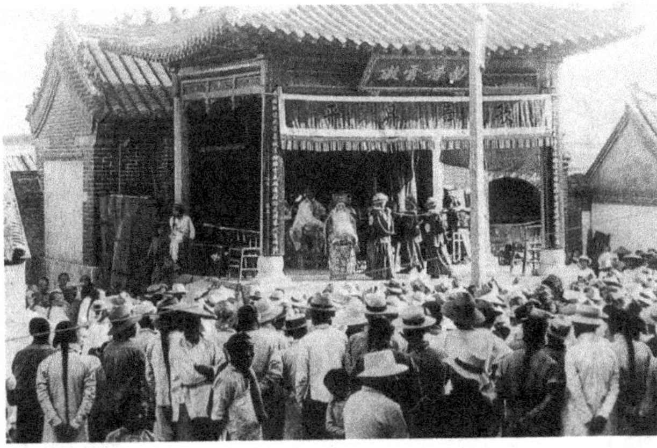

Photo 2: The theater in the city of Weihai, with a performance in progress.

Beiyang Navy, probably in the 1870s or 1880s.[20] The front stage of the theater is 5.8 meters wide and 5.6 meters deep, and the front of the stage is precisely square. In those days every time the Temple of Li the Dragon King had a festival, or the Beiyang admiral returned from a period at sea, the scholar and merchant leaders on the island would organize performances in celebration.

The size of Weihai theaters was not very large, but they were very well done. The bracketing under the eaves was colorfully decorated, the animals above the eaves were arrayed in fine rows, and the front of the stage was absolutely square. At the base of the stage were four sturdy stone columns reaching up to the eaves, each with portrayals of myths and flowers, and the illustrations were beautiful. On the main columns, poetic couplets were carved or hung on scrolls. The area behind the stage was made of brick and stone.

There are some folk sayings about these stone columns in opera theaters. In building a theater, it is most important to set the four base columns correctly. Other parts may be built in daylight. Only the individual columns must be set after dark. Some people say this is to guard against disclosing the secrets of construction. Others simply say that in the daylight hours so many people are there watching the performance that it's easy to have some kind of accident.

---

20. The Beiyang Navy was one of the four modernized Chinese navies established during the "Self-strengthening Movement" in the late nineteenth century. It was led by Li Hongzhang (1823–1901), one of the most powerful leaders in late Qing China.

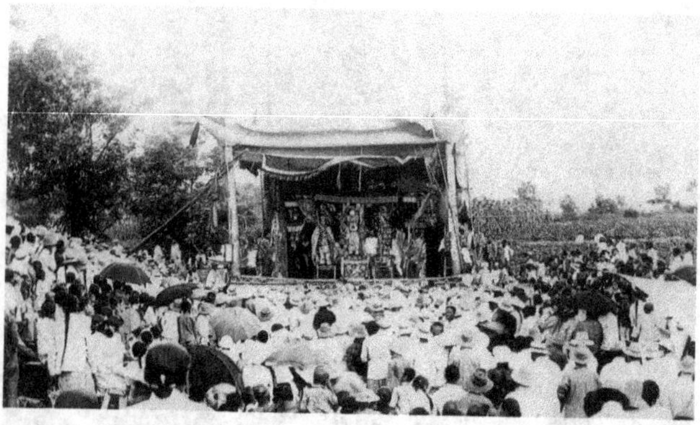

Photo 3: The district theater, also with a performance in progress.

Before the beginning of the Qing Dynasty, theaters in Weihai mostly performed Yuan-period opera. Beijing opera came in during the Daoguang period (1820–1850), when it was called "greater opera" and came into general favor. Whenever there was a temple festival or an important traditional festival, performances would be arranged. When the fishermen went out to sea, groups of performers would be invited to put on plays for three to five days. People called this "invited opera." The great performer Ma Shaotong wrote a book called *The Early Years of Beijing Opera in Weihai*, in which he observed:

> In the latter part of the Guangxu period (1875–1908) people said there were "outside troupes" who would sometimes perform clapper plays and Beijing opera at temple festivals in Weihai. Weihai people called performances by these "outside troupes" "rustic stage opera." In Beijing, it was called "grass stage opera," and in Hebei it was called "shed opera." If you think about these terms, the point is that these opera troupes didn't have a regular place to perform. Whether it was a city or out in the country, any temple festival, anytime someone wanted to beseech the gods or pray for rain, to interact with a god, the idle winter days in farming villages, or wishing fishermen a bounteous catch, an "outside troupe" would be invited to perform for a few days.

Gradually, Weihai developed its own local theater. Performers there included both local actors and some well-known performers

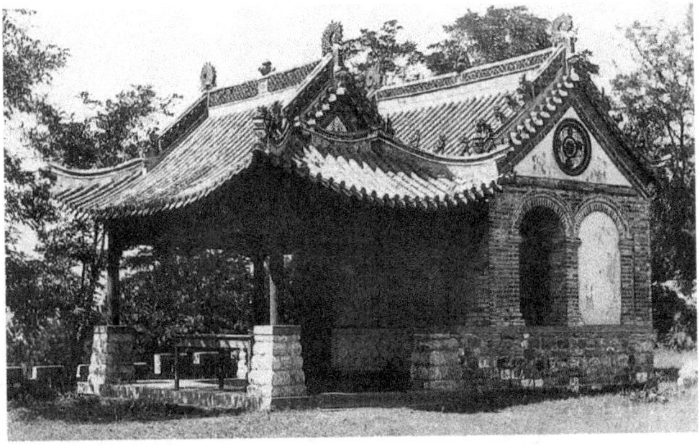

Photo 4: The theater at the Admiral's Academy on Liu Gong Island.

from other places. And all along, the "outside troupes" were also invited to perform in Weihai. This continued until the Japanese reached this area in 1938. After this, "outsiders" could no longer be seen.

On the eve of Liberation, performance of traditional opera had almost died out.

The theaters got older, fell into disrepair, and gradually were destroyed. Some were leveled as part of the "Four Olds" during the Cultural Revolution, and some were razed because of urban expansion.[21]

This is by Liu Weifeng. It appeared in Old Photos *no. 67, pp. 168–71.*

21. The term "Four Olds" refers to old customs, old culture, old habits, and old ideas. One of the stated goals of the Cultural Revolution was to destroy the Four Olds.

TEN

Staying Alive and Doing Your Best

My Older Brother's First Radio

Introduction

This is the story of an enterprising young man in a time of great difficulty, the point at which Japanese armies moved further inland—into Shanxi Province in this case. His hobby in early radio-building made him helpful to his community, providing the latest news of hostilities. When the news led many people to consider relocating to escape the Japanese advance, this young man made his way, first riding a bicycle, toward the relatively safe area of Sichuan, where he continued his education and started his career.

The Text

My older brother Sun Puyi graduated from Mingyi Middle School in Fenyang, Shanxi, and examined into the medical school at Jilu University in Jinan (Shandong). He had a hobby—making radios.

During his first year at Jilu University my older brother made his first radio. This was a crystal radio receiver, a square board with two batteries attached, and several coils of wire that he wound himself. It looked simple, but it provided good reception. This was the first radio I had ever seen, and it might have been the first radio in Fenyang.

My brother came back to Fenyang by train for summer vacation in 1937, and right after he passed through Beiping [Beijing], the Marco Polo Bridge Incident occurred. It was said that when the passengers on the train heard this news, they couldn't hold back the feelings of hatred that had begun with the "September Eighteenth

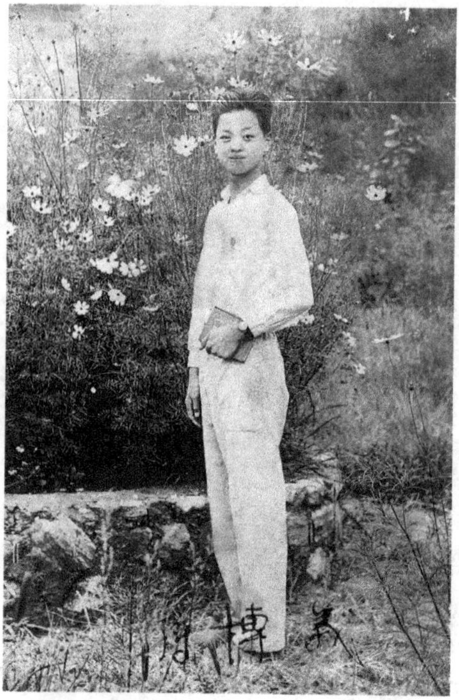

Photo 1: My older brother as a student at Mingyi Middle School in Fenyang in the early 1930s.

Incident."[1] They started shouting slogans and singing anti-Japanese songs as their feelings erupted. The train police came through the cars to get things under control, but they couldn't quite keep order.

When my brother arrived in Fenyang, the people there were also angered by the renewal of Japanese aggression, and they wanted to get news of the situation as quickly as possible. At that time Fenyang didn't have its own newspaper, and it took two or three days for newspapers from Beiping or Taiyuan, the provincial capital, to get there, so everyone was always anxious about the situation. Our family lived at the Fenyang Women's School. We were in a house just inside the school gate, with a parlor and a bedroom. When my brother arrived that day, right after he put down his luggage he set up his radio in the parlor.

The news got around Fenyang very quickly that Sun Puyi had

1. The September Eighteenth Incident occurred in 1931. It is also known as the "Mukden Incident." This was a staged action on the China Northeastern Railway in which Japanese forces placed dynamite along the railroad track. The explosion did not actually disrupt use of the line at that point. The Japanese claimed that Chinese troops had set off the explosion and used the incident to extend their authority in Manchuria. By March 1932 the Japanese created the puppet state of Manchuguo and installed Puyi, the last emperor of the Qing dynasty, as figurehead ruler of Manchuguo.

The Marco Polo Bridge incident," commonly referred to as the "7-7 incident" among Chinese, occurred outside Beiping on July 7, 1937, and marked the beginning of World War II in Asia. Again, Beiping was the name given to Beijing during the period 1927 to 1949 when the Nationalist government moved the capital to Nanjing.

a radio. Day after day the numbers grew of those who came to our house to hear news of the fighting, and soon there was no way for everyone to sit down in the room. The only thing to do was to move the radio out into the courtyard. Every evening my brother would set up the radio out in the yard, and we helped him by carrying a table out to serve as a platform for the radio. Every evening huge crowds of people gathered around the radio. When the news broadcast began, there wasn't any need to try to create order. Everyone became very quiet, and only the announcer's voice was heard reverberating through the evening quiet. The listeners' reactions rose or fell according to news of the action. Our fighters against the Japanese did not fear death. Heroism surged in the bloody fighting, but there was a string of losses between Beiping and Tianjin.

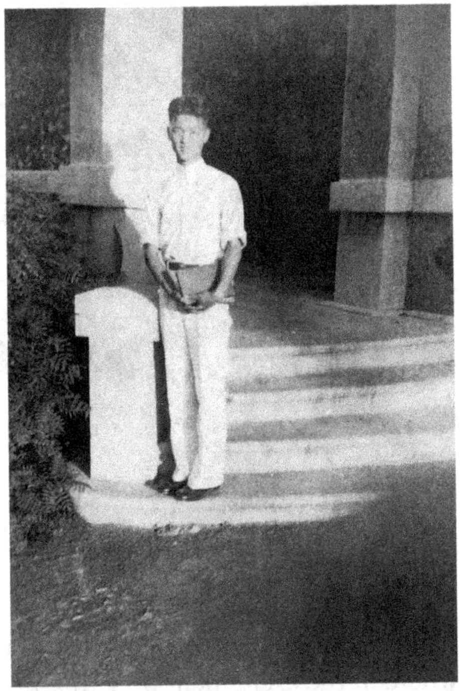

Photo 2: My brother as a student at Jilu University in Jinan (Shandong) in 1936–1937.

An old Mingyi School classmate, Deng Quanhai, recalls that he came to our house every evening to hear the news. He listened and wrote it down, and then the next morning he would go to the gate of the Guangzhi Academy on East Main Street, write a big-character poster and put it up at the gate of the academy. A crowd gathered around to read his poster, and there was always someone who copied the poster and spread the news further. One time there was good news about the fighting at Fengtai, near Beiping, a victory for us. So Deng Quanhai prepared an "extra," and when he posted it there was great rejoicing. An elderly gentleman, Ren Liu, knelt on the ground

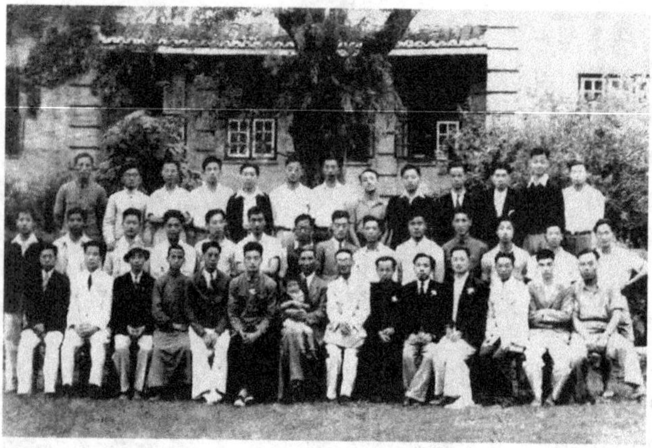

Photo 3: My brother (second from right in the upper row) as a student at West China University in Chengdu during the wartime years.

to copy the "extra." A worker at the Hanmo Bookstore, opposite the academy, was deeply moved by the news and took it upon himself to print a hundred copies of the "extra" free of charge so that the news could be spread further. That evening in Fenyang was just like having a festival, with people flying flags and firecrackers popping everywhere in the excitement over this victory.

But the war situation did not go well. With the Beiping-Tianjin region in hand, the Japanese devils set out to get control of Shanxi's coal resources. They divided their forces into two lines of advance, and conditions got worse day by day. By October two critical passes into the province had fallen to the Japanese. Taiyuan was in danger, and Fenyang also fell under "a dark cloud." Those who came to hear the news every evening became more and more concerned. Should we stay in Fenyang, or should we leave now?

My older brother was firmly resolved. He pledged with his life that he wouldn't become a slave in a defeated country. At the beginning of November, he got on an old bicycle and rode, carrying his radio, climbing mountains and crossing rivers. He rode more than a thousand *li* (more than three hundred miles). From Fengling he crossed the Yellow River on a wooden ferry boat and went on to Xi'an. Then he rode a bus from Xi'an, crossing the Qinling Mountains in southern Shaanxi and on to Chengdu. After arriving in Chengdu he lived alone, and in order to get work earlier and in hopes of providing help to his family, he gave up his original idea of studying medicine and entered the pharmacy school at West China University.[2] But his hobby didn't change, and he made his second and third radios.

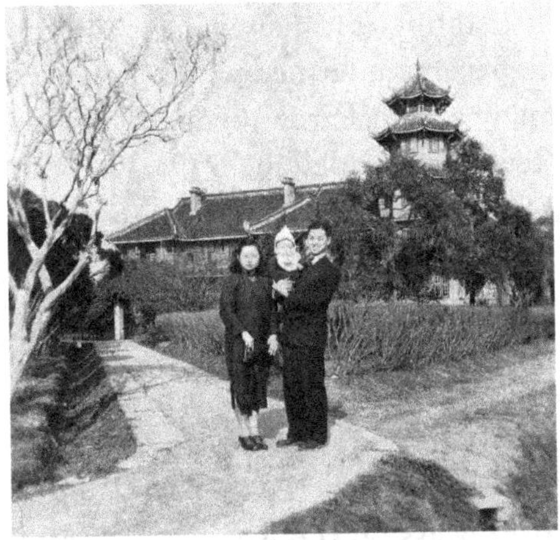

Photo 4: My brother and his wife with their first son, Zhengyang, on the campus in 1947.

After the successful end of the Anti-Japanese War, my brother not only had his radios, he also made a broadcast device and was active in nation-wide amateur radio station activities.

Looking at it from the present, I hope my brother has kept his first homemade radio as a memento because with it he made efforts on behalf of China during the Anti-Japanese War. We do not want to forget any human or material effort that went into the success of that war.

This is by Sun Yude. It appeared in Old Photos *no. 72, pp. 132–36.*

A Village Doctor

Introduction

Feng Shikai, the father here, practiced a traditional profession and was good at it. He and other members of his family who shared this profession, and other associates in it, gradually incorporated some modern techniques. They served their communities well and were respected members of those communities. Yet when the Cultural Revolution began, Father Feng was subjected to criticism and forced to discontinue his practice. Like so many other cases, his turned on an insignificant detail from earlier years. Yet the results proved

2. This is the same university where the woman professor Xiao Yiju taught, beginning at about this same time. See "The Life of a Woman Scientist" elsewhere in this collection.

devastating. Ultimately his medical practice was taken from him and he became a farmer, which he also did successfully. Again as for so many others, Father Feng did not get redress until after the Cultural Revolution ended in 1979.

Among the details of this account, we may observe some of the ways used to continue a family when, for whatever reason, a couple does not produce a male heir. Father Feng became the heir of an uncle so this uncle's family line would be maintained. The surprise ending here was also a surprise to the writer of this piece, Mr. Feng's son. It indicates yet another way for a couple to get an heir when they do not succeed in producing a son of their own. And as seen in "Byways on the Road to Success," a family line can be carried on when a man from outside a family marries into another family and takes on its family name.

The Text

My father's name is Feng Shikai. He was born in 1921 in Shigu Village, Jiejiayuan District, Xiangying County, Shanxi Province. My grandfather was the third oldest child in his generation, and he had two sons, my father's older brother and then my father. Because his own father's elder brother didn't have any sons, my father became that uncle's heir and carried on his name. My uncle had a son, who joined the army and went missing in action. When Father and his elder brother grew up, like their uncle they left home and moved to neighboring counties to practice medicine. Then on the eve of New China, they had settled in Dingfan Village of Gaocun District in Wanrong, and they couldn't return home. By this time they both had become their uncle's sons. Back home, both their father and their other uncle died. But because of the difficulties of communication, they weren't able to go home to take care of the funeral, so they got help from their relatives who had stayed there.

My uncle was a well-known eye doctor who had learned from his family. His predecessors had become widely known for their mastery of eye diseases, not just because of their knowledge of the special prescriptions for eye problems but also because, at a time when Western medical knowledge had not yet become common, he was able to use medical instruments to operate on cataracts. They were extremely

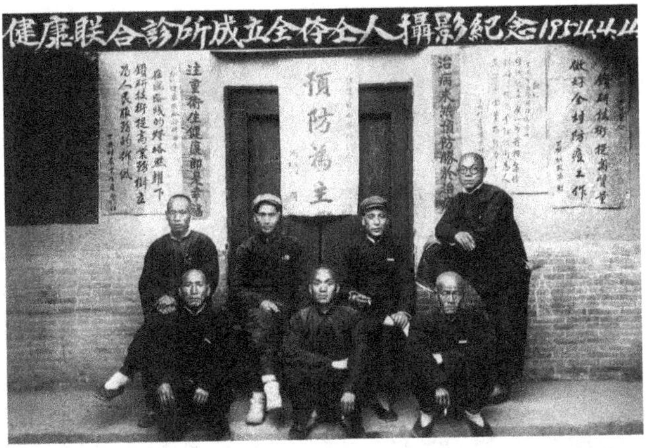

Photo 1

well regarded. The emperor had presented them with a signboard recognizing them as "Famous Doctors of Shanxi." Although I didn't find anything about this in local history materials, I went back to their old village of Shigu, some distance from the county seat in a mountain valley. There I found an impressive set of buildings in a stone courtyard. The main building was a two-story home with decorated pillars and beams, and although it was badly deteriorated, it would have been impossible to build such a home without great resources. Taught by their uncle, my own uncle, my father, an uncle by marriage, several more distant relatives, and still others, all became well-known doctors capable of operating a clinic.

Photo 1 was taken on April 4, 1954, in a courtyard of the kind commonly found in southern Shanxi, where Father and several colleagues from nearby villages opened a clinic. Father is second from the right in the back row. He was relatively young among the seven colleagues, and there's a little smile on his face, reflecting his confidence and hope in the newly opened clinic. And things did indeed turn out that way! Just as the bright possibilities of the New China were beginning to unfold, Father reached out for great achievements.

I have heard Mother say that during the period of land reform in Dingfan, the farmers' committee often met at our house, and under their influence and teaching, Father had soon been able to join the Chinese Communist Party. In this photo you can see a sign and a couplet in the background. The sign says "Prevention is most important," and the couplet says, "In healing sickness, prevention is better

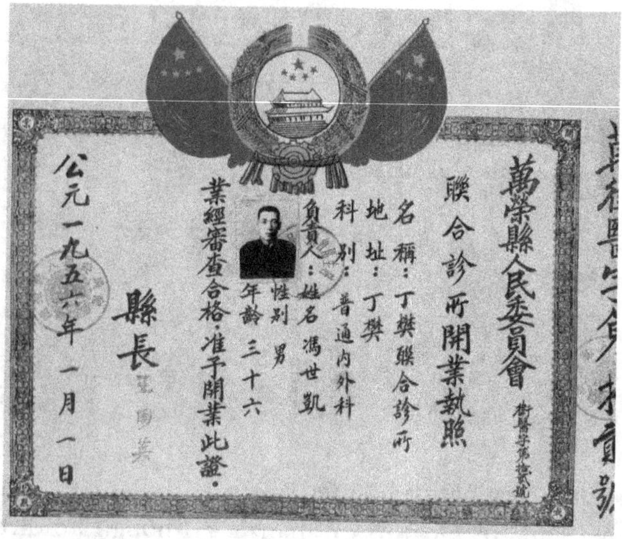
萬榮縣人民委員會

聯合診所開業執照

名稱：丁樊聯合診所

地址：丁樊

科別：普通內外科

負責人：姓名 馮世凱

性別 男

年齡 三十六

業經審查合格，准予開業此證。

縣長

公元一九五六年一月一日

Photo 2

than treatment. Emphasis on hygiene and fitness brings happiness." No doubt the united clinic was Father's response to the Party's call, and his enthusiastic relationships with his colleagues helped to make it all possible. In that place a clinic was something new in people's lives, so they invited a photographer to record its establishment.

It is impossible to know whether the clinic was operated on investment by shares, or just how it was financed. More than a year later, Father's united clinic had moved along well. On January 1, 1956, the Wanrong County Head, Wang Guoying, affixed his seal to the "United Clinic's Business License." The areas of responsibility are shown, and for "ordinary internal and external matters" this person is Feng Shikai. Father was thirty-six that year. That this license to practice medicine remains after fifty years indicates that Father kept up with new developments and studied new techniques. He kept his distance from old-time conservative traditions. Photo 3 shows Father with his fellow students in the Southern Shanxi acupuncture and moxibustion training class. Further, County Head Wang Guoying always spoke well of Father, and Mr. Wang's word meant something because he himself was praised by many for his work on behalf of the people.

Photo 4 was taken at the time of the Mid-Autumn Festival in 1959. This is a photo of the Gao village hospital in Wangya

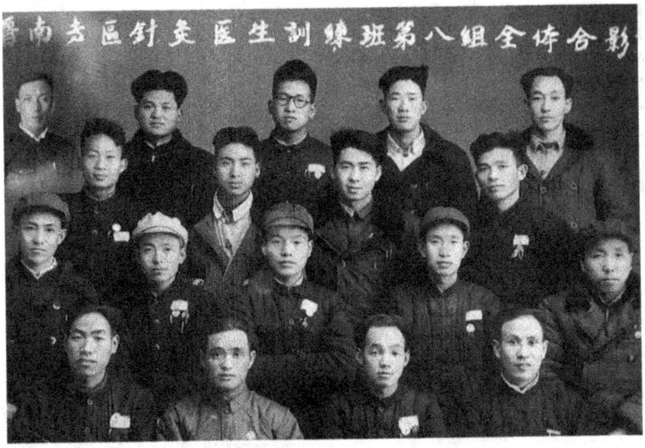

Photo 3

Commune. This was probably the best period in Father's life. Professionally, he had been made head of the hospital. In his career, he had become adept at both Chinese and Western medicine. His areas of special competence were ophthalmology, pediatrics, and acupuncture and moxibustion, and he was regarded as "the best surgeon" in the county. Things were going well with his family too. Mother was managing things well and was always thoughtful and hardworking. Relations with all our neighbors were good, and progress was good for my older sister and me. Everything was harmonious. As he appears in this photo (at right in front), Father parted his hair very stylishly, his eyes are lively, and he looks very self-confident. He is the only person in this photo wearing white shoes. This probably was a gesture of filiality as his grandfather had died shortly before this photo was made.

Photo 5 is the entire staff of the Wangya Commune Hospital at a meeting to see off a doctor who was leaving. By this time the staff had grown to twenty people. The photo suggests the hospital was operating well under Father's leadership, and he was successful in his profession. Father, first on the right in the second row, wears a stylish collared white shirt. There's a little smile in the corners of his eyes, and again he looks confident. He has the best expression of anyone in this photo.

But the world would soon come apart for Father. In 1964 the "Four Cleans" Movement was launched throughout the country. The medical and hygiene personnel of Wanrong County were to be

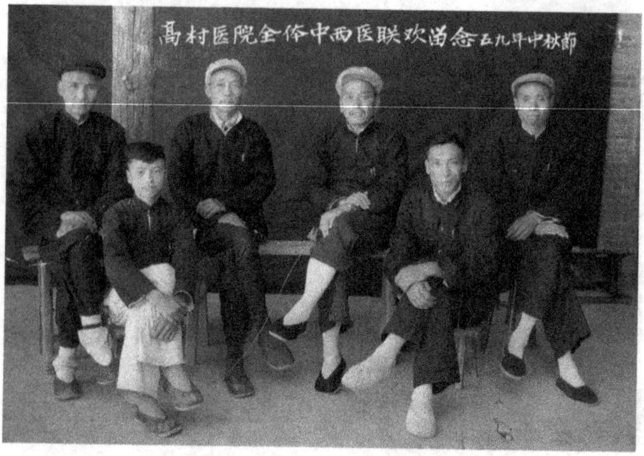

Photo 4

investigated as to their thought, their economic life, their work style, and also as to their life's history. Everyone had to pass the test, and everyone was in danger. Father was a Party member and was head of the hospital. He would be one of those to bear the brunt of this kind of scrutiny. Who knows what kind of official went to look up records from the years of Yan Xishan's rule?[3] Those records showed that while practicing medicine in Beitong Town of Xinjiang county before Liberation, Father had served for three months as a company commander. This duty was part of Yan Xishan's structure in village organization and probably was very similar to that of company commander in a large village's militia in the years after Liberation.

Father's "service" in this duty came about in a conversation that he'd had with a local official whom he was treating back then. As it happened, the position was vacant at the time, and the official had thanked Father for filling the vacancy. Father didn't want to hurt the man's feelings and found it difficult to decide. Actually he did not respond to the official, and certainly he did not know that someone had put his name down for the position. To say that he "filled in for" suggests that it wasn't clear whether he could actually perform

3. Yan Xishan was the provincial governor of Shanxi from soon after the Revolution of 1911 until 1949. Although he has been deemed a "warlord" because of his efforts to remain independent from Jiang Kaishek, Yan was a progressive leader and sought to develop Shanxi economically and to advance education. Yan was known for using a different track gauge for the province's railroads, a decision intended to reinforce its independence but a policy of mixed effect.

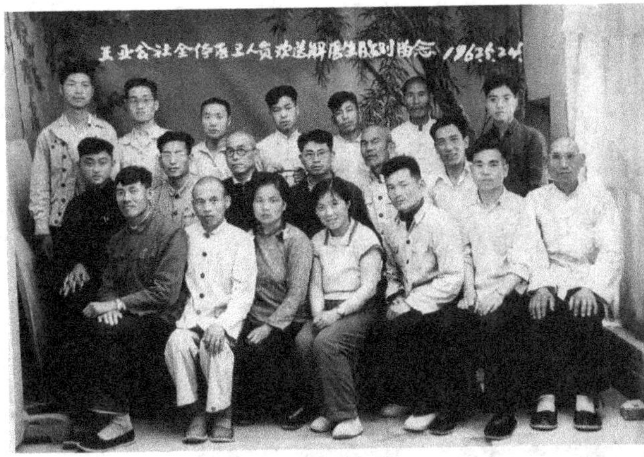

Photo 5

that service. Surprisingly, three months later the People's Liberation Army moved into the Xinjiang county seat. The full story on this "service" had never been revealed to Father, and it became "history not revealed to the Party"—a disaster for Father.

The later development of the "Four Cleans" Movement tended ever further to "extreme leftism," which in Father's case meant that he would be removed from the Party and from his duties in it. He would not receive any punishment, but he was sent down, returning to his former job at the Dingfan clinic. You can readily imagine the effect of this sudden disaster on Father, yet he didn't show any resentment but returned to the old clinic and quietly worked hard to look after the medical needs of the local people. He was well received by officials and ordinary people.

When the Cultural Revolution began, the entire country went into chaos. The original Dingfan party secretary was brought down by the rebel group. Then after the rebels gained power, they created continuous problems for Father and attacked him as an "anti-Party history" person. They expelled him from the clinic and sent him to brigade labor reform, the same treatment as for the "rich landlords and bad right-wing elements" of the village. Father too was paraded through the streets and suffered criticism. He was punished with dirty, difficult or even dangerous labor. He was not exonerated until 1979. Photo 6 was taken during the 1970s, when Father had already become simply an ordinary peasant, and he also had been relegated to the position of "dirty fifth element."

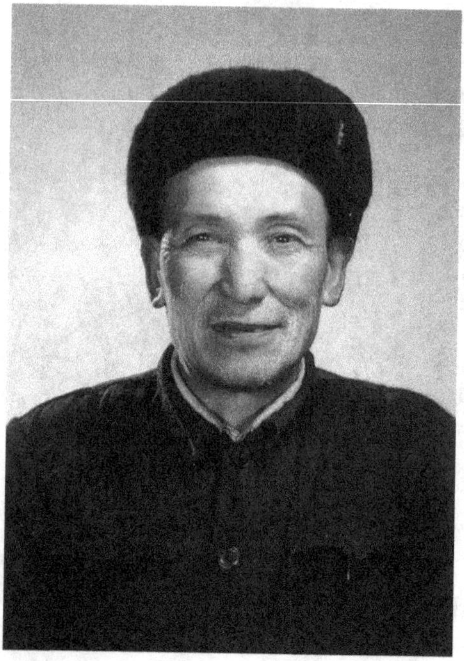

Photo 6

By that time I was already grown, and I saw that after work, almost every day some commune member from our village or the larger area would come and ask Father to treat him or her. Father was not allowed to write prescriptions, but he would use acupuncture or moxibustion if that were relevant, and he would give them advice about what medications to buy. He treated countless people in this way without receiving a single *fen* in payment. Some of those he treated were embarrassed that payment wasn't allowed, and at New Year or other festival times they would bring some high-quality *mantou* or some fruit from their own homes.

In 1970, the "Attack One and Oppose Three Movement" was launched (attack anti-revolutionary elements, oppose corruption and bribery, oppose speculation and profiteering, and oppose extravagance and waste), and as part of that our home underwent a surprise inspection by the local militia. They rummaged through everything and dug underneath the house and took everything. Finally they got Father's licenses, his needles and moxibustion equipment, and his instruments for treating eye diseases. All of these were shown in an exhibit of items confiscated in home searches for the revolutionary masses to see. The criminal name given to Father's items was "From the underground hospital of Feng Shikai, the anti-revolutionary with historical questions." At the time I had to visit this exhibit in an outing organized by my school. I couldn't stand to see it, and it would be difficult to express the pain I felt.

Father's "historical questions" were like a great mountain weighing on our family of four that previously had been quite fortunate.

Father had never had the chance to go to school, and he had experienced difficulties because of this lack of education. He had great respect for those in the family who had been able to study. In southern Shanxi males were important and females not. Girls were not allowed to go to school at all, or at the most they might go through three or four years of elementary school. My big sister was the only girl in Dingfan who had finished upper middle school. She had sat for the university admissions exam just before the Cultural Revolution, but because of Father's situation did not pass and was kept from attending university.

I also finished upper middle school, with good grades, but again because of Father's situation I couldn't be recommended as a lower or middle class peasant. Father took my failure as another blow and was greatly distressed that his problems hurt his son's future. Working with others, joining the Party, and hiring workers had actually hurt our prospects. Mother continued to manage the household, always worried and watchful. Father endured the great mental and physical pressure that came to him. He learned quickly about agricultural life, studying things he had never had any contact with, and even though he was a "black fifth element," he was able become a project leader in the production brigade. He used his work points and then a bit of land that he could control and furtively produced some by-products to sell. Thus, he could meet the household expenses and raise his daughter and son. This was not a matter of one or two years' effort but took ten years of hard work and strong will.

When I finished upper middle school, I went back to the village and did farm work for two and a half years. When the entrance exams were restored in 1978, I was able to enter Fudan University's journalism department. Father held up the letter notifying us of my admission. Both his hands were trembling, and his eyes filled with tears of joy. Well-wishers from the village brought strings of firecrackers that they set off continuously for two days to show their happiness, and Father and Mother bought cake, fruit, and many canned goods to entertain guests who had come to our house over the years. The next year, Father's case was redressed, and he went back to the village clinic to practice medicine again. That next summer I used some of my school funds that I had saved, and Father and I traveled to Shanghai

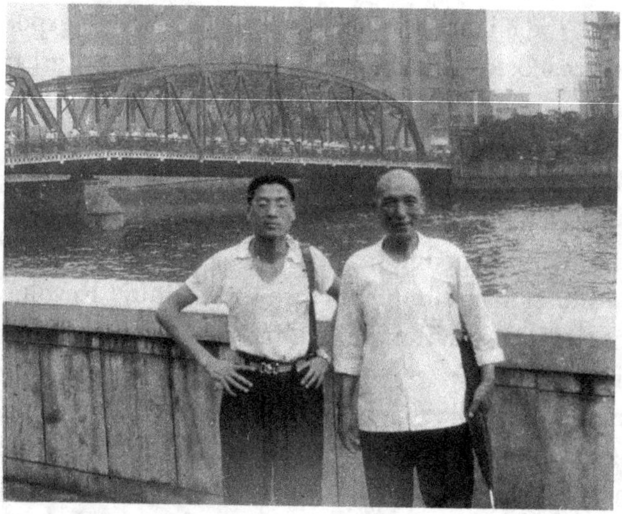

Photo 7

so he could see the city. We also took the riverboat on up the Yangzi River to Nanjing.

Photo 7 is of Father and me at Pudong Park in Shanghai, with the familiar sights of the Waibaidu Bridge and Shanghai Mansion in the background. Father's kindly little smile in this photo reflects his liberation from those earlier burdens and his heartfelt appreciation of the new policies of reform and opening.

Father continued to work at the Dingfan Village clinic until the early 1990s. We knew he was getting old and urged him to retire. But at home he never really stopped his activities. He took it upon himself to plant fruit trees, and if someone came who needed help with a medical problem, he treated them without any fee. In 1998 Father developed his own medical problem—lung cancer. He was able to live for one more year and died at the age of seventy-eight. The next year Mother developed stomach cancer. Before she died, she told me a very important secret. From the beginning I was their adopted son!

This memoir is by Feng Yinpu. It was published in Old Photos *no. 52, pp. 40–47.*

A Peasant Couple with Twelve Children

Introduction

This simple story of a peasant family is repeated many times over in the pages of *Old Photos* on life in modern China. This one is set in the northeastern province of Heilongjiang. Those who work the land do have somewhat less representation than other classes of people in photographic reminiscences on life in modern China. Still, we hear their voices through this archive. Their generally simple stories are usually very moving, as is this one. Neither Father nor Mother was literate, and Mother didn't even have a name until she was in her forties. But they schemed and bled to advance the welfare and opportunities for their children.

Before the most recent ten or fifteen years, it was rare for a farm family to have most of its offspring relocate in a city, as happened for this family. Education was indeed the key. It was probably an important factor that this was Heilongjiang, in the frigid north, where the cities were somewhat less crowded and educated younger people could gain opportunities in the urban economy.

A point to note in this narrative is that Father's ambition for his family backfired when a new wave of anti-rightism broke over the rural society. The father's ambition to produce more and work as hard as ever clashed with the village's need for class enemies. Already past seventy, Father worried himself to death over this problem. It was often true that injustices did not originate with Chairman Mao or central authorities, but instead with local leaders jealous of such hard-working people, or with some other petty problem.

The Text

My father and mother have already passed away.

According to the estimates put forward at the time of their deaths, my father was probably born in 1899 and my mother in 1913. My father was called Sun Zhiming (see Photo 1) and my mother (Photo 2) did not have a name for most of her life. During the period of "communization," a large dining hall was opened in the village, and all the members of the commune ate there. When they began to put together a register of names and the commune accountant got to our family, he said to my mother, "Now it's no longer alright not to have a name, so I'll give you a name. May we call you Feng Rong?" Mother said, "Anything will be alright." So this was how my mother got the name Feng Rong at the age of forty, a name that neither her father and mother nor other members of her family knew. (Feng was her family name; Rong may mean "glory" or "honor." Of course, she deserved both.)

Father and Mother were farmers all their lives and never traveled. The largest place they visited was Jiamusi. That was during the Great Cultural Revolution when my fourth elder brother was studying at the teachers' college in Jiamusi. He had joined the Red Guard and was a leader. The organization he belonged to actually was organized by a representative of the army as an opposition Red Guard, so he had been sent to a branch military district.[4] Our parents went to Jiamusi to visit him, but they didn't get to see him! Photos 1 and 2 were taken later by Fourth Elder Brother and were the only photos made of them in their entire lives. The photo of my mother was taken in the vegetable garden behind our family's house in the village of Yongjiu, Tangyuan County, Heilongjiang Province. I am not sure where Father's photo was taken. Photo 3 was taken in 1965, the graduation photo as I completed elementary school in Yongjiu. Before Fourth Elder Brother passed the examination to go to the teachers' college, he taught at this school for a year as a local teacher.[5] My brother, Sun Yongcai, is the first person on the left in the front row. I am third from the left in the back row.

My father and mother had twelve children, six boys and six girls. Only one of us, an older sister who died of smallpox, did not live to adulthood. The rest of us had our own families and professions, and ten of the twelve children left the village, got work, and lived in a city.

4. During the Cultural Revolution, it was not unusual for rival groups of Red Guards to develop. As suggested here, sometimes local authorities became involved. A student from a peasant background might readily be influenced to join a group connected to the army.

5. Some teachers were recruited locally to supplement a school's teaching staff. They could be paid less than fully trained teachers. They were given the opportunity to take the entrance examinations for more formal training. On this also see "Is It Worthwhile for Village Kids to Go to School?" in this collection.

This was what our village most admired about our family and also what our parents regarded as their best achievement.

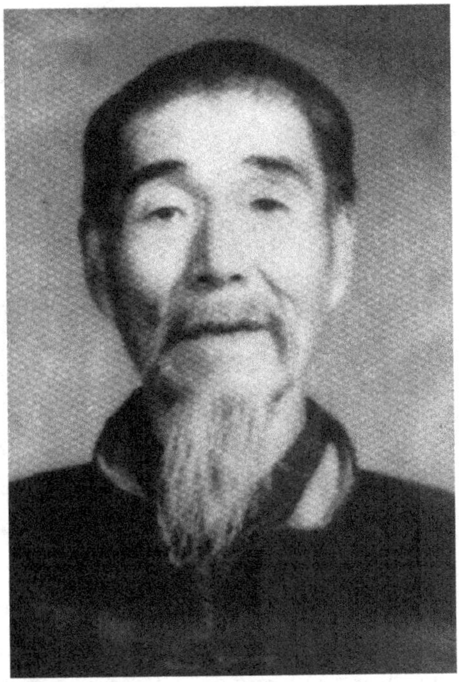

Photo 1

Father was illiterate and didn't even know the characters for his own name. But he was not conservative. He knew the importance of study and made sure we studied well. At each New Year he killed a hog or got some fish, and each time he invited our teachers from school to our home to eat and drink. So he gained respect from the teachers and asked them to make sure we were learning well. We all depended on study to make our way out of the village.

Father was always direct and honest, and always wanted to excel. He didn't have any bad habits. Thus he made his way out of the old society. As far back as I can remember, we heard him talk about his experiences. During one winter he had taken care of the landlord's household, threshing the grain. The soft grass used to stuff the work shoes went bad, so Father spent the whole night re-stuffing and re-sewing the shoes. The landlord saw Father doing this and called the other workers and told them to see how quickly he worked and how he was up working before daylight. Father had not closed his eyes that whole night, and after he finished that repair job he got all the workers at it on the threshing floor.

Father worked hard to build up some family property, and when land reform got started, he contracted for the holdings of a middle class peasant. During the Cultural Revolution, according to the policies of the time, middle class peasants were targets of class struggle, but the landlords and rich peasants of our village all moved

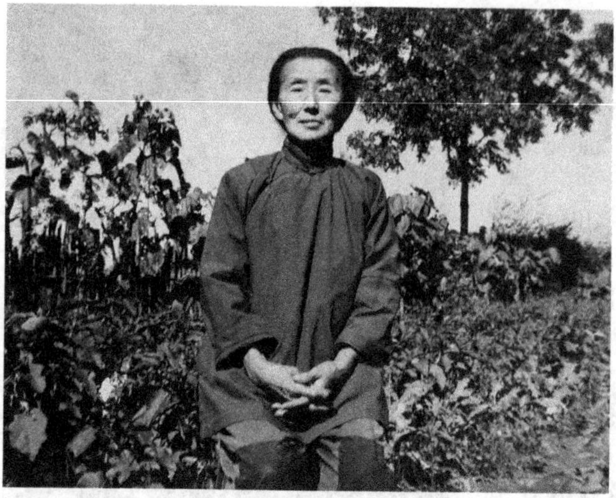

Photo 2

away. Class struggle cannot be without an enemy, so our family became a target of discrimination. The "revolutionary committee" of the brigade concluded that we were not covered by any items from the fourth volume of "Selections from Mao Zedong's Writings" and also said we were not included in anything about the poster entitled "Mao Zedong Goes to Anyuan." But they could claim we were targets for class struggle, based on the fact that the younger brothers in our family had not gone to the military. My older brother's work in the military was not enough. Father got very worried, lying on the *kang* and venting his feelings. One day he suddenly said he had made the wrong decision at the time of land reform, and he should give the land back. So after this he often went over to Shuanghe Village, where our land was registered during the land reform program, and sought out the person who had the records. He never was able to get this worked out. I knew that the question of how he was classified was not important to Father himself, but that he was simply concerned about the future of his children. By 1972 he suffered a brain hemorrhage and died.

My mother was a good-hearted, hard-working village woman. In those times the productive ability and the standard of living were so low it's easy to imagine how hard it was to raise eleven children. Besides our older brother and sister who were already working elsewhere, our family still had thirteen people and was the biggest

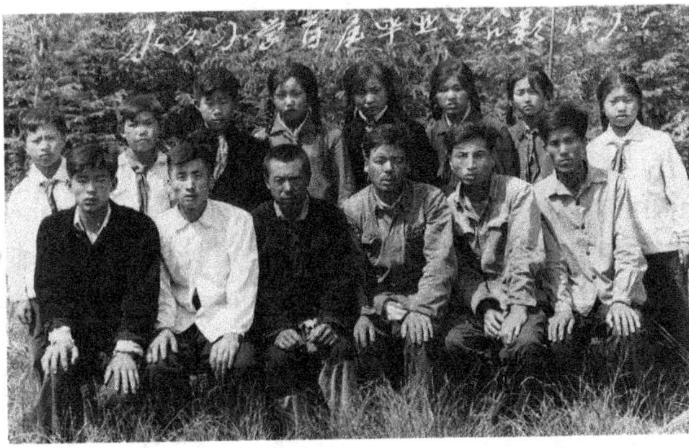

Photo 3

household in the village. Yet only our oldest brother was counted as a work-hand, and his ten work-points only brought a few *jiao* each month. So year after year we had what was called "three *jiao* debt" to the brigade.[6] The clothes we wore and the shoes we had no money to buy were all made stitch by stitch by our mother. And she still had to do the cooking and take care of the pigs. Do you know how many layers there are in the cotton clothes Heilongjiang people wear? To be able to wear a "top-stitched" cap, you had to put a needle through the crown of the cap countless times. Many times I saw that Mother's fingers were bleeding from needle pricks. When you add to her ongoing hard work the recoveries from her many pregnancies—and without enough supplementary food or medicine in those critical times—Mother developed arthritis in her fingers and would feel pain just before it was going to rain.

After I graduated from the teachers' college and started teaching at the county's First Middle School, my salary was 36.50, and after the cost of meals I could only save a few *yuan* each month. Often I would borrow a bicycle from my older brother's unit and pedal madly the ten miles home to see my mother. At first I would buy a little fruit or pastry, but later I discovered that Mother was not

---

6. Evaluation of the productive capacity of a household was left to the local leadership, and it would seem they had chosen this as another way to penalize this large, relatively prosperous family, which surely had more than one mature worker. Their modest but ongoing debt was common enough to have been given its own name. Given that some items were purchased through the collective, the agricultural workers in this "revolutionary" system shared some problems faced by tenant farmers in the old society.

eating these things herself but was passing them on to her grandchildren or to the children of relatives, so I began simply giving her money. But I could only afford eight or ten *yuan* each time.

In 1980, when I was thirty, I got married. The wedding was as simple as you could imagine. There were no photos, and we didn't invite any guests. Some of my colleagues helped me take my bride over to the gathering on a bike, and we ate some sweets and fruit and drank a cup of tea. We regarded this as our wedding. Before we were married, Mother told me to come back to the village. When I got home, my mother bent down to the lower drawer of an old wooden chest, took out a decorative quilt cover and a set of pillows and pillowcases, and gave them to me. So I asked her, "How were you able to buy these?" My mother replied, "I never spent any of that money you brought home. It was all to be saved so I could give something nice to you and your young wife." All I could do was to cry right on the spot!

Until now, every time I think about that deed I shed more tears. My loving mother's heart, lighting the days and months! Her love is as great as Mt. Tai!

My mother died in 1982, on the Sixth of the Third Month on the old calendar—the same day my father had died, ten years earlier.

This was written by Sun Yongsheng. It appeared in Old Photos *no. 42, pp. 106–109.*

Qiu Sanbao's Story

Introduction

Readers will quickly see the connection between this and the selection by Qiu Sanbao, "A Photo of My Father," in the first section of this book. The editors of *Old Photos* tried repeatedly to locate Qiu after the photo, but the payment they sent was returned by the post office. As this story by a journalist in Wuhan tells us, Qiu's whereabouts in that city came to light after he visited *Old Photos* editor Feng Keli hundreds of miles away in Jinan, Shandong. The publication of this article in Number 90 seems an anticipation of the magazine's celebration of Number 100 about a year later.

The Text

In 1998 Qiu Sanbao, a clothing factory worker in Wuhan, sent a brief manuscript to *Old Photos.* It was a photo of their father, who had worked on the Yangzi River Bridge at Wuhan, and it told how the father wasn't comfortable back home working as a farmer. (This was published in *Old Photos,* no. 10.) After the piece was published, the editors mailed the materials back to the address given by Qiu Sanbao—including the fee for the article and the father's photo. But strangely, the photograph and the postal money order for the fee were returned by the post office with a notice saying "This Person Has Left."

More than ten years passed, and Qiu Sanbao still could not be found. It was as if this person was no longer in the world.

Speaking logically, you would think that Qiu Sanbao would certainly have paid attention to see if the article got published, and at least would ask the editors about returning the father's precious photo. "Maybe he had an accident? Every time we thought about this, we became very uneasy." Every time the chief editor Feng Keli had a media interview he would ask journalists for their help in locating Qiu Sanbao.

Most recently, because Feng Keli published a new book called *History as We Can See It,* I as a reporter for the *Yangzi River Daily* went to interview him.[7] During my interview it was natural that we would talk about Qiu Sanbao and his father. Their story happened in Wuhan, where our newspaper is published.

The day my interview with Feng Keli was published in our newspaper, a worker at the Mt. Mulan Scenic Area in the Huangling Disrict of Wuhan saw the article and told Qiu Sanbao, who also works at the Scenic Area as a tour guide. Qiu Sanbao immediately called the *Yangzi River Daily* hotline.

Who would ever have imagined that Qiu Sanbao is a woman!

---

7. This book is Feng Keli's retrospective on *Old Photos,* with reflections on many of the journal's articles from its beginning in 1996 until 2013, when the book was published by the highly respected Guangxi Normal University Press in Guilin.

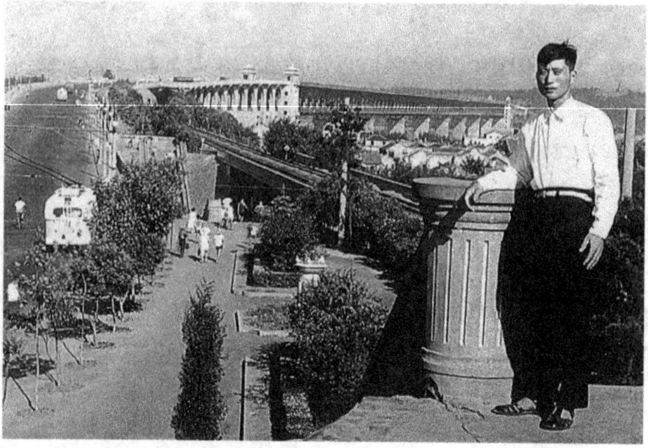

The photo of Qiu Hongyuan published in Old Photos *no. 4 in 1998.*

Why Didn't You Follow Up on the Result of Your Article?

On July 25, I went to the Mt. Mulan Scenic Area on Changling Street of Hanfan village in the Lishu section and met Qiu Sanbao and her father.

In 1996, Qiu Sanbao was twenty years old and saw her first issue of *Old Photos.* In 1998 she decided to give it a try and sent a manuscript to *Old Photos*. Because the job she had at the time was so busy and tense, and also because she soon found a position as a guide at Mt. Mulan Scenic Area, she went back to her home district of Huangling.

"You submitted the article, so why didn't you follow up to see what happened about it? And why didn't you buy *Old Photos* again? And what about that precious photo of your father? Why didn't you try to get that back?"

I just kept asking her questions, one after another.

"At first, my work situation was too busy. I had to move from one job to another. I would move on to a different factory every half a year or so. I just drifted from place to place and didn't even have a set place to live. Really, there's no way you could imagine what my life was like then. Then I moved to Huangling, a suburban county, a village, where you couldn't buy *Old Photos*. And so gradually I just forgot about it."

Qiu Sanbao was wearing a full-length yellow dress, bright enough in the sunlight that it was a bit glaring.

She said she hadn't met with any unexpected changes of fortune.

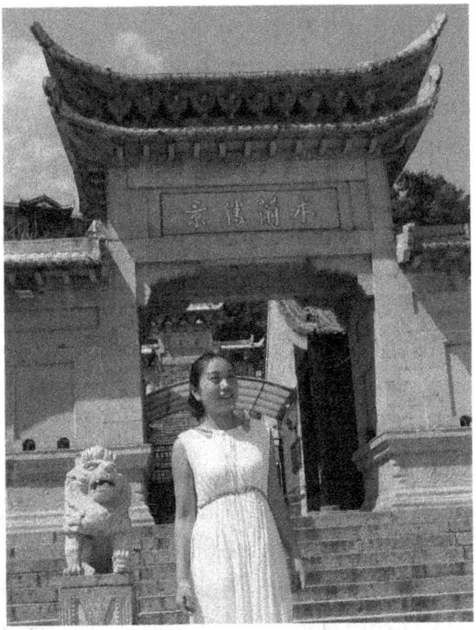

Qiu Sanbao during her interview with the journalist Liu Gonghu, writer of this article.

If there were any, they had tended to move her in a good direction. She was no longer dripping sweat every day, working all night and finishing in the morning. She had taken a less demanding job.

Qiu Sanbao's experiences over fifteen years can be summarized like this. First she worked as a seamstress in clothing factories on Liji Road, Minsheng Road, and Tongyi Road. Then in 2000 she became a guide in the Mt. Mulan Scenic Area. In 2003 she got pregnant and gave birth to her child, and then for ten years after that was completely consumed in her role as a housewife and mother. She also got opportunities to travel to some famous places around the country.

Then this year she returned to work as a guide at the scenic area.

"Do You Want Me to Tell the Truth?"

Maybe it was because this was the first time she had talked with a reporter that Qiu Sanbao was a bit guarded and nervous. When I asked about details of her life, she often said, "Do you want me to tell the truth?"

At about three in the afternoon, as she requested, I went to Changling Street in Hanfan and met her father, who is seventy-five years old—the person in that photo, Qiu Hongyuan. As the old man began recalling the hardships of his work life, his daughter would often interrupt him: "You often see such things in movies or on television. People on the lower level of society don't think this is strange or uncommon."

The old man and I talked pleasantly. And before this, she would remind me that he could be a little strange and if I asked more, it might be a bother to him. After I finished the interview and went back to my office, suddenly there was a telephone call from Qiu Sanbao, and she spoke emotionally, "Surely you weren't pleased with my behavior today? I must have seemed strange. I should tell you that yesterday when I knew you'd be coming over, my older brother and I and some friends discussed it, and they thought I should be on guard and not tell you everything. But when I talked with you, I discovered that you really are an honest reporter. Have I thought too much on this?"

Then she continued quietly and told me that after she got settled in at Mt. Mulan, she had married a man who is a policeman and had a beautiful little girl, and she really doesn't like to recall the difficulties of that earlier time.

And indeed, those times were truly difficult for her.

From 1996 until 2000 she went from one sewing workshop to another. She could only get four hours of sleep, and the pay was low. Sometimes when they got hungry at about three in the morning, her workmates would ask her to go out and get hot noodles. She would find a shop that was open and carry ten or more bowls of noodles back to the factory. The hot noodles would help them keep going until about five in the morning. When they got off work, they went back to their dormitory to rest for those few hours and went back to work again.

Even under these conditions, she liked to read books and newspapers, and sometimes to write. She happened to discover *Old Photos* in a bookstore on Minsheng Road, and sent in her article. But she didn't dare to think it would be published. She just thought she would do such a thing as a break from her difficult routine.

"I really would never have thought the editors at *Old Photos* would follow up on this for such a long time. Receiving the payment wasn't important to me. What is important is that they have such a heart, and their attitude. I haven't slept very well these past few days. This has drastically changed many of my attitudes about society."

She told me all of this during our conversation.

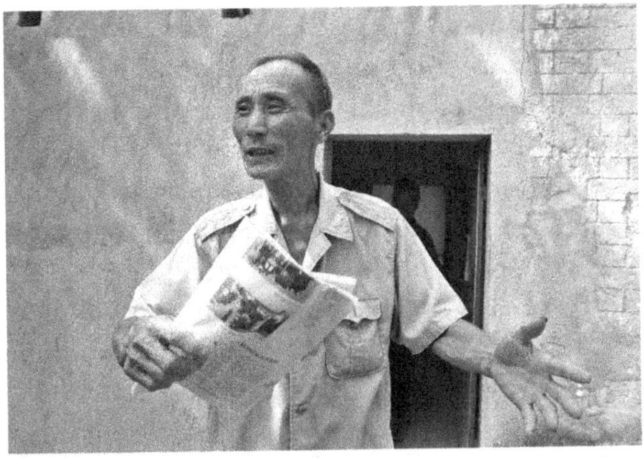

Qiu Hongyuan in recent years as a retiree, age 75.

The Father Had Many Difficulties in His Own Life

Going back to the beginning, Qiu Hongyuan had already left the construction brigade when the photo was taken in 1965. He told me he was a strong worker in the village, and when he transferred to East-West Lake farm in Wuhan and specialized in cross-breeding and transplanting rice seedlings, he could make four or five *yuan*. After sending part of this to the village he could still have two *yuan* for his living expenses.

One time, he and about ten of the colleagues in the village decided to go to the Yangzi Bridge on an outing. They walked from East-West Lake to Gusao station to catch a bus to the bridge. There was a photo service at the bridge, so each of them spent a few *mao* (30–40 cents) to get a photo made. A few days later you would receive your photo in the mail. They were very pleased with themselves about these photos.

Besides his ID photo and the photos on other certifications, Qiu Hongyuan didn't have any other photos from life.

"I worked away from home for many years. My three children were born but I wasn't there to help raise them or enjoy them. I was away from home beginning in the 1960s and working in Wuhan for more than twenty years. I could only get back home for the New Year and at the busiest work times back home, like the days when we harvested the crops."

He was a highly skilled stonemason. He worked as a bricklayer and stonemason in the construction of the Wuhan Steel Mill, the Wuhan Heavy Machinery Plant, and the Navy Engineering School. When the Mt. Mulan Scenic Area began construction, he worked there for several years and was one of the stonemasons who engraved several stonework landmarks in the area.

When he turned sixty, Qiu Hongyuan returned home and never left again. His oldest son has a rather large-scale vegetable farm not far out of the city. His younger son works at the local inspection office. His daughter married a policeman, and through her ability as a writer was able to get a rather relaxed job.

Qiu Sanbao also serves as the group host for an Internet group on "QQ."[8] Her essays and songs appear on her own QQ space and are seen and discussed by many "friends" in the group. A little more than ten years ago she was a sentimental young woman, and with the pressures of living she understood early that we age easily and that an individual might have frailties. As she contemplated that old photo of her father, she expressed her feelings of wistfulness and admiration. "Father can never be young again, and I am left to carry on his youthful years. Everyone who is old was once young. And everyone who is young will gradually become old and weak. A day will come when he leaves this earth."

Now she is a sturdy mother, who no longer holds a pessimistic view of life.

This is by Liu Gonghu. It was published in Old Photos *no. 90, pp. 182–87.*

Byways on the Road to Success

Introduction

This family story covers three generations, from grandparents to the present younger generation. Clearly the father of this present generation is the main figure. As in several other such memoirs, the

8. QQ is a Chinese Internet group, an equivalent of Facebook. Qiu Sanbao wrote herself into the role of a group host.

hardening of class background after 1949 became a major determinant of this man's fortunes. After the last of several sojourns far from home, he found a way to begin to change his circumstances. In a traditional practice to enable a family to maintain its name, he married into a family that was without a male heir, taking that family's name. This action, along with the reopening of university entrance exams at the end of the Cultural Revolution, opened the way for this man to apply his talent. The writer is pleased that she and her siblings were able to graduate from university and join the Communist Party.

The Text

This photo was taken in 1984. That's my grandfather in the middle of the back row with my father and mother on either side of him. Those in the front, from right to left, are my first younger brother, my younger sister who's the third one of us, then the youngest brother, and me. That year I was in first grade.

During his lifetime my grandfather had the experience of both the old and the new society. He had both "red" and "black" in his background. Before Liberation, he was a supplier of gunpowder, an armed merchant-landlord, and an stalwart family head in Yilong County, Sichuan. When the Red Army passed through Yilong County all those years ago, Grandfather gave them several fat pigs, and he was also invited to become an underground Party member. Because of this special aspect of his background, whenever problems came up after Liberation, he was able to avoid difficulties. But unfortunately, although the "red umbrella" sheltered Grandfather, it couldn't protect Grandmother, nor could it protect Father and his four children. Grandmother had to wear the dunce cap of a landlord's wife. She couldn't take the criticism and open humiliation and killed herself.[9]

Father and his three brothers had been given personal names that, taken together, read "an official position, rising high," from which you can see Grandfather's high hopes for them. But his elder brother, who at the time was in primary school, could not escape

9. In cases like this that refer to incidents occurring many years before 1949, it is not clear how local Party officials would know about the grandfather's role in earlier Red Army activities. Evidently some local organizations have long-term records. Perhaps it was a recollection of an early member that saved the grandfather.

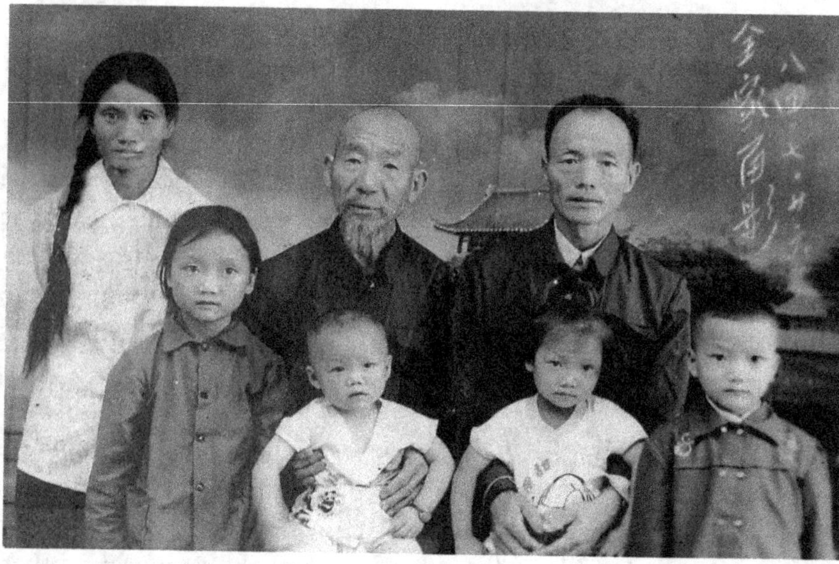

the misfortune of kneeling beside his mother in struggle meetings. Father was the second child, and because he was still young was exempted from such mistreatment. Grandfather turned his clever younger child over to Grandmother's younger brother. Father traveled with his uncle doing various kinds of work, including making *doufu*, and eventually they wound up in Chongqing. After graduating from a specialized middle school, because of his family background, Father had to give up his wish to become a policeman and decided to work in a factory. After starting at the Chongqing Steel Plant, Father's talent quickly showed itself. They invited him to become a member of the Communist Youth League, but when Father revealed his true family background they said he had to leave. He returned to his hometown again.[10]

After going back home, Father realized there wasn't any opportunity in that backward area, and once again strapped on his backpack and went out on his own to seek his fortune. He took on jobs as a stone-worker, as a coolie, and as a woodcutter deep in the mountains. Finally he went to Dechang County in the Liang Mountain

10. "The organization" refers to the Chinese Communist Party or to cadre, many of whom were not actually Party members. Clearly the young man's background was deemed sufficiently "rightist" that he could not be accepted even as an entry-level company official.

region, where through someone's introduction he met a family that had been without a male heir for three generations. Father married in as an outside male heir for the family. He was now far away from his original home where nobody knew his background so he was able to carefully give himself a "straw hat," that is, to become a person of "poor and lower-middle class peasant" background. Even his very shrewd mother-in-law never learned about his actual class background.[11]

After the university entrance examinations were resumed, Father could rely on his talents to get into Sichuan Architectural College.[12] Following graduation, he was employed by an architectural firm and worked his way up to become senior engineer and company manager. Later, he retired from the company and began doing his own design projects. His projects were often evaluated as excellent, so he sometimes got opportunities to appear on television to talk about his work.

One of Father's greatest regrets was that he was unable to join the Party because of his class background. He passed this long-held hope on to his children and three of the four of us became Party members. What satisfied him most was not his work but the fact that all four of his children were able to go to university.

Father was always concerned to watch over our education. The happiest times in my memories were at supper time when we all sat around the table and Father would ask us to tell what we had done in school. We all were pleased to tell our parents what new things we had learned that day.

Even more time was spent with Grandfather, who had been a *xiucai* ("flowering talent") scholar in late Qing times, so he too took an interest in our schoolwork. In the leisure of evening time, in the twilight, he would lean back in his cane chair with his eyes

---

11. Dechang is closer to Kunming, capital city of Yunnan, than to either Chengdu or Chongqing—his home—in Sichuan. Dechang lies between the south-flowing Yalong River and the Jinsha River ("River of Golden Sand") about two hundred miles north of Kunming. This was the setting in which the father got married and also gained class acceptability.

12. University entrance examinations were resumed in 1978 after over a decade of suspension. At this point the father was able to begin his college education.

closed listening to me read *The Complete Biography of Yue Fei.* I always enjoyed having Grandfather as my "living dictionary" to learn unusual terms.

Although the age-old thinking that males are superior was deeply set in Grandfather's thinking, as the oldest child I was inclined to follow Father's enthusiastic nature, and the brother just next to me took after Mother's more yielding temperament. So Father had little choice but to take me under his wing in the role of an eldest son. When we had guests at our modest home, Father would have me make tea for them. When such times came, my brother would retreat upstairs to read. He was simply not comfortable interacting with people. During school vacations Father would take me to work with him to spend some time at the construction site, and as I got older I would sometimes carry sand or stone to where they were needed or just be an odd-job worker on the construction site and make a day's pay of eight *yuan*.

Because I was delayed by sickness, my brother, who was a little less than two years younger, and I took the university entrance examination together. Father was still in the middle of a project in Chongqing and couldn't come home, but he called us on the telephone and said, "You two do your best and choose the school and the major you want to get into. Just be sure to do your best." I got into the Law Department at Southwest College of Politics and Law, and my brother, following my advice, went to Sichuan Agricultural University to study food science. At that time, Father went together with Mother's uncle to open an apple drying factory in Lanyuan County. I thought my brother could go there after he graduated and be of help to Father. Also, as food is the essential in life, with a degree in food science, my brother would not be unemployed. After graduation he returned to our home area and became a deputy village head.

Our third sister went to West China University and majored in Chinese. After graduating she returned to our home district and worked in cultural affairs. Our youngest brother went to Jinyang County, where he is an optometrist.

When vacations come, many friends get together and travel to different places. But I take my husband and children back to the old home with so many memories and help Father in his work in the construction field, go buy food with Mother, and help with the cooking. Mother often says, "When you get a vacation, I can take a few days off too!"

My greatest hope in life is that when I retire, I can return home and do the best I can to look after my parents as all children should.

This is by Luo Qiong. It was published in Old Photos *no. 64, pp. 112–14.*

Further Reading

Barbey, Bruno. *China: From Mao to Modernity*, 1973-2013. Singapore: Didier Millet, 2015.

Beers, Burton F. *China in Old Photographs 1860–1910*. New York, NY: Dorset Press, 1981.

Bertholet, Ferdinand M. and Lambert van der Aalsvoort. *Among the Celestials: China in Early Photographs*. Brussels: Mercatorfonds, 2014.

Boerschmann, Ernst. *Old China in Historic Photographs*. Mineola, NY: Dover Publications, 1982.

Bulfoni, Clara and Anna Pozzi. *Lost China: The Photographs of Leone Nani*. New York: NY: Skira, 2004.

Cabot, Mabel. *Vanished Kingdoms: Tibet, China, and Mongolia, 1921–1925*. New York, NY: Aperture, 2005.

Cameron, Nigel. *The Face of China: As Seen by Photographers and Travelers, 1860–1912*. New York, NY: Aperture, 1978.

Carter, Tom. *China: Portrait of a People*. Hong Kong: Blacksmith Books, 2010.

Cody, Jeffrey W. and Frances Terpak, eds. *Brush and Shutter: Early Photography in China*. Los Angeles, CA: Getty Research Institute, 2011.

Cook, James A., Joshua Goldstein, Matthew D. Johnson, and Sigrid Schmalzer, eds. *Visualizing Modern China: Image, History, and Memory, 1750–Present*. Lanham, MD: Lexington Books, 2014.

French, Howard W. and Qiu Xiaolong. *Disappearing Shanghai: Photographs and Poems of an Intimate Way of Life*. Paramus, NJ: Homa & Sekey Books, 2012.

Guo, Guang, ed. *China: A Photographic Journey Through the Middle Kingdom*. New York, NY: Abbeville Press, 2010.

Harris, David. *Of Battle and Beauty: Felice Beato's Photographs of China*. Santa Barbara, CA: Santa Barbara Museum of Art, 1999.

Henriot, Christian and Wen-hsin Yeh, eds., *Visualising China, 1845-1965: Moving and Still Images in Historical Narratives*. Leiden, Netherlands: Brill, 2012.

Jenner, W. J. F., *China: A Photography, 1937–1987*. New York, NY: Pantheon, 1988.

Lu, Hanchao. *The Birth of a Republic: Francis Stafford's Photographs of China's 1911 Revolution and Beyond*. Seattle: University of Washington Press, 2010.

Morrison, Hedda. *Travels of a Photographer in China, 1933–1946*. Oxford, UK: Oxford University Press, 1987.

——. *A Photographer in Old Peking*. Oxford, UK: Oxford University Press, 1986.

Naumkin, V. V. *China: Caught in Time*. Berkshire, UK: Garnet Publishing, 2009.

Neville, Peter, et al. *China: People Place Culture History*. London: DK Publishing, 2007.

Page, Brian. *Beijing Then and Now*. San Diego, CA: Thunder Bay Press, 2007.

Pang, Laikwan. *The Distorting Mirror: Visual Modernity in China*. Honolulu: University of Hawaii Press, 2007.

Parr, Martin and Wassink Lundgren, eds. *The Chinese Photobook: From the 1900s to the Present*. New York, NY: Aperture, 2015.

Riboud, Marc. *The Three Banners of China*. New York, NY: Macmillan Company, 1966.

——. *Visions of China: Photographs, 1957–1980*. New York, NY: Pantheon Books, 1981.

Spence, Jonathan and Annping Chin, eds. *The Chinese Century: A Photographic History of the Last Hundred Years*. New York, NY: Random House, 1996.

——. *Imperial China: Photographs 1850–1912*. Milton, ON, Canada: Penwick Publishing, 1978.

Thomson, John. *China, the Land and Its People: Early Photographs*. Hong Kong: John Warner Publications, 1977.

——. *China and Its People in Early Photographs: An Unabridged Reprint of the Classic 1873/4 Work*. New York, NY: Dover Publications, 1983.

Wakeman, Carolyn and Ken Light, eds. *Assignment: Shanghai: Photographs on the Eve of Revolution*. Berkeley: University of California Press, 2003.

Wu, Hung. *Zooming In: Histories of Photography in China*. London, UK: Reaktion Books, 2016.

Yao, Betty, ed. *China: Through the Lens of John Thomson, 1868–1872*. Bangkok Thailand: River Books, 2010.

Index

This index is organized into categories. Readers may quickly locate a desired entry by becoming familiar with the several headings.

Written works and publication / Movies / Pictures / Songs

Groups

People, inc. fictional characters

Places (inc. institutions & structures)